UNDERWATER ACOUSTIC SYSTEM ANALYSIS

PRENTICE-HALL SIGNAL PROCESSING SERIES

Alan V. Oppenheim, Editor

UNDERWATER ACOUSTIC SYSTEM ANALYSIS

William S. Burdic

Autonetics Marine Systems Division
Rockwell International

PRENTICE-HALL, INC.   Englewood Cliffs, NJ   07632

Library of Congress Cataloging in Publication Data

Burdic, William S.
  Underwater acoustic system analysis.

  Includes index.
  1. Underwater acoustics. 2. System analysis.
I. Title.
QC242.2.B87   1984   620.2′5   83-9489
ISBN 0-13-936716-0

Editorial/production supervision and
   interior design: *Shari Ingerman*
Cover design: *Marvin Warshaw*
Manufacturing buyer: *Tony Caruso*

Printed in the United States of America

10 9 8 7 6 5 4 3 2

ISBN 0-13-936716-0

Prentice-Hall International, Inc., *London*
Prentice-Hall of Australia Pty. Limited, *Sydney*
Editora Prentice-Hall do Brasil, Ltda., *Rio de Janeiro*
Prentice-Hall Canada Inc., *Toronto*
Prentice-Hall of India Private Limited, *New Delhi*
Prentice-Hall of Japan, Inc., *Tokyo*
Prentice-Hall of Southeast Asia Pte. Ltd., *Singapore*
Whitehall Books Limited, *Wellington, New Zealand*

To our beloved grandchildren
Trevor, Brian, and Justin

Contents

CHAPTER SEVEN
Discrete Fourier Methods   201

CHAPTER EIGHT
Correlation and Correlation Functions   214

CHAPTER NINE
Random Processes   244

CHAPTER TEN
Ambient Noise in the Ocean   297

CHAPTER ELEVEN
Spatial Filtering: Beamforming   322

CHAPTER TWELVE
Acoustic Characteristics of Targets   361

CHAPTER THIRTEEN
Statistical Basis for Performance Analysis   380

CHAPTER FOURTEEN
System Performance Analysis: Examples

411

Index

439

Preface

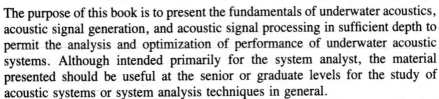

The purpose of this book is to present the fundamentals of underwater acoustics, acoustic signal generation, and acoustic signal processing in sufficient depth to permit the analysis and optimization of performance of underwater acoustic systems. Although intended primarily for the system analyst, the material presented should be useful at the senior or graduate levels for the study of acoustic systems or system analysis techniques in general.

The subject of underwater acoustic systems encompasses a variety of applications including acoustic communication, the detection and location of surface and sub-surface objects, depth sounders, and sub-bottom profiling for seismic exploration. Although each application has its own special problems, the basic physical phenomena are the same. The required system analysis skills are therefore similar.

The analysis of underwater acoustic systems requires an understanding of the generation and propagation of compressional acoustic waves in the ocean, including the complex effects that occur at the surface and bottom boundaries. These subjects are covered in Chapters 2 through 5, following a historical review in Chapter 1.

Chapters 6 through 9 present a review of Fourier methods, correlation techniques, and random processes. Those readers already familiar with the analysis of signals and noise may need only a cursory review of these chapters to familiarize themselves with notation to be used in the remainder of the book.

The sources and characteristics of ambient noise in the ocean are covered in Chapter 10. Special attention is given to the directional characteristics of the ambient noise field and the resulting effect on the noise spatial correlation function.

Chapter 11 considers the subject of acoustic beamforming as a spatial filtering operation, designed to improve the detection of spatially concentrated signals in the presence of spatially distributed ambient noise. Using the results from Chapter 10, the effect of both uniform and nonuniform noise fields on spatial filter performance is discussed.

The detection of objects in the water depends on their radiated acoustic energy for passive systems or on their acoustic reflectivity for active systems. Chapter 12 presents the target characteristics of importance for both passive and active detection systems. Also discussed in this chapter is the undesired signal called *reverberation* with which the received target echo must compete in an active system.

In Chapter 13 the basis for calculating the system detection performance and for estimating target parameters is established using the methods of statistical hypothesis testing. Finally, Chapter 14 presents selected examples of overall system performance analysis using many of the principles developed in earlier chapters. Examples include detection prediction and parameter estimation for both passive and active systems.

The material for this book was developed over a period of years for company-sponsored courses in the Autonetics Marine Systems Division of Rockwell International. I would like to thank my colleagues at Rockwell for their helpful suggestions and direct help in preparing the manuscript, and the management at Rockwell for their support of the project.

W. S. Burdic

UNDERWATER ACOUSTIC SYSTEM ANALYSIS

1

Historical Background

1.0 EARLY BEGINNINGS

Most modern technologies are able to trace their roots back several centuries to the fundamental discoveries and accomplishments of scientists in many diverse fields. It is a tribute to these early scientific pioneers that they often were able to make important contributions in several seemingly unrelated areas of technology.

Few men can match Leonardo da Vinci for the sheer scope of his activities and thought processes. More of an engineer than a scientist, he anticipated a remarkable number of practical applications of then existing and future technology. There is a striking example in his notes on acoustics that is of particular interest to those involved with underwater acoustic systems. Toward the end of the fifteenth century he wrote [1]:

> If you cause your ship to stop, and place the head of a long tube in the water and place the outer extremity to your ear, you will hear ships at a great distance from you.

This remarkable disclosure includes all the essential elements of a modern passive sonar system. It recognizes that moving ships, even sail- or oar-driven ships, generate sound in the water that may propagate to considerable distances. It describes a receiving device, in this case an air-filled tube to convert water-borne sound to airborne sound, coupled with the ear–brain combination, resulting in the perception and identification of distant ships. It even recognizes the fact that better results are obtained if an effort is made to reduce self-noise (by causing your own ship to stop) that competes with the sounds of distant ships.

1

The progress made since Leonardo's time is primarily in the sophistication (and cost) of the receiving equipment and in the level of understanding of the factors affecting the propagation of sound in the water.

Leonardo recognized that sound traveled at a finite speed as evidenced by the following passage from his notes:

> It is possible to recognize by ear the distance of a clap of thunder, on first seeing its flash. . . .

In 1635, a French philosopher, Pierre Gassendi, used this idea in reverse to measure the speed of sound. He timed the interval between the observed flash of a firearm and the reception of the audible report to obtain an estimate of 1569 ft/sec [2]. Although this is not very close to the true value of about 1100 ft/sec, it is one of the earliest attempts at quantitative measurement of acoustic parameters. At about the same time, another Frenchman, Marin Mersenne, measured the speed of sound in air by timing the return of an echo and achieved an accuracy of better than 10% [3].

In 1687, Sir Isaac Newton published his *Mathematical Principles of Natural Philosophy,* which included the first mathematical treatment of the theory of sound. He was able to relate the propagation of sound in fluids to measurable physical quantities such as density and elasticity. He derived from theory the speed of sound in air as being proportional to the square root of the ratio of atmospheric pressure to air density. This actually gave a value that was too low until corrected later by Laplace, who included the specific heat ratio in the expression [2]. In addition to this pioneering work on the theory of sound, Newton's development of the calculus was of course very important in the formulation and solution of the mathematical expressions describing wave phenomena.

Much of the theoretical work on sound in the eighteenth and nineteenth centuries was related to the study of pitch and quality of sound produced by musical instruments. Investigators in the field included such notables as d'Alembert, Lagrange, Daniel Bernoulli, and Euler. Actually, the harmonic analysis of sound has its beginnings in the solution of certain problems in the study of heat. Baron Jean Baptiste Joseph Fourier used the infinite trigonometric series that now bears his name to represent more complicated functions in his treatise *The Analytical Theory of Heat* (1822). In 1843, the German physicist Georg Simon Ohm recognized that complex sounds can be decomposed into a series of simple tones and expressed mathematically as a Fourier series. This was an important event not only in the study of acoustic signals but of other types of signals as well.

The speed of sound in water was measured in 1827 by Daniel Colladon, a Swiss physicist, and Charles François Sturm, a French mathematician. At Lake Geneva they used a light flash coupled with the sounding of an underwater bell to obtain a speed of 4707 ft/sec for a water temperature of 8°C. This value is surprisingly close to the accepted value today for fresh water [2].

In 1877, Lord Rayleigh (John William Strutt) published his book *Theory of Sound* [4]. This monumental work covered the generation, propagation, and reception of sound in a rigorous fashion. He covered the elastic behavior of solids, liquids, and gases, establishing the basis for acoustic theory even as it exists today.

1.1 PRE-WORLD WAR I COMPONENT TECHNOLOGY

With some notable exceptions during World War I, the development of underwater acoustic systems relied heavily on the development of electrical and electronic components and technology. Devices were required for the generation of underwater acoustic signals, and for the conversion of acoustic signals to the more convenient electrical analogs. The transmission, amplification, and other signal processing tasks required in acoustic systems make extensive use of the components and analysis methods of electronic communication systems.

Of particular interest in the generation of acoustic signals is the fact that mechanical forces exist in the presence of electric or magnetic fields. The force exerted on ferromagnetic material by a magnetic field has been known since very early times. In 1819, Hans Christian Oersted demonstrated that a magnetic field exists around a wire carrying electric current, and in 1831, Michael Faraday demonstrated the induction of electric current in one wire by current flowing in a nearby coil of wire. In the 1840s, James Joule quantified the magnetostrictive effect by measuring the change in dimensions of a magnetic material in the presence of a magnetic field. In 1880, Pierre Curie and his brother discovered the piezoelectric effect, which results in an electric charge on the faces of certain types of crystals when subjected to a mechanical strain [5]. Conversely, these crystals undergo mechanical deformation in the presence of an electric field. Magnetostriction, the piezoelectric effect, and the force exerted by an electromagnet on an iron armature provide the basis for the design of most underwater transducers.

In 1912, R. A. Fessenden developed the first high-power underwater source [6]. This device, the Fessenden oscillator, was driven electrically at a single frequency and operated on much the same principle as the electrodynamic speaker. Operating in the range 500 to 1000 Hz, it was capable of acting as an underwater receiver as well as a transmitter. In 1914, stimulated by the sinking of the *Titanic* in 1912, Fessenden used his device to demonstrate "echo ranging" on an iceberg at a range of 2 miles.[1] Because of its simplicity and reliability, the Fessenden oscillator remained in use as a source of underwater sinusoidal signals until relatively modern times.

Fessenden was also a consultant to the Submarine Signal Company, which developed the first commercial application for underwater acoustic devices. By

---

[1] L. F. Richardson filed for a British patent on an underwater acoustic echo ranging device five days after the sinking of the *Titanic*. However, Fessenden was apparently the first to reduce the idea to practice.

simultaneously sounding an underwater bell and an above-water foghorn, both located on a lightship, this company provided ships with a method of determining range to the lightship by measuring the difference in time of arrival of the airborne and waterborne sound. Although this technique did not gain wide acceptance because of the emergence of radio as an aid to navigation, the parent company remains intact as the Submarine Signal Division of Raytheon.

The development of telegraph and telephone systems of communication in the nineteenth century provided a reservoir of component and system analysis technology that would eventually aid in the development of underwater systems. Morse's telegraph was put in public use in 1844 and provided valuable experience in the transmission of signals over considerable distances. Methods of multiplexing signals were developed to reduce costs by, among others, Thomas Alva Edison.

In 1876, Alexander G. Bell was granted a patent for the telephone. Although others had anticipated the basic idea, Bell was the first to make a practical instrument for converting acoustic speech to electrical signals and then back to speech at the receiving end. Bell's transmitter and receiver were identical, each consisting of a metallic diaphragm in the field of an electromagnet. Sound waves striking the diaphragm caused it to vibrate, thereby causing the current in the coil of the electromagnet to vary. At the receiving end, this current variation caused the magnetic field to vary, inducing vibrations in the diaphragm, thereby reproducing the original sound.

Telephone technology stimulated an understanding of the concepts of signal and network analysis. The analysis of electric circuits originated with the work of Georg Ohm, who specified the relationship between current and voltage associated with a simple impedance element such as a resistor. This work was extended by Gustav Kirchhoff's famous laws relating to voltage and current distribution in more complicated electrical networks.

The invention of the vacuum triode by Lee De Forest in 1907 signaled the beginning of the modern electronics industry. Vacuum-tube amplifiers permitted the amplification of weak signals so that underwater acoustic systems no longer had to rely on the remarkable sensitivity of the unaided human ear.

1.2 WORLD WAR I ACOUSTIC TECHNOLOGY[2]

The initial application of acoustic technology in World War I was not under the water against the submarine threat, but in the air against the threat of enemy bombers and zeppelins. Much progress was made by the British and French in the early war years in the detection and location of enemy aircraft using their radiated acoustic signals. These signals were detected either directly, using

[2] The history of underwater acoustic technology development from World War I to 1950 is covered in an article by Marvin Lasky in the *Journal of the Acoustical Society of America* [7]. This article also contains a useful bibliography for those with more than a passing interest in the history of the subject.

horns and air tubes coupled to the operator's ears, or electrically, using carbon button microphones and telephone receiver systems. This technology is of importance to our present subject because it was applied directly to the underwater environment when the submarine threatened to starve the Allies into submission.

Lord Rayleigh, through his earlier work and as a consultant, was able to contribute significantly to the British effort in acoustics. In the nineteenth century it was generally thought that the binaural capability by which we determine sound direction was the result of the difference in sound intensity received at each ear when the sound does not arrive from directly ahead or behind us. This intensity difference was supposedly caused by the acoustic shielding effect of the skull. Rayleigh [8] reasoned that this would not explain our ability to perceive the direction of low-frequency sound with a wavelength large compared to our skull dimensions. For instance, at a frequency of 100 Hz, the acoustic wavelength in air is approximately 10 ft, so that the skull cannot provide appreciable acoustic shielding. Rayleigh concluded that, in addition to intensity differences, the ear–brain combination is sensitive to the phase, or time, difference between signals received at our two ears. If direction can be estimated by using the phase difference caused by the separation of our ears, the sensitivity should obviously be improved by increasing the separation between the primary sensors. This observation gave rise to the development of binaural listening devices for determining bearing.

The typical aircraft listening device consisted of a pair of large acoustic horns, each connected to one of the operator's ears by means of a tube and stethoscope earpiece. By rotating the pair of horns until the sound seemed equal in both ears, the operator determined bearing.

It was noted early in the development that aircraft engines generated a number of predictable tonal components. The predominant tonals were in the range 80 to 130 Hz. Tuned and tunable filters were developed to cover this range and thereby enhance performance by rejecting interfering acoustic signals at other frequencies. For the nonelectrical listening systems, these filters consisted of Helmholtz resonators.

The acoustic horns for detecting aircraft were 4 ft long and separated by 7 ft. This separation represents one-half wavelength at about 80 Hz. Thus, for a target at right angles to the direction of maximum response, the signals in the two horns would exactly cancel at a frequency of 80 Hz. In addition to the two horns separated in the azimuth plane, two horns separated in elevation provided elevation direction.

To avoid the necessity of training the large four-horn structure mechanically, a device was developed to permit the operator to adjust the effective length of tubing to each ear. In this manner the effective acoustic path length was adjusted to make the sound appear equal in both ears. This acoustic path "compensator" was calibrated so that bearings could be read directly from the control.

In 1915, concern with the German submarine attack on shipping intensified efforts to develop means for detecting submerged submarines. Optical, thermal, magnetic, and electromagnetic means were considered as well as acoustical means. It was concluded that acoustic sensing devices offered the only practical solution.

The work on underwater acoustic systems was carried on in Great Britain by A. B. Wood and his associates at the Board of Inventions and Research in the British Admiralty. In the United States, Harvey Hayes pioneered in the field of passive sonar arrays at the New London Experimental Station (now the Naval Underwater Systems Center).

An early successful passive detection and localization system was the American SC device. Shown in schematic form in Figure 1-1, this device was a direct descendant of Leonardo's original listening tube. Instead of placing one tube in the water, two tubes were used with their sensitive bulbs separated by approximately 5 ft. The bulbs were then connected to the ears by air tubes that terminated in stethoscope-type earpieces. This device gave the operator a binaural capability that was especially effective in the frequency range centered around 500 Hz. The SC device was operated successfully from both surface ships and submarines.

A natural extension of the SC listening device was the MB Tube shown in Figure 1-2. This device had six rubber bulbs on each side of a rotatable tube. The six bulbs on each side were connected through a tubing manifold that was acoustically "impedance matched" and directed to one side of a stethoscope, as with the SC device. Direction finding was again accomplished by manual rotation of the structure to equalize the sound in both ears. The MB Tube gave improved sensitivity and better angular resolution than the SC device.

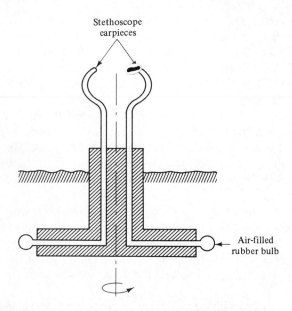

Figure 1-1   SC tube binaural airtube listening device.

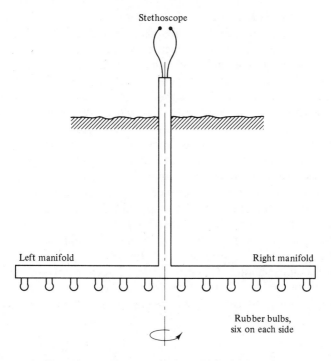

Figure 1-2   MB tube manually steered airtube array.

The rotatable binaural listening devices were deployed external to the ship's hull and generally could only be used at low speed. To avoid this problem, the MV Tube consisted of a flush-mounted array of bulbs on the hull. Steering was accomplished by variable acoustic length "compensator" lines in each of the bulb channels. Six elements on each side of the array were coupled together, after proper steering compensation, to apply to each ear. A training wheel on the compensator provided the operator with a scale to indicate target direction. The MV Tube came in lengths from 18 to 40 ft. A typical installation used two MV devices, one on each side of the ship's bow. The acoustic shielding provided by the ship's hull permitted resolution of the left–right ambiguity normally associated with a single-line array. The MV Tube was the most sophisticated non-electrical listening device used in World War I and reportedly permitted tracking submarines at ranges of 2000 yards while traveling at 20 knots.

The adverse effect on performance caused by the self-noise of a moving ship carrying an acoustic sensor was well known during World War I. This led to the development of a number of towed hydrophone devices in an attempt to move the sensor as far as possible from the self-noise field. The most successful of the towed systems was the U-3 Tube developed in 1918. Twelve equally spaced hydrophones (carbon button microphones) were housed in a flexible rubber tube 40 ft long. The ends were sealed and the tube filled with fresh water to achieve an overall neutrally bouyant system in salt water. This line array of hydrophones was called an "eel." The U-3 Tube system consisted of two eels towed approximately 300 to 500 ft behind the ship with about 12 ft of separation between the horizontal arrays. The electrical signals were brought onboard by means of a multiconductor cable and compensated for different signal arrival angles by means of an electrical compensator. The compensator introduced the proper electrical delay into each hydrophone channel before combining. As with the airtube arrays, the array was split into two halves, with six hydrophone signals combined from each half to apply to the left and right operator earphones. As before, the operator rotated the compensator to equalize the sound and determine target direction. The purpose of towing two eels with horizontal separation was to resolve the left–right ambiguity. One group of six hydrophones from each of the two eels was connected to a compensator, which was then adjusted to equalize the phase of the two signals. This gave a separate bearing measurement, relative to a different baseline direction, from which the ambiguity could be resolved. The geometry of this technique is illustrated in Figure 1-3. In this example the angle α, measured with the starboard eel, could be in either the first or second quadrant. Using the forward sections of both the port and starboard eels,the measured angle β could be in either the first or fourth quadrant. By using both α and β, the target direction is identified as being in the first quadrant.

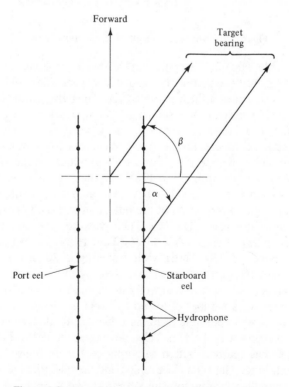

Figure 1-3 Use of two eels to resolve bearing ambiguity.

To improve the signal-to-noise ratio, the U-3 Tube systems were equipped with two electrical high-pass filters, one with a 450-Hz corner frequency and the other with a 900-Hz corner. The higher-frequency filter was presumably used at higher towing speeds to compensate for the increased low-frequency flow noise.

The U-3 Tube system was operated effectively in combination with the hull-mounted MV system. By obtaining target bearings on both the towed and hull-mounted systems, and knowing their separation, the approximate target range was obtained by triangulation. A simple nomograph chart was provided to the operator for reading range by entering bearings from the two systems.

Passive systems such as the SC, MV, and U-3 Tube systems had reasonable success in detecting submerged submarines. However, it was still difficult to localize with sufficient accuracy for weapon delivery. This led to interest in active echo ranging systems.

The noted French physicist Paul Langevin, after several years' work, demonstrated the detection of a submarine with an active system in 1917. He used radio transmitting equipment, operating at 38 kHz, to drive a piezoelectric transducer. The transducer was large enough to create a narrow beam of energy in the water so that both range and bearing to the target could be determined. As a result of Langevin's work, the British began the development of active echo ranging equipment. However, the war ended before any of these active devices became operational.

1.3 ACOUSTIC DEVELOPMENT BETWEEN THE WARS

At the end of World War I, the pace slowed considerably in the development of underwater acoustic systems. In the United States the Acoustics Division of the Naval Research Laboratory was established in 1923 with Harvey Hayes as its first chief. By 1927, Hayes and four others constituted the entire staff of the department.

Toward the end of the war, both sides were expending considerable effort to reduce the intensity of acoustic signals radiated by submarines. Success in this effort, together with the positive results of Langevin's experiments, led both the United States and Great Britain to devote most of their postwar effort to the development of active sonar. In addition to being independent of the radiated noise of the target, active sonar is capable of rapid target localization.

For active sonar systems in this period, the frequency region of primary interest was 10 to 30 kHz. This frequency range is above most of the acoustic self-noise of a ship, and permits the generation of narrow beams of acoustic energy with moderate-size sensors. The British developed transducers using piezoelectric devices similar to that used by Langevin, while the United States used magnetostrictive devices to generate the ultrasonic pulses. In other respects the U.S. and British systems were very similar in principle to Langevin's original echo ranging system. Of course, the rapid development of vacuum-tube technology, spurred on by the emerging radio industry, resulted in improved reliability and performance of the electronic portions of the systems.

The Treaty of Versailles restricted Germany to a small naval force without submarines or airplanes. Under these conditions they viewed the role of the sonar as primarily defensive rather than offensive. As a result, their postwar concentration was on the development of passive listening devices rather than active systems. In keeping with this defensive posture, they first studied carefully the noise sources on-board ship, including propeller, machinery noise, and bow wave noise, and used this information in the design of new ships. The location and design of the passive sonar sensor elements were carefully coordinated with the ship design to obtain the best possible performance. The electrically steered conformal array technology developed by Harvey Hayes was then applied, and improved, to produce an excellent conformal passive array for their capital ships. This system, called the GHG, was installed on the German cruiser *Prinz Eugen* and, according to German accounts, was instrumental in the survival of the ship during extensive torpedo attacks.

The GHG conformal array was elliptical in shape and located in the bow region. On the *Prinz Eugen* it had as many as 60 hydrophones on each side. Beamforming was accomplished by using a broadband tapped electrical delay line connected to a mechanical device that was essentially a small geometric replica of the hydrophone array. A set of rotatable brushes picked up the connections to the delay line and inserted the proper compensating delay into each hydrophone channel to form the receiving beam in the desired direction. The design of this system was studied carefully by sonar experts in the United States following World War II and had considerable influence on the subsequent design of passive conformal systems.

An improved understanding of the physical processes involved in sound propagation in the sea began to emerge in the 1930s. The absorption of sound by seawater was measured at ultrasonic frequencies at the Naval Research Laboratory, introducing a field of investigation that is still active in quantifying and explaining the absorption phenomena over the entire frequency range of interest.

The seemingly unpredictable behavior of the ocean as a sound transmission medium began to be understood with the invention of the bathythermograph by Athelstan Spilhaus in 1937. This device permitted rapid and convenient measurement of the water temperature versus depth, from which the sound speed–depth profile could be calculated. From this it was learned that even small temperature gradients result in sound refraction that dramatically affects the propagation characteristics.

By the beginning of World War II, U.S. ships were equipped with bathythermographs (or BTs) so that the sound propagation characteristics could be assessed. This same capability was given to submarines, significantly adding to the information available for making tactical decisions.

Improved understanding of the medium was a necessary condition for the proper development of a systems approach to the design and analysis of underwater acoustic systems. During the interwar period the groundwork was being laid for the understanding of informational processing systems by such men as

Nyquist and Hartley at Bell Telephone Laboratories. Estimation theory and methods of hypothesis testing were beginning to be applied to the problems of identifying useful signals embedded in unwanted noise in communications systems. These beginnings of complete system understanding were among the more important advances made in the interval between the wars.

1.4 WORLD WAR II AND BEYOND

After the entry of the United States into World War II, the National Defense Research Committee (NDRC) was formed under the leadership of Vannevar Bush. This organization sponsored areas of research and development that were deemed vital to the successful prosecution of the war. Among these efforts were the development of radar at the MIT Radiation Laboratory, the atom bomb Manhattan Project, and an extensive program in underwater acoustics organized by Division 6 of NDRC. The underwater studies were carried out by the University of California Division of War Research, the Naval Electronics Laboratory at San Diego, and the Woods Hole Oceanographic Institution. Valuable contributions were also made by laboratories associated with Columbia University, Harvard, the Massachusetts Institute of Technology, and other naval laboratories and industrial organizations.

The purpose of the work of these groups was to improve the design and use of underwater acoustic systems. Because of this short-range practical objective, the experimental programs were not always designed to answer fundamental questions in the scientific sense. Thus almost all the experiments were conducted in the ultrasonic region then of interest for active sonar systems. However, the quality of the personnel involved in these studies was such that a great deal of progress was made in the fundamental understanding of underwater sound transmission. Fortunately, at the end of the war this knowledge was gathered together and published as a series of summary reports, collectively titled *Physics of Sound in the Sea* [9]. These reports formulated the theoretical bases for describing sound propagation in the sea, reverberation, and the reflection of sound from submarines, ships, and wakes in much the same form that it exists today. The instruments and techniques of measurement developed during the war set the stage for ocean measurement programs in subsequent years that would fill the gaps created by the limited objectives of the wartime program.

By the end of the war, the factors affecting sound speed in the ocean were identified and understood. Experimental measurements of sound speed were available from all over the world, covering a wide variety of seasons, weather, and water depths. Using primarily the methods of ray acoustics, the effect of the sound speed profile on propagation paths was investigated and the different path types identified and given names that are still in use. Woolard, Ewing, and Worzel investigated long-range low-frequency sound transmission in deep water using explosive charges, verifying the existence of the deep sound channel that permits the reception of sound at ranges of 1000 miles or more.

The radar technology program at the Radiation Laboratory of MIT developed the component and circuit technology and the signal processing techniques required for the generation, transmission, reception, and detection of pulsed microwave signals. The problem of processing weak radar signals in the presence of noise is essentially the same as the sonar signal processing problem. The work accomplished in this program was also well documented after the war in a remarkable set of 28 volumes known as the Massachusetts Institute of Technology *Radiation Laboratory Series,* sponsored by NDRC. Of particular interest to the general subject of processing signals in noise is Volume 24 of this series, titled *Threshold Signals,* by Lawson and Uhlenbeck. This book was one of the first to present a unified treatment of the statistical and spectral properties of signals and noise. By a combination of experimental procedures and analysis, the relationship between the output signal-to-noise ratio and the receiver filter characteristic was investigated, and an approximation to an optimum filter was identified. The detection process was described as a hypothesis test based on the statistical properties of the output waveform with and without the target signal present. Although the particular criterion used (called the ideal observer criterion) was later changed, this was an important step in the development of a statistical model characterizing the performance of a signal processing system in combination with an output decision-making device (often a human observer).

Independent of the work at the Radiation Laboratory, important contributions to understanding the processing of signals and noise were made by Norbert Wiener and S. O. Rice. In 1942, Wiener produced his classic work "The Extrapolation, Interpolation and Smoothing of Stationary Time Series," originally issued as an NDRC report [10]. He rigorously derived by estimation theory the form of a filter that provides the *best* separation of a desired signal from undesired noise. Rice published his articles "Mathematical Analysis of Random Noise" in the *Bell System Technical Journal* in 1944–1945 [11]. This work dealt extensively with the statistical properties of noise and the effect on these statistical properties of passing noise through various circuit operations.

From approximately 1945 through 1955 the work of Shannon [12] in the United States, and Gabor [13] and Woodward [14] in England, established the subject of information theory as a mature discipline. Gabor and Woodward in particular provided new insight into the properties of signal waveform design that affect system capability in the areas of target resolution and estimation of target range and velocity.

At the conclusion of World War II, the pace of technical activity did not slow to the extent that it had after World War I. The pressure of the cold war, plus the stimulation of the remarkable advances in science and technology during the war, resulted in an explosion of activity on a broad front. The disbanding of key organizations such as the Radiation Laboratory at MIT, the University of California Division of War Research, and others actually spread this capability throughout the country, resulting eventually in a dramatically larger technological base. The 15 years immediately following the war saw the beginning and rapid development of solid-state technology, digital computer systems, rocketry, precision long-range guidance systems, and the application of nuclear technology to power-generation devices suitable for primary ship propulsion systems.

In the field of underwater acoustics, the experimental and theoretical work begun during the war was continued and expanded to cover the entire sonic as well as ultrasonic regions. Knudsen [15], Wenz [16], Marsh [17], Urick [18], and others identified the various sources and characteristics of ambient noise in the ocean. Additional understanding of the causes of absorption of sound in the sea was provided by Lieberman [19] and Leonard et al. [20], and experimentally determined absorption coefficients were available from below 100 Hz to above 1 MHz [21–23]. Massive amounts of acoustic propagation data were obtained at different sites around the world. The increasing availability of general-purpose computers made it possible to analyze these data and to develop a statistical picture of the relative frequency of occurrence of various propagation conditions. The computer also made possible the solution of equations describing acoustic propagation on a routine basis so that theory and experiment could be easily compared. This led to more accurate and complete mathematical models to be used in system synthesis and performance calculations.

An important postwar development was the introduction of digital techniques to the field of signal processing. The first application in underwater acoustics was the digital multibeam steering system (DIMUS) [24]. This system used hard limiting on the hydrophone signals, followed by digital shift registers, to provide the necessary delays for beamforming. The introduction of fast transform techniques [25], together with the rapidly decreasing cost and increased capability of solid-state digital components, resulted in the design of all-digital signal processing systems by the late 1960s and early 1970s.

The existence of strategic nuclear weapons in the years following World War II gradually changed the role and design of underwater acoustic systems. The typical World War II sonar was required to localize a submarine in close proximity to a convoy to aid in convoy defense. Short-range active systems served this purpose well. With the development of nuclear submarines with a capability to launch long-range ballistic missiles, the situation changed considerably. The area to be defended now included whole continents, with potential missile launch areas covering whole oceans. It became mandatory to be able to detect and classify underwater targets at long range. This led to renewed interest in passive systems that do not suffer from the round-trip transmission loss of echo ranging systems.

The increased range requirement resulted in the use of lower frequencies to avoid the high absorption losses at ultrasonic frequencies. Lower frequencies, in turn, required larger acoustic apertures to maintain the ability to determine target direction. In general, shipborne installations tend to use the largest available on-board aperture or use a long towed-line array to obtain the required range performance.

Basic experimental programs were conducted to measure low-frequency propagation over long ranges and to measure the correlation properties of signals and noise at the lower frequencies. Detection and processing of narrowband tonal components of the target signals were spurred on by the advances in high-speed digital technology. All of these activities were of course countered by the efforts of ship designers to reduce the radiated acoustic energy from submerged submarines. Considering the total energy generated by a nuclear submarine, a remarkably small amount is radiated as acoustic energy.

In terms of the classic problem of detecting a single-target signal in a background of purely random, well-behaved noise, underwater acoustic systems were capable of very near ideal performance by the 1970s. Performance under these conditions is limited only by the available spatial aperture, time aperture, frequency aperture, and the characteristics of the medium. However, as the sensitivity to detect weak signals improved, the simple model of the competing noise field became less appropriate. The ocean is full of target-like noise sources, both human-made and biologic, and increasing sensitivity increases the number of these sources with which the desired target signal must compete. The problem then becomes one of classifying and separating the signals received from many targets. The volume of data generated under these conditions by a modern multibeam sonar system is very large. The key technical issue suddenly becomes the methods by which this information will be processed to eliminate the extraneous and present only a manageable amount to the human operator.

The attempt to solve these problems of extracting useful information from large volumes of data has led to numerous attempts to detect and classify targets automatically and to control system parameters adaptively to improve performance in particular environments. These developments make use of the ever-increasing capability of solid-state digital processing hardware. The story is not yet complete in this area, and it will undoubtedly continue to constitute an important subject for investigation in the development of underwater acoustic systems.

1.5 CONCLUSION

The modern worker in the field of underwater acoustics must be impressed with the accomplishments of the past. Harvey Hayes would feel quite at home in a modern-day discussion on the design of acoustic arrays, and quite probably would be able to provide some valuable insights. From Leonardo da Vinci's original declaration that you can hear the sounds of ships in the water, we are still working on answers to the questions: How many, what direction, how far, and of what type?

On the other hand, a great deal of information and understanding has been added to the level of knowledge that existed a hundred years ago. We now understand *why* and *under what circumstances* the sounds of ships are heard at a distance. We can characterize the medium and our systems so that performance can be predicted, at least in simple circumstances. Finally, the remaining problems that limit performance are understood and possible approaches to obtain further improvements are being investigated.

SUGGESTED READING

1. MacCurdy, E., *The Notebooks of Leonardo da Vinci*. Garden City, N.Y.: Garden City Publishing Co., Inc., 1942, Chap. X.

2. "Sound," *The New Encyclopaedia Britannica*, Vol. 17. Chicago: Encyclopaedia Britannica, Inc., 1974, p. 19.

3. "Sound," *Funk and Wagnalls New Encyclopedia*, Vol. 22. New York: Funk & Wagnalls, Inc.. 1975, p. 24.

4. Strutt, John W. (Lord Rayleigh), *Theory of Sound*, Vols. I and II, New York: Dover Publications, Inc., 1945.

5. Albers, V. M., *Underwater Acoustics Handbook—II*. University Park, Pa.: The Pennsylvania State University Press, 1965, Chap. 10.

6. Urick, R. J., *Principles of Underwater Sound for Engineers*. New York: McGraw-Hill Book Company, 1967, Chap. 1.

7. Lasky, M., "Review of Undersea Acoustics to 1950," *J. Acoust. Soc. Am.*, Vol. 61, No. 2, p. 283, Feb. 1977.

8. Rayleigh, Lord, "On the Perception of the Direction of Sound," *Proc. R. Soc. A*, Vol. 83, pp. 61–64 (1909); or *Scientific Papers by Lord Rayleigh*, Vol. V. New York: Dover Publications, Inc., 1964, p. 522.

9. *Physics of Sound in the Sea*, Parts I–IV; originally issued as Division 6, Vol. 8, NDRC Summary Technical Reports; reprinted in 1969 by the Government Printing Office, Washington, D.C.

10. Wiener, N., "The Extrapolation, Interpolation and Smoothing of Stationary Time Series," NDRC Progress Report No. 19 to the Services, MIT, Feb. 1, 1942.

11. Rice, S. O., "Mathematical Analysis of Random Noise," *Bell Syst. Tech. J.*, Vols. 23 and 24 (1944–1945).

12. Shannon, C. E., "The Mathematical Theory of Communications," *Bell Syst. Tech. J.*, Vol. 27, pp. 379–424 and 623–637 (July and Oct. 1948).

13. Gabor, D., "Theory of Communication," *J. Inst. Electr. Eng.*, Vol. 93(III), p. 429 (1946).

14. Woodward, P. M., *Probability and Information Theory with Applications to Radar*. Elmsford, N.Y.: Pergamon Press, Inc., 1955.

15. Knudsen, V. O., R. S. Alford, and J. W. Emling, "Underwater Ambient Noise," *J. Mar. Res.*, Vol. 7, p. 410 (1948).

16. Wenz, G. M., "Acoustic Ambient Noise in the Ocean: Spectra and Sources," *J. Acoust. Soc. Am.*, Vol. 34, p. 1936 (1962).

17. Marsh, H. W., "Origin of the Knudsen Spectra," *J. Acoust. Soc. Am.*, Vol. 35, p. 409 (1963).

18. Urick, R. J., "Some Directional Properties of Deep Water Ambient Noise," Naval Research Laboratory Report 3796, 1951.

19. Lieberman, L. N., "Origin of Sound Absorption in Water and Sea Water," *J. Acoust. Soc. Am.,* Vol. 20, p. 868 (1948).

20. Leonard, R. W., P. C. Combs, and L. R. Skidmore, "Attenuation of Sound in Synthetic Sea Water," *J. Acoust. Soc. Am.,* Vol. 21, p. 63 (1949).

21. Urick, R. J., "Low Frequency Sound Attenuation in the Deep Ocean," *J. Acoust. Soc. Am.,* Vol. 35, p. 1413 (1963).

22. Thorp, W. H., "Deep Ocean Sound Attenuation in the Sub- and Low-Kilocycle-per-Second Region," *J. Acoust. Soc. Am.,* Vol. 38, p. 648 (1965).

23. Sheehy, M. J., and R. Haley, "Measurement of Attenuation of the Low-Frequency Underwater Sound," *J. Acoust. Soc. Am.,* Vol. 29, p. 464 (1957).

24. Anderson, V. C., "Digital Array Phasing," *J. Acoust. Soc. Am.,* Vol. 32, p. 867 (1960).

25. Cooley, J. W., and J. W. Tukey, "An Algorithm for the Machine Computation of Complex Fourier Series," *Math. Comp.,* Vol. 19, pp. 297–301 (Apr. 1965).

26. Hunt, F. V., *Electroacoustics.* New York: John Wiley & Sons, Inc., 1954, Chap. 1.

2

Acoustic Waves in a Homogeneous Medium

~~~~~~

2.0 INTRODUCTION

In this chapter we consider the nature of acoustic waves in a highly idealized medium. In the ocean, sound transmission is affected by such factors as temperature, pressure, chemical composition, and details of the surface and bottom boundaries. In the idealized homogeneous medium, we assume that all such properties of the medium are constant throughout and that the boundaries are sufficiently remote that they may be ignored. Finally, all mechanical loss mechanisms such as viscous effects are assumed to be zero. These restrictions are eventually removed or modified, but in the beginning they permit important simplifications in the derivation of acoustic wave equations.

The propagation of sound in water is a mechanical phenomenon and depends on the mechanical properties of the medium. In particular, the inertial and elastic properties of an elemental volume are of interest. A net force across a volume element results in an acceleration opposed by inertial properties, and mechanical strain is created in the element related to applied force and the elastic properties of the medium. The total energy involved in these mechanical effects includes the kinetic energy of motion and the stored potential energy represented by internal strain.

A localized variable source of mechanical force imposes unbalanced forces on neighboring volume elements. The propagation of the resulting motion-strain effects away from the source results in a *longitudinal compression wave* that transmits mechanical, or acoustic, energy away from the source.

For an oscillating source, the wave consists of regions of compression, where the pressure exceeds the original equilibrium value, and regions of rar-

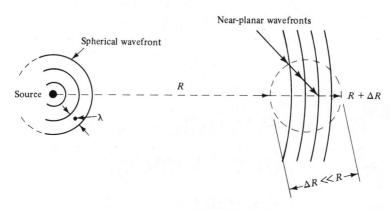

Figure 2-1   Propagation from a compact source in a homogeneous medium.

efaction with pressure less than the original value. These regions move, or propagate, away from the source at a constant rate determined by the properties of the medium. In Figure 2-1 the pressure peaks are shown at some instant of time following the start of the source oscillations. For a homogeneous medium these peaks will appear as concentric spherical shells surrounding the source. For a single-frequency sinusoidal source motion, the spheres are equally spaced with a separation equal to the *acoustic wavelength,* λ, in water.

Waves propagating with spherical symmetry, as shown in Figure 2-1, are called *spherical waves* and the equations describing their characteristics in time and space are *spherical wave equations*.

Now consider a small region in space, far removed from the source, with dimensions small compared with the range to the source. As shown in Figure 2-1, the curvature of the wavefronts is slight within this region. At sufficiently long ranges, the spherical wavefront may be approximated as a plane surface provided that the region of interest is appropriately limited. The spherical wave equations then simplify to *plane wave equations* describing the transmission of *acoustic plane waves*.

Obviously, the plane wave approximation is not suitable for all problems involving acoustic propagation. However, in many practical situations the approximation is valid, and the plane wave equations lead to simple relationships among the various acoustic parameters.

Normal procedure at this point is to derive the generalized wave equation. The plane wave equation is then obtained as a special case of the general wave equation. This approach is well documented in numerous excellent texts [1–3]. In this chapter we are interested primarily in developing simple relationships among the physical properties of the medium and parameters describing the acoustic wave. This is best accomplished by direct consideration of the solution of the plane wave equation. This approach has the added advantage of having a simple and well-known electrical analog, permitting identification of analogous relationships among the various acoustic and electrical parameters.

Following consideration of plane wave equations, the general wave equation is obtained by logical extension. This chapter concludes with a discussion of units, reference standards, and the decibel notation commonly used in underwater acoustics.

2.1 ELECTRICAL ANALOG OF PLANE WAVE TRANSMISSION

Rather than derive the acoustic plane wave equation directly, it is instructive to consider an electrical analog. A lossless electrical transmission line with distributed series-inductive and parallel-capacitive elements is shown in Figure 2-2. The quantities L and C represent the incremental inductance and capacitance per unit length of line.

The sending end of the line is located at $x = 0$ and the receiving end is at $x = \ell$. The voltage and current at any point on the line are functions of both x and t. From elementary circuit theory we relate the partial derivative of $v(t, x)$ with respect to x to the partial of $i(t, x)$ with respect to t and the incremental inductance. Thus

$$\frac{\partial v(t, x)}{\partial x} = -L \frac{\partial i(t, x)}{\partial t} \tag{2-1}$$

Equation (2-1) describes the well-known fact that a voltage is developed across an inductive element that opposes, or impedes, a time rate of change of current in the element.

In a similar manner, the parallel capacitance opposes a change in voltage by absorbing current from the line. The change in line current as a function of distance along the line is

$$\frac{\partial i(t, x)}{\partial x} = -C \frac{\partial v(t, x)}{\partial t} \tag{2-2}$$

The simultaneous partial differential equations (2-1) and (2-2) can be solved for either the voltage or the current. For instance, take the partial of (2-1)

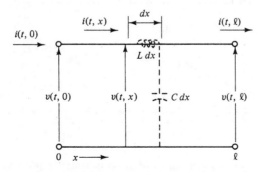

Figure 2-2   Lossless electrical transmission line with distributed constants.

with respect to x and (2-2) with respect to t to obtain

$$\frac{\partial^2 v}{\partial x^2} = -L\frac{\partial}{\partial x}\left(\frac{\partial i}{\partial t}\right)$$

$$\frac{\partial^2 v}{\partial t^2} = -\frac{1}{C}\frac{\partial}{\partial t}\left(\frac{\partial i}{\partial x}\right) = -\frac{1}{C}\frac{\partial}{\partial x}\left(\frac{\partial i}{\partial t}\right)$$

from which

$$\frac{\partial^2 v}{\partial t^2} = \frac{1}{LC}\frac{\partial^2 v}{\partial x^2} \qquad (2\text{-}3)$$

The arguments of v and i in these equations have been dropped for simplicity. Equation (2-3) is a well-known differential equation with solutions of the form

$$v(t, x) = v_1[t - \sqrt{LC}(x + k_1)] + v_2[t + \sqrt{LC}(x + k_2)] \qquad (2\text{-}4)$$

To examine the nature of v_1, we let k_1 equal zero and sketch the waveform as a function of t for fixed values of x, and then as a function of x for fixed values of t. Assume that v_1 is zero for nonpositive values of its argument and has the shape shown in Figure 2-3 (a) at $x = 0$. At some other location, $x = x_0$, v_1 has the same shape but is shifted in time by an amount $t_0 = x_0\sqrt{LC}$ [Figure 2-3(b)].

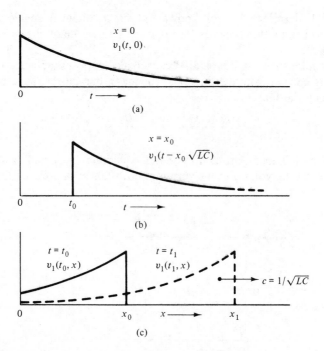

Figure 2-3   Voltage waveforms on a lossless transmission line as functions of time and distance.

Now consider the voltage as a function of *distance* along the transmission line at a fixed instant of *time* t_0. The value of $v_1(t_0, x)$ is shown in Figure 2-3(c). For $x\sqrt{LC} > t_0$, the voltage on the line is zero. At a later instant of time, t_1, all points on the waveform are shifted to the right by the amount $(t_1 - t_0)/\sqrt{LC}$. The waveform *travels* to the right, or *propagates* with a *speed of propagation* given by

$$c = \frac{1}{\sqrt{LC}} \tag{2-5}$$

The constant k_1 in (2-4) determines the origin for v_1. For $k_1 = 0$, v_1 can be thought of as a voltage originating at $x = 0$, traveling to the right with speed c. This is called the *forward traveling wave* component of the total voltage $v(t, x)$.

By similar reasoning we could show that $v_2(t, x)$ is a waveform traveling in the *backward* or negative x-direction. The transmission line in Figure 2-2 does not exist except for positive values of x. Therefore, v_2 cannot exist unless it originates to the right of the point $x = 0$. This is possible provided that k_2 is negative and nonzero. The relative amplitudes and polarities of v_1 and v_2 and the value of k_2 are determined by the transmission line boundary conditions.

The relationship between the voltage and current associated with either a forward or a backward traveling wave is a fundamental parameter of the transmission line called the *characteristic impedance*. From (2-1) and (2-2) we can write

$$\left(\frac{\partial v}{\partial i}\right)^2 = \frac{L}{C} \tag{2-6}$$

For the forward traveling wave originating at $x = 0$, this has the solution

$$v_1\left(t - \frac{x}{c}\right) = +\sqrt{\frac{L}{C}} i_1\left(t - \frac{x}{c}\right) \tag{2-7}$$

and for the backward traveling wave

$$v_2\left(t + \frac{x + k_2}{c}\right) = -\sqrt{\frac{L}{C}} i_2\left(t + \frac{x + k_2}{c}\right) \tag{2-8}$$

The characteristic impedance is now defined as

$$Z_0 = \sqrt{\frac{L}{C}} \tag{2-9}$$

For the lossless transmission line, Z_0 is a pure resistance.

Consider now a lossless transmission line extending to infinity in the positive x-direction, with a single voltage source at $x = 0$. A forward traveling wave originates at the source and travels in the positive x-direction. Because the line is lossless, the energy associated with the forward traveling wave does not diminish with time or distance. Since no other energy source exists in this example, the backward wave solution will have zero amplitude.

Because only the forward wave is present on the infinite line, the ratio of total voltage to total current at any point on the line equals the characteristic impedance. This is also true at $x = 0$, resulting in an input impedance equal to Z_0. This leads to the interesting conclusion that a finite length of transmission line, terminated in a load impedance equal to Z_0, behaves exactly as an infinite transmission line. A forward wave launched at the sending end will travel the length of the line and be absorbed in the characteristic impedance. Because the energy in the forward wave is completely dissipated in the load, no backward wave is created.

In contrast to the behavior of either the infinite line or the finite line terminated in Z_0, we now examine the behavior of a finite-length line terminated in a short circuit. The short circuit at some location $x = \ell$ is a boundary condition that requires the voltage at the boundary to be zero at all times. This condition is met by assuming a backward traveling wave exactly equal and opposite to the forward wave at the location of the short circuit. Referring to Figure 2-4, the backward wave is shown as originating in an imaginary source, or *image,* at the end of an imaginary line to the right of the short circuit. The length of the imaginary line is equal to the length of the real line, so that

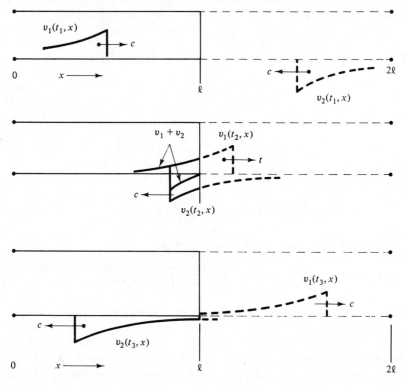

Figure 2-4 Voltage waveforms on a short-circuited electrical transmission line.

$k_2 = -2\ell$. Thus, for $0 \le x \le \ell$,

$$v(t, x) = v_1(t, x) + v_2(t, x)$$

$$= v_1\left(t - \frac{x}{c}\right) - v_1\left(t + \frac{x - 2\ell}{c}\right)$$

Notice that for $x = \ell$, $v(t, x)$ is zero for all values of t.

The forward traveling wave proceeds past the short circuit into the imaginary transmission line, while the backward wave, v_2, appears as a real waveform moving toward the original sending end of the line. The boundary has actually converted the forward wave into a backward wave of opposite polarity.

For obvious reasons, the backward traveling wave in the real transmission line is said to be caused by the *reflection* of the forward, or *incident*, wave at the short-circuit boundary. If the voltage source in this example has a source impedance equal to Z_0, there will be no reflection of v_2 at the source location. If the source impedance is not equal to Z_0, multiple reflections will occur, with the waveform traveling back and forth between the source and the short circuit. From the relationships in (2-7) and (2-8), it is easy to show that the current in the short circuit is exactly double the current in the forward or backward wave at that point.

By interchanging the roles of voltage and current, the effect of an open-circuit termination can be visualized. The backward voltage wave has the same polarity as the forward wave in this case, resulting in a doubling of the voltage at the open circuit. The current is of course zero into the open circuit.

To investigate the effect of other terminations, let the source signal be sinusoidal with a source impedance equal to Z_0, and the terminating load be Z_ℓ, possibly complex. It is convenient to represent a real sinusoidal signal as a complex exponential. The complex voltages at any point for the forward or backward waves are, for $t \ge 0$,

$$v_1(t, x) = a_1 \exp\left[j\omega\left(t - \frac{x}{c}\right)\right] \tag{2-10}$$

$$v_2(t, x) = a_2 \exp\left[j\omega\left(t + \frac{x + k_2}{c}\right)\right] \tag{2-11}$$

where a_1, a_2, and k_2 are real constants. Keep in mind that the actual waveforms are the real parts of (2-10) and (2-11).

The terminating impedance, Z_ℓ, is the ratio of the total voltage to the total current at $x = \ell$. Using (2-7) and (2-8), we write (for $x = \ell$)

$$Z_\ell = \left[\frac{v_1 + v_2}{i_1 + i_2}\right]$$

$$= Z_0\left[\frac{v_1 + v_2}{v_1 - v_2}\right] \tag{2-12}$$

Solving for v_2, we obtain

$$v_2 = v_1 \left[\frac{Z_\ell - Z_0}{Z_\ell + Z_0} \right] \qquad \text{(at } x = \ell) \tag{2-13}$$

or

$$v_2(t, \ell) = a_1 \left[\frac{Z_\ell - Z_0}{Z_\ell + Z_0} \right] \exp \left[j\omega \left(t - \frac{\ell}{c} \right) \right] \tag{2-14}$$

In the general case with Z_ℓ complex, the bracketed term in (2-13) is complex. For R real, define

$$\left[\frac{Z_\ell - Z_0}{Z_\ell + Z_0} \right] = R \exp (j\phi)$$

Substitution in (2-14) gives

$$
\begin{aligned}
v_2(t, \ell) &= Ra_1 \exp \left[j\omega \left(t - \frac{\ell}{c} + \frac{\phi}{\omega} \right) \right] \\
&= a_2 \exp \left[j\omega \left(t + \frac{\ell + k_2}{c} \right) \right]
\end{aligned}
\tag{2-15}
$$

From the requirement that a_1, a_2, and R are real, we get

$$a_2 = Ra_1 \tag{2-16}$$

$$k_2 = -2\ell + \frac{\phi c}{\omega}$$

The voltage at any other point on the line is obtained from (2-4), (2-15) and (2-16).

$$
\begin{aligned}
v(t, x) &= a_1 \exp \left[j\omega \left(t - \frac{x}{c} \right) \right] + Ra_1 \exp \left[j\omega \left(t + \frac{x - 2\ell + \phi c / \omega}{c} \right) \right] \\
&= a_1 \exp (j\omega t) \left\{ \exp \left(-j \frac{\omega x}{c} \right) + R \exp \left[j\omega \left(\frac{x - 2\ell}{c} + \frac{\phi}{\omega} \right) \right] \right\}
\end{aligned}
$$

With some rearranging, this can be put in the form

$$
\begin{aligned}
v(t, x) = a_1 \exp \left(-j \frac{\omega \ell}{c} \right) \Big\{ &\exp \left[+j\omega \left(\frac{\ell - x}{c} \right) \right] + \\
&R \exp \left[-j\omega \left(\frac{\ell - x}{c} \right) + j\phi \right] \Big\} \exp (j\omega t) \tag{2-17}
\end{aligned}
$$

The voltage on the line as given by (2-17) is a complex exponential time function with a complex amplitude that varies in magnitude and phase as a function of x. The voltage is separated into a product of a time function and a distance function. Thus

$$v(t, x) = V(x) \exp (j\omega t) \tag{2-18}$$

where $V(x)$ is the complex amplitude as a function of distance. Because the amplitude, $V(x)$, does not vary with time (assuming steady-state conditions), the waveform described by (2-18) is said to define a *stationary* or *standing wave* of voltage on the transmission line.

As an example of a standing wave, we return to the short circuit termination described previously, with $Z_\ell = 0$, $R = -1$ and $\phi = 0$. Equation (2-17) then reduces to

$$v(t, x) = a_1 \exp \left(-j\frac{\omega\ell}{c} \right) \left\{ \exp \left[+j\omega\left(\frac{\ell - x}{c}\right) \right] - \right.$$

$$\left. \exp \left[j\omega\left(\frac{\ell - x}{c}\right) \right] \right\} \exp (j\omega t)$$

From the basic Euler relationships relating complex exponentials and sinusoids, this may also be written as

$$v(t, x) = a_1 \exp \left(-j\frac{\omega\ell}{c} \right) \left[2j \sin \omega\left(\frac{\ell - x}{c}\right) \right] \exp (j\omega t) \tag{2-19}$$

Finally, the real part of (2-19) is

$$\text{Re}[v(t, x)] = -2a_1 \sin \left[\omega\left(\frac{\ell - x}{c}\right) \right] \sin \left[\omega\left(t - \frac{\ell}{c} \right) \right] \tag{2-20}$$

At $x = \ell$, the voltage is zero as required by the short-circuit condition. The magnitude of the voltage is plotted as a function of x in Figure 2-5. The

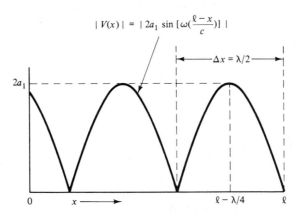

Figure 2-5 Voltage standing wave on a short-circuited lossless electrical transmission line.

maximum voltage magnitude is $2a_1$ at a distance from the short circuit determined from

$$\omega\left(\frac{\ell - x}{c}\right) = \frac{2\pi f}{c}(\ell - x) = \frac{\pi}{2} \tag{2-21}$$

Defining the electrical wavelength λ as the ratio c/f, we obtain

$$x = \ell - \frac{\lambda}{4}$$

as the location of the voltage maximum closest to the termination. This peak repeats at half-wavelength intervals. The voltage nulls are also separated by half-wavelength intervals.

A current standing wave exists on the short-circuited line also. Obviously, the current will be a maximum at the short circuit. This maximum repeats at half-wave intervals resulting in a standing wave that is spatially in quadrature with the voltage standing wave.

The voltage and current are also in time quadrature at all points on the line. The impedance at any point must, therefore, be purely reactive, or zero (at points of voltage null), or infinite (at points of current null). A little thought will show that the impedance behaves as an inductive reactance for distances less than $\lambda/4$ from the short circuit. For $\lambda/2 > \ell - x > \lambda/4$, the impedance appears capacitive. The reactance alternates between inductive and capacitive at quarter-wave intervals along the line. Notice that at the locations of the voltage maxima, the line appears to be terminated in an open circuit.

As another example, let Z_ℓ be purely resistive and equal to $Z_0/2$. For this case $R = -\frac{1}{3}$ and $\phi = 0$. Substituting these values in (2-17) and solving for the real part, we obtain

$$\text{Re}[v(t, x)] = \frac{2a_1}{3}(\cos^2 \alpha + 4 \sin^2 \alpha)^{1/2} \cos[\beta + \tan^{-1}(2 \tan \alpha)] \tag{2-22}$$

where

$$\alpha = \frac{2\pi}{\lambda}(\ell - x)$$

$$\beta = \omega\left(t - \frac{\ell}{c}\right)$$

Notice that in this example the phase, $\tan^{-1}(2 \tan \alpha)$, associated with the time-varying component is a function of distance. The magnitude of the voltage standing wave described by (2-22) is shown in Figure 2-6 together with the accompanying current standing wave. With a finite terminating resistance, neither the voltage nor the current standing wave magnitude is zero at any point on the line. With a purely resistive termination, the voltage minimum coincides with the current maximum, and vice versa, with the maxima and minima occurring at quarter-wave intervals measured from the termination.

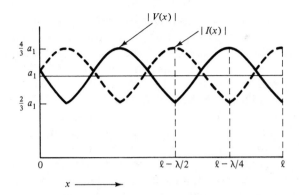

Figure 2-6 Standing waves on lossless electrical transmission line for $Z_\ell = Z_0/2$.

For the finite terminating impedance, the impedance measured at any point on the line may, in general, be complex, even for a purely resistive termination. The general expression for impedance at any location x for the lossless line is

$$Z(x) = Z_0 \left[\frac{Z_\ell \cos \alpha + jZ_0 \sin \alpha}{Z_0 \cos \alpha + jZ_\ell \sin \alpha} \right] \qquad (2\text{-}23)$$

where α is as defined with (2-22). Equation (2-23) of course reduces to Z_0, independent of x, if $Z_\ell = Z_0$.

With a constant average power source at the sending end, the power flow on a lossless transmission line is particularly simple. Because there are no losses in the line, the power delivered to the input equals the power delivered to the load. It follows that the power flow is the same at all points. This is true for any fixed termination.

For a line terminated in the characteristic impedance or for an infinite line, only the forward wave exists, and the power flow at any point is simply

$$P = \frac{\langle v_1^2 \rangle}{Z_0} = \langle i_1^2 \rangle Z_0 \qquad (2\text{-}24)$$

where $\langle \bullet \rangle$ indicates a time average.

If the line termination does not equal Z_0, the power flow in terms of the voltage or current at any location x is

$$P = \langle v_x^2 \rangle \operatorname{Re}[Y_x] = \langle i_x^2 \rangle \operatorname{Re}[Z_x] \qquad (2\text{-}25)$$

where $\operatorname{Re}[Y_x] = \operatorname{Re}[1/Z_x]$ = real part of admittance at x

$\operatorname{Re}[Z_x]$ = real part of impedance at x

This power is the same at all locations even though v_x, i_x, and Z are all functions of x.

It follows from (2-25) that the power delivered to a line terminated in either a short circuit or an open circuit is zero since the real part of Z_x or Y_x is zero at all locations.

To summarize, the properties of a lossless electrical transmission line are determined by its incremental inductance and capacitance per unit length, and by the line terminations. An input distrubance may result in waves traveling in opposite directions with speed of propagation $c = 1/\sqrt{LC}$. The impedance to a traveling wave in either direction is the characteristic impedance $Z_0 = \sqrt{L/C}$.

The relationship between the forward and backward waves is determined by the line boundary conditions. For an infinitely long line, or a line terminated in Z_0, only the forward wave will exist. For any other termination, waves in both directions are possible. The voltage and current at any point on the line are obtained as the sum of the forward and backward components. For a continuous sinusoidal source with internal impedance Z_0, the sum of the forward and backward waves results in standing waves of voltage and current on the line.

2.2 ACOUSTIC PLANE WAVES IN A HOMOGENEOUS MEDIUM

Now we return to the problem of sound propagation in water. Assume an infinite homogeneous medium in equilibrium, and a coordinate system as shown in Figure 2-7. A force is applied in the x-direction resulting in a uniform pressure $p(x_0, t)$ in the vertical infinite plane parallel to the y-z plane, at a distance x_0 from the origin.

As an aid to identifying terms and concepts, first consider a static case. Assume a rigid plane boundary to the right of the applied force, and assume that the force is not varying with time. The excess pressure (relative to the equilibrium condition) is everywhere the same, so that there is no net force across any small volume element of the water between the plane at x_0 and the rigid

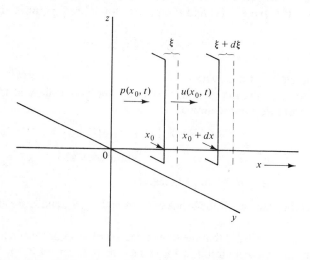

Figure 2-7   Infinite homogeneous medium with plane applied pressure.

boundary. Therefore, there is no particle motion and the particle speed $u(x, t)$ is everywhere zero. However, the compressional force, acting in the positive x-direction against the rigid boundary, will result in a particle displacement, in the x-direction, at the location x_0. This displacement decreases linearly, as x increases, to zero at the rigid boundary.

Consider the small volume element with original length dx and unit area in the plane of the applied force. The equilibrium volume, V, of the element is therefore equal to dx. The application of the compressional force changes the volume of the element to

$$V' = dx + d\xi = dx + \frac{\partial \xi}{\partial x} dx$$

The change in volume is

$$dV = V' - V = \frac{\partial \xi}{\partial x} dx$$

The *strain* produced in the volume element is defined as the ratio of the volume change to the original volume:

$$\text{strain} = \frac{dV}{V} = \frac{(\partial \xi / \partial x)\, dx}{dx} = \frac{\partial \xi}{\partial x} \tag{2-26}$$

Hooke's law states that the ratio of stress to strain in an elastic medium is a constant. In this case the stress is the static applied pressure and the constant is the *bulk modulus of elasticity, B*. For a positive pressure in the x-direction, the strain is negative. Therefore,

$$\frac{p(x)}{\partial \xi / \partial x} = -B \tag{2-27}$$

or

$$p(x) = -B \frac{\partial \xi}{\partial x} \tag{2-28}$$

We now return to the more general case with the applied pressure varying in time, rather than static. With a time-varying pressure the pressure magnitude will be a function of distance, x, as well as time, t. There will be in general a pressure differential across a volume element of length dx, defined by

$$dp = \frac{\partial p(x, t)}{\partial x} dx$$

This net pressure results in an acceleration of the element, or particle, defined by Newton's second law of motion. Thus

$$\frac{\partial p(x, t)}{\partial x} dx = -(\rho\, dx) \frac{\partial u(x, t)}{\partial t} \tag{2-29}$$

The negative sign is required because a net acceleration to the right requires a negative spatial pressure gradient. In this equation ρ is the density of the medium, and ρdx is, therefore, the mass of the volume element. The particle acceleration is given by the partial derivative of the particle speed $u(x, t)$ with respect to time. Since dx appears on both sides of (2-29), we can also write

$$\frac{\partial p}{\partial x} = -\rho \frac{\partial u}{\partial t} \qquad (2\text{-}30)$$

where the arguments of p and x are dropped for simplicity. This equation expresses the well-known fact that an *inertial reaction*, resulting from particle mass, resists a change in speed under the influence of an applied force. Compare (2-30) with (2-1) for the electrical transmission line. In the electrical system, an *inductive reaction* opposes a change in current under the influence of an applied voltage.

With a time-varying applied pressure, the strain produced in a volume element is also time varying. A second differential equation is obtained from (2-28) by taking the partial derivative with respect to time.

$$\frac{\partial p(x, t)}{\partial t} = -B \frac{\partial}{\partial t}\left(\frac{\partial \xi(x, t)}{\partial x}\right) = -B \frac{\partial}{\partial x}\left(\frac{\partial \xi(x, t)}{\partial t}\right) \qquad (2\text{-}31)$$

But the particle speed, $u(x, t)$, is defined as the time rate of change of particle displacement. That is,

$$u(x, t) = \frac{\partial \xi(x, t)}{\partial t} \qquad (2\text{-}32)$$

Substitution of this relationship in (2-31) and rearranging gives

$$\frac{\partial u}{\partial x} = -\frac{1}{B} \frac{\partial p}{\partial t} \qquad (2\text{-}33)$$

Equation (2-33) relates the time rate of change of pressure to the differential speed across an element by means of the elastic properties of the medium. This is similar to (2-2), which relates time rate of change of voltage to differential current.

Equations (2-30) and (2-33) may be solved simultaneously for either pressure or particle velocity in exactly the same manner as used for the electrical transmission line. Thus

$$\frac{\partial^2 p}{\partial t^2} = \frac{B}{\rho} \frac{\partial^2 p}{\partial x^2} \qquad (2\text{-}34)$$

This is one form of the differential acoustic plane wave equation. The acoustic or mechanical system described by (2-30), (2-33), and (2-34) is obviously the mechanical analog of the electrical lossless transmission line. The analogous relationships among the electrical and mechanical parameters are as follows:

Electrical Parameter		Mechanical Parameter	
Voltage	v	Pressure	p
Current	i	Particle speed	u
Inductance/unit length	L	Density	ρ
Capacitance/unit length	C	Inverse bulk modulus of elasticity	B^{-1}

Because of these relationships, we may transfer the concepts and solutions directly from the electrical system to the acoustical system. The solutions to (2-34) have the form

$$p(x, t) = p_1\left[t - (x + k_1)\sqrt{\frac{\rho}{B}}\right] + p_2\left[t + (x + k_2)\sqrt{\frac{\rho}{B}}\right] \quad (2\text{-}35)$$

The functions p_1 and p_2 represent forward and backward traveling waves, respectively, propagating with a speed in the water $c = \sqrt{B/\rho}$. Please note that the propagation speed of the pressure wave must not be confused with the *particle speed*, $u(t, x)$. In a longitudinal compression wave, such as an acoustic wave, there is no net translation of the particle or elemental volume elements, assuming a lossless linear medium. The average displacement of the particles is zero and the particle speed averaged over time is zero. The propagation speed c is the speed of a pressure maximum (or minimum) as the wave moves in the x-direction. As the wave passes a given point, the particles of the medium move first one direction and then the other, eventually returning to their equilibrium position.

By analogy with the electrical system, a characteristic acoustic impedance is defined for the medium as the ratio of the pressure to particle speed for either the forward or backward waves. This impedance involves the inertial and elastic properties of the medium, and is given by

$$Z_0 = \sqrt{\rho B} \quad (2\text{-}36)$$

For a lossless medium and the plane wave assumption, the characteristic impedance is purely real.

An alternative expression for Z_0 is obtained by recognizing that

$$B = \rho c^2$$

Therefore,

$$Z_0 = \sqrt{\rho(\rho c^2)} = \rho c \quad (2\text{-}37)$$

The characteristic impedance is identified in (2-37) as the product of density and propagation speed.

The relationship between pressure and particle speed for the forward wave is

$$p_1\left(t - \frac{x + k_1}{c}\right) = Z_0 u_1\left(t - \frac{x + k_1}{c}\right) \quad (2\text{-}38)$$

and for the backward wave

$$p_2\left(t + \frac{x + k_2}{c}\right) = -Z_0 u_2\left(t + \frac{x + k_2}{c}\right) \tag{2-39}$$

2.2.1 Plane Wave Acoustic Intensity

Recalling the analogies of pressure with voltage and particle velocity with current, (2-38) and (2-39) may be thought of as statements of an acoustic Ohm's law. Carrying the analogy one step further, we can define the rate of energy flow in the acoustic systems. At a particular location, the kinetic energy per unit volume is proportional to the square of the particle speed. This corresponds to the stored magnetic energy (proportional to current squared) in an electrical system. Potential energy is stored in a volume element because of the strain and is proportional to square of pressure. This corresponds to stored electric field energy (proportional to voltage squared) in the electrical system.

The average rate of energy flow past a given location is proportional to the mean value of the product of pressure and particle speed. The average rate of energy flow in a unit area is equal to the average power density. Thus

$$\frac{\text{acoustic power}}{\text{unit area}} = \frac{P}{A} = \langle up \rangle \tag{2-40}$$

where $\langle \bullet \rangle$ denotes a time average. For a forward plane wave, we use (2-38) to obtain

$$\frac{P}{A} = \langle u(Z_0 u) \rangle = Z_0 \langle u^2 \rangle \tag{2-41}$$

or

$$\frac{P}{A} = \left\langle \left(\frac{p}{Z_0}\right)p \right\rangle = \frac{\langle p^2 \rangle}{Z_0} \tag{2-42}$$

The power per unit area in an acoustic wave is called the *acoustic intensity, I*. Thus

$$I = \frac{\langle p^2 \rangle}{Z_0} = Z_0 \langle u^2 \rangle \tag{2-43}$$

In a lossless infinite medium only the forward wave will exist in the region to the right of the location of the pressure source in Figure 2-7. As with the electrical transmission line, this same conclusion is valid if the medium is terminated in a plane boundary with an acoustic impedance such that the ratio of pressure to particle speed at the boundary is equal to Z_0. In either case, the power flow at any point between the source and boundary is given by (2-43). Notice that for the plane wave in a lossless homogeneous medium, the power density, or intensity, does not diminish with range.

Cartesian coordinates may be expressed as

$$\text{gradient } p = \mathbf{i}\frac{\partial p}{\partial x} + \mathbf{j}\frac{\partial p}{\partial y} + \mathbf{k}\frac{\partial p}{\partial z} \tag{2-45}$$

The plane wave case, with the direction of motion aligned with the x-axis, is a special case of (2-45) with $\partial p/\partial y$ and $\partial p/\partial z$ both zero.

The plane wave differential equation, as defined in (2-34), involves the *change* in the *gradient* of pressure as a function of x. This is represented by the second partial derivative of pressure with respect to x. In the general case we must describe the change in the vector (gradient of pressure) with respect to a three-dimensional coordinate system. This change is called the divergence of the vector and in Cartesian coordinates is defined as

$$\text{divergence (gradient of } p) = \frac{\partial^2 p}{\partial x^2} + \frac{\partial^2 p}{\partial y^2} + \frac{\partial^2 p}{\partial z^2} \tag{2-46}$$

The operation defined by (2-46) is also called the *Laplacian* of the scalar pressure field $p(x, y, z)$.

The plane wave equation is a special case of the following statement:

$$\text{Laplacian } (p) = \frac{1}{c^2}\frac{\partial^2 p}{\partial t^2} \tag{2-47}$$

with $\partial^2 p/\partial y^2$ and $\partial^2 p/\partial z^2$ equal to zero. Equation (2-47) is a valid statement of the *general wave equation* and does not require the disappearance of the derivatives with respect to z and y.

Now consider a concentrated mechanical force at the origin of the coordinate system in Figure 2-8. The force is assumed to act uniformly in all radial directions. That is, at any distance r from the origin, the pressure is independent

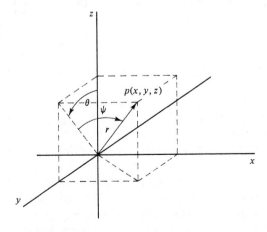

Figure 2-8   Coordinate system for spherically symmetric acoustic field.

of the polar angles ψ and θ. Thus, with a time-varying source at the origin, the pressure at any point in space is a function only of radial distance r from the origin, and time. Similarly, particle displacements and velocities are functions of r and t. With these qualifications the Laplacian of the pressure field can be shown to be

$$\text{Laplacian } (p) = \frac{\partial^2 p}{\partial r^2} + \frac{2}{r}\frac{\partial p}{\partial r} \tag{2-48}$$

The *spherical wave equation* now becomes

$$\frac{\partial^2 p}{\partial r^2} + \frac{2}{r}\frac{\partial p}{\partial r} = \frac{1}{c^2}\frac{\partial^2 p}{\partial t^2} \tag{2-49}$$

Equation (2-48) is simplified by recognizing that

$$\frac{1}{r}\frac{\partial^2 (rp)}{\partial r^2} = \frac{\partial^2 p}{\partial r^2} + \frac{2}{r}\frac{\partial p}{\partial r}$$

and

$$\frac{1}{r}\frac{\partial^2 (rp)}{\partial t^2} = \frac{\partial^2 p}{\partial t^2}$$

from which

$$\frac{\partial^2 (rp)}{\partial r^2} = \frac{1}{c^2}\frac{\partial^2 (rp)}{\partial t^2} \tag{2-50}$$

This equation is identical in form to plane wave equation (2-34) with p replaced by rp. The solution has the form

$$rp(t,r) = f_1\left(t - \frac{r + k_1}{c}\right) + f_2\left(t + \frac{r + k_2}{c}\right)$$

or

$$p(t,r) = \frac{1}{r}f_1\left(t - \frac{r + k_1}{c}\right) + \frac{1}{r}f_2\left(t + \frac{r + k_2}{c}\right) \tag{2-51}$$

The first term on the right represents a spherical wave diverging from the origin with velocity c. In contrast to the plane wave, the pressure amplitude decreases inversely with the distance from the origin.

The second term in (2-51) is a traveling wave that is converging toward the origin, with amplitude increasing toward infinity as r approaches zero. Although this portion of the solution is formally possible, it represents such a special circumstance that it will not be considered in this section on acoustic propagation in a homogeneous medium. The subject of converging acoustic waves arises more naturally in a discussion of acoustic lenses or propagation in a nonhomogeneous medium where refractive effects cause convergence.

Even with the diverging wave in (2-51), difficulty is experienced for r close to zero, because the pressure amplitude can be arbitrarily large. Actually, basic assumptions made in deriving the wave equations are violated as the pressure increases without bound. For this reason we shall normally confine our attention to situations such that r is finite and the acoustic displacements at r are small compared with r.

2.3.1 Pressure, Particle Velocity, and Displacement in Spherical Waves

Particle acceleration in spherical waves is related to pressure gradient by Newton's second law of motion in the same manner as for plane waves. Thus

$$\frac{\partial p}{\partial r} = -\rho \frac{\partial u}{\partial t} \tag{2-52}$$

where u is the particle velocity in the radial direction. This relationship is used to obtain particle velocity from pressure as follows:

$$du = \frac{\partial u}{\partial t} dt + \frac{\partial u}{\partial r} dr$$

or

$$\frac{du}{dt} = \frac{\partial u}{\partial t} + \frac{\partial u}{\partial r} \frac{dr}{dt}$$

Assuming small particle displacements and velocities, we may neglect the second-order product term. Therefore,

$$du \simeq \frac{\partial u}{\partial t} dt = -\frac{1}{\rho} \left(\frac{\partial p}{\partial r} \right) dt \tag{2-53}$$

and

$$u(t, r) = -\frac{1}{\rho} \int_{-\infty}^{t} \frac{\partial p(\tau, r)}{\partial r} d\tau \tag{2-54}$$

Particle displacement in the radial direction is obtained from particle velocity in much the same manner. By definition

$$u(t, r) = \frac{\partial \xi(t, r)}{\partial t} \simeq \frac{d \xi(t, r)}{dt}$$

from which

$$\xi(t, r) = \int_{-\infty}^{t} u(t, r) d\tau \tag{2-55}$$

Relationships (2-54) and (2-55) are easily demonstrated assuming a complex sinusoidal pressure function. Considering only the diverging pressure wave, let

$$p(t,r) = \frac{p_m}{r} \exp\left[j\omega\left(t - \frac{r}{c}\right)\right] \tag{2-56}$$

where p_m is the peak pressure amplitude at the range $r = 1$.

To determine the particle velocity, we require the partial derivative of (2-56) with respect to r. From elementary calculus we write

$$\frac{\partial p}{\partial r} = -\left(j\frac{\omega}{c} + \frac{1}{r}\right)p(t,r) \tag{2-57}$$

Substitution of (2-56) and (2-57) in (2-54) gives

$$u(t,r) = +\left(j\frac{\omega}{c} + \frac{1}{r}\right)\left(\frac{p_m}{\rho r}\right) \exp\left(-j\frac{\omega r}{c}\right) \int_{-\infty}^{t} \exp(j\omega\tau)d\tau$$

or

$$u(t,r) = +\left(\frac{1}{j\omega\rho}\right)\left(j\frac{\omega}{c} + \frac{1}{r}\right)p(t,r)$$

$$= \frac{1}{\rho c}\left(1 - j\frac{c}{\omega r}\right)p(t,r) \tag{2-58}$$

2.3.2 Specific Acoustic Impedance in Spherical Waves

Now let $\omega r/c = 2\pi r/\lambda$ and define *specific acoustic impedance* as the ratio of pressure and particle velocity.

$$Z = \frac{\rho c}{1 - j\lambda/2\pi r} = |Z| \exp(j\theta) \tag{2-59}$$

where

$$|Z| = \rho c[1 + (\lambda/2\pi r)^2]^{-1/2}$$

$$\theta = \tan^{-1}(\lambda/2\pi r)$$

Using (2-59), (2-58) is rewritten as

$$p(t,r) = Zu(t,r) \tag{2-60}$$

It is convenient to separate Z into its real and imaginary parts:

$$\text{Re}[Z] = \rho c\left[1 + \left(\frac{\lambda}{2\pi r}\right)^2\right]^{-1} \tag{2-61}$$

$$\text{Im}[Z] = \rho c\left(\frac{\lambda}{2\pi r}\right)\left[1 + \left(\frac{\lambda}{2\pi r}\right)^2\right]^{-1} \tag{2-62}$$

Now consider the two extreme conditions represented by

$$\text{Region 1:} \quad \frac{2\pi r}{\lambda} \gg 1$$

$$\text{Region 2:} \quad \frac{2\pi r}{\lambda} \ll 1$$

In region 1 the range is long compared with the wavelength (long range and/or short wavelength), and in region 2, the range is short compared with the wavelength.

In region 1, (2-61) and (2-62) predict that the real component of Z approaches ρc (the plane wave characteristic impedance), and the imaginary part of Z approaches zero. In region 1, the spherical wave takes on many of the characteristics of a plane wave, at least when considered over a limited range interval.

In region 2, the real part of Z approaches zero with the square of r (or inversely with the square of λ). The imaginary term approaches zero linearly with r and inversely with λ. Notice that the sign of the imaginary term is positive, indicating that the reaction is caused by inertial effects in the medium. The resistive and reactive parts of Z are shown in Figure 2-9 over the full range of the variable $2\pi r/\lambda$. The reactive term is a maximum at $2\pi r/\lambda = 1$. In the region $2\pi r/\lambda < 1$, the impedance is dominated by the reactive term. For large values of the abscissa the ratio of resistance to reactance increases as the reactance falls to zero.

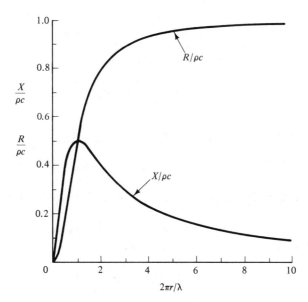

Figure 2-9   Real and imaginary parts of specific acoustic impedance for spherical wave.

Equations (2-55) and (2-60) may be used to obtain particle displacement with a complex sinusoidal pressure wave.

$$\xi(t, r) = \frac{1}{Z} \int_{-\infty}^{t} p(\tau, r) d\tau$$

$$= \frac{1}{j\omega Z} p(t, r) = \frac{u(t, r)}{j\omega} \tag{2-63}$$

2.3.3 Acoustic Intensity in Spherical Waves

The acoustic intensity of the spherical wave is the time average of the product of pressure and particle velocity, just as for the plane wave. Using the complex representation for pressure and velocity, we must be careful to form the product of the *real parts* of these quantities when calculating intensity for the real acoustic wave. Thus

$$I = \frac{1}{T} \int_T \text{Re}[p] \, \text{Re}[u] dt \tag{2-64}$$

But this time average is easily shown to be equal to

$$I = \frac{1}{2} \text{Re}[pu^*] \tag{2-65}$$

where $(\bullet)^*$ denotes the complex conjugate. Using (2-56), (2-59), and (2-60) we obtain

$$I = \frac{[1 + (\lambda/2\pi r)^2]^{1/2}}{2\rho c} \left(\frac{p_m^2}{r^2}\right) \cos \theta \tag{2-66}$$

But $\cos \theta = [1 + (\lambda/2\pi r)^2]^{-1/2}$. Therefore,

$$I = \frac{1}{r^2} \left(\frac{p_m^2}{2\rho c}\right) = \frac{1}{r^2} \left(\frac{p_{rms}^2}{\rho c}\right) \tag{2-67}$$

where p_{rms} is the root-mean-square amplitude of the real pressure waveform at $r = 1$.

Equation (2-67) is similar to the plane wave equation for intensity (2-42), except for the fact that the intensity of the spherical wave decreases inversely as the square of range. This type of intensity loss with range is called *spherical spreading loss*.

The total outward power flow at range r is obtained as the product of intensity and the area of a sphere with radius r.

$$P(\text{total}) = 4\pi r^2 I = 4\pi \left(\frac{p_m^2}{2\rho c}\right) \tag{2-68}$$

Since the medium is lossless, this total power flow is independent of range.

2.4 PULSATING SPHERICAL SOURCE

To demonstrate the use of relationships developed in the preceding section, consider a pulsating spherical source centered at the origin as shown in Figure 2-10. Although a spherical device is not a particularly common source of acoustic power, it results in simple relationships that provide insight into the performance of more complicated shapes.

2.4.1 Pressure, Velocity, and Displacement at Spherical Surface

Let the spherical radius vary in a sinusoidal manner from the equilibrium value, a. The acoustic pressure at a in complex notation is

$$p(t, a) = \frac{p_m}{a} \exp\left(j\omega t\right)$$

From (2-60) and (2-63), the particle radial speed and displacement at the surface of the sphere are

$$u(t, a) = \frac{p_m}{aZ(a)} \exp\left(j\omega t\right) = u_m(a) \exp\left(j\omega t - \theta\right)$$

$$\xi(t, a) = \frac{p_m}{j\omega a Z(a)} \exp\left(j\omega t\right) = \xi_m(a) \exp\left(j\omega t - \theta - \frac{\pi}{2}\right)$$

where $u_m(a)$ and $\xi_m(a)$ are the peak radial particle speed and displacement, respectively, at the surface of the sphere, and $Z(a)$ is the specific acoustic impedance. The total acoustic power radiated by this source is

$$P = 4\pi\left(\frac{p_m^2}{2\rho c}\right)$$

For a fixed source radius and power requirement, the peak particle speed and

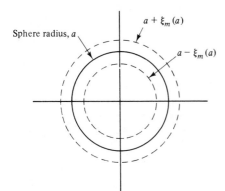

Sphere radius, a

$a + \xi_m(a)$

$a - \xi_m(a)$

Figure 2-10   Pulsating spherical acoustic source.

displacement vary as a function of wavelength. In particular,

$$\frac{u_m(a)}{p_m} = \frac{1}{a|Z(a)|} = \frac{[1 + (\lambda/2\pi a)^2]^{1/2}}{a\rho c} \tag{2-69}$$

$$\frac{\xi_m(a)}{p_m} = \frac{1}{\omega a|Z(a)|} = \frac{\lambda[1 + (\lambda/2\pi a)^2]^{1/2}}{2\pi a\rho c^2} \tag{2-70}$$

For $2\pi a/\lambda \gg 1$, the velocity amplitude at a is approximately the presure amplitude at a divided by the characteristic impedance, ρc. This relationship is
independent of frequency in region 1. The displacement amplitude at a is
approximately the pressure amplitude divided by $\omega\rho c$. Thus in region 1 the
displacement amplitude must increase as frequency decreases (increasing λ) to
maintain the same pressure and velocity amplitudes, and therefore the same
power flow. The relationships in region 1 may be summarized as

$$\left.\begin{array}{l} u_m(a) \simeq \left(\dfrac{p_m}{a}\right)\dfrac{1}{\rho c} \\[2ex] \xi_m(a) \simeq \left(\dfrac{p_m}{a}\right)\dfrac{1}{\omega\rho c} \end{array}\right\} \quad \begin{array}{l} \text{Region 1} \\[1ex] 2\pi a/\lambda \gg 1 \end{array} \tag{2-71}$$

Region 2, where $2\pi a/\lambda \ll 1$, is also of practical interest. For instance,
suppose that we wish to build a spherical acoustic source at a frequency of 50
Hz. Assuming a sound speed of 1500 m/sec, the wavelength at this frequency
is 30 m. Operation in region 1 implies that $a \gg \lambda/2\pi = 4.78$ m. Thus the
spherical diameter must be large compared with 9.56 m or 31.35 ft to behave
in accordance with (2-71). Because of the impracticality of spherical sources of
this size, single-element acoustic sources do not operate in region 1 in the low
acoustic frequency band.

If $2\pi a/\lambda \ll 1$, (2-69) and (2-70) simplify to

$$\left.\begin{array}{l} u_m(a) = \left(\dfrac{p_m}{a}\right)\dfrac{\lambda}{2\pi a\rho c} \\[2ex] \xi_m(a) = \left(\dfrac{p_m}{a}\right)\dfrac{\lambda}{2\pi a\omega\rho c} \end{array}\right\} \quad \begin{array}{l} \text{Region 2} \\[1ex] 2\pi a/\lambda \ll 1 \end{array} \tag{2-72}$$

2.4.2 Acoustic Intensity at Spherical Surface

In region 1, the pressure and particle velocity are in time phase, with the
result that the intensity is the simple product of rms pressure and speed. In region
2, the speed is no longer in phase with pressure, so that the intensity must be
defined as the product of rms pressure with the rms value of the velocity
component that is in phase with the pressure. Thus

$$I = \left(\frac{p_{\text{rms}}}{a}\right)u_{\text{rms}}(a)\cos\theta = \frac{p_{\text{rms}}^2}{a^2\rho c} \tag{2-73}$$

where θ is the phase angle associated with the specific acoustic impedance as defined before. The factor $\cos\theta$ may be thought of as a *power factor*, having the same meaning as the power factor associated with electrical networks containing both resistive and reactive components. To maintain a constant intensity, the particle speed must increase as the power factor decreases. However, the increasing component of speed is in time quadrature with the pressure waveform, and therefore does not contribute to useful power output of the device. The in-phase component of speed must be constant and independent of power factor to maintain a constant intensity.

The relationship between particle speed and pressure amplitude for a fixed ratio $p_m/a\rho c$ is shown in Figure 2-11 as a function of $2\pi a/\lambda$. Also shown in this figure is the power factor, $\cos\theta$, as a function of the same variable.

Regions 1 and 2 are identified, arbitrarily, as the regions where $2\pi a/\lambda > 10$ or < 0.1, respectively. In region 1 the power factor exceeds 0.99, and in region 2 the power factor is less than 0.1. Obviously, the intermediate region between regions 1 and 2 is of great practical interest when compromises must be made between size and efficiency.

2.4.3 Volume Flow Rate

The rapid decrease in power factor as $2\pi a/\lambda$ approaches zero places practical constraints on the size of devices designed to produce appreciable

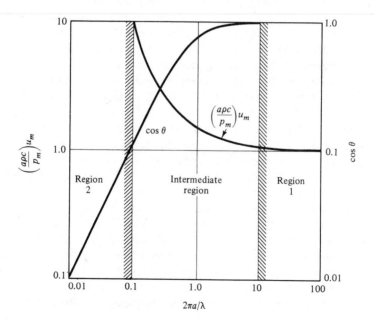

Figure 2-11   Relationship between pressure and velocity amplitude for a spherical source.

amounts of acoustic power. In this region a large volume of water must be made to flow back and forth as the sphere pulsates, and yet only a small fraction of this effort results in useful output power. The peak rate of water volume flow is obtained as the product of the peak particle speed (assumed equal over the surface of the sphere) and the area of the spherical surface. Thus the peak volume rate, Q, is

$$Q = 4\pi a^2 u_m(a) = \frac{4\pi a p_m}{\rho c}\left[1 + \left(\frac{\lambda}{2\pi a}\right)^2\right]^{1/2} \tag{2-74}$$

In region 2 this becomes approximately

$$Q \simeq \frac{2\lambda p_m}{\rho c} \tag{2-75}$$

and approaches infinity as frequency approaches zero.

It is tempting to try to establish design constraints relating source diameter, total output power, and operating frequency for a source such as the pulsating sphere. Because of the wide variety of applications and constraints, this can be done only very crudely. For instance, if only a very small acoustic power output is required, the efficiency of the source may not be a significant consideration and the designer may settle for a very small source operating in or near region 2.

For a very high power requirement, source efficiency is important, resulting in the selection of a larger source size operating near the boundary of region 1. For a source intended to radiate power equally in all directions, operation far into region 1 is usually avoided because of the difficulty in forcing all surface elements of the sphere to move in phase.

Operation very far into region 2 rapidly results in completely incompatible conditions. For instance, with $a = 0.01$ m, the required peak displacement at 100 Hz for a 1-W output is equal to the radius, and the required volume flow rate is more than 2000 times the equilibrium volume of the sphere. Long before these conditions are reached, all the basic assumptions used to derive the wave equations (such as small acoustic disturbances relative to equilibrium) are violated.

The peak displacements and velocities that can be sustained in a pulsating source are limited by such practical considerations as the strength of materials, thermal limitations of the driving mechanism, and electrical breakdown limitations. The medium itself also imposes a limitation on the intensity at the spherical surface. The peak pressure amplitude cannot exceed the ambient pressure of the medium without creating a void on the negative half-cycle, thereby preventing a further increase in pressure amplitude. For a source near the water surface, the maximum pressure amplitude is near the value of atmospheric pressure. This limitation on source power by the medium is discussed further in Chapter 3.

2.5 UNITS, REFERENCE STANDARDS, AND DECIBEL NOTATION

Over the years the literature dealing with underwater acoustic systems evolved a mixed system of units. Literature dealing with the physics of sound historically used the metric cgs system of units until 1960 when the mks, or meter–kilogram–second, system was officially adopted by the Eleventh General Conference on Weights and Measures.

The users of underwater acoustic systems measure range in such diverse units as yards, nautical miles, and meters. Prior to 1971, the most common combination of units included the cgs system for acoustic parameters such as pressure, with range measured in yards. Sound speed could be found in either English or metric units.

In 1971, the U.S. Navy officially adopted the mks system, as recommended by the 1960 conference. This consistent system is now in general use for both acoustic and system-level parameters. Because much important literature exists using the earlier systems of units, we shall describe both, together with methods of conversion from one to the other.

2.5.1 Mixed CGS System

We shall first consider the system of units used almost exclusively prior to 1971. The equation for acoustic intensity is

$$I = \frac{\langle p^2 \rangle}{\rho c}$$

where, in the cgs system, p = pressure = force/unit area, dynes/cm^2
ρ = water density = mass/unit volume, grams/cm^3
c = sound speed = distance/unit time, cm/sec
I = intensity = power/unit area, ergs/sec/cm^2

The unit of pressure is also called a microbar (or μbar) and is one millionth of a standard atmosphere.

A more convenient unit of intensity is the watt/cm^2, where one watt is equal to 1 joule/sec or 10^7 ergs/sec. Thus

$$I(\text{W}/\text{cm}^2) = 10^{-7}\frac{\langle p^2 \rangle}{\rho c} \qquad (2\text{-}76)$$

In the cgs system the density of sea water is approximately 1 g/cm^3 and the sound speed is approximately 1.5×10^5 cm/sec. The numerical value for characteristic impedance is, therefore,

$$Z_0 = \rho c = 1.5 \times 10^5 \text{ g/cm}^2\text{-sec} \qquad (2\text{-}77)$$

As a numerical example, let the rms acoustic pressure be 1 μbar. The resulting intensity of the acoustic wave is

$$I = \frac{(10^{-7})(1)^2}{1.5 \times 10^5} = 0.667 \times 10^{-12} \text{ W/cm}^2 \qquad (2\text{-}78)$$

Assuming a lossless homogeneous medium, the intensity at range r-cm from a compact source of total power P-watts is given by

$$I(\text{W/cm}^2) = \frac{P}{4\pi r^2}$$

Converting the units of range from centimeters to yards, we have

$$I(\text{W/cm}^2) = \frac{P}{4\pi(r_{yd} \times 91.4 \text{ cm/yd})^2} = \frac{0.952 \times 10^{-5}P}{r^2_{yd}} \qquad (2\text{-}79)$$

For a 1-W compact source, the intensity at 1-yd is 0.952×10^{-5} W/cm^2. At a range of 10,000 yd this same source would produce an intensity of 0.952×10^{-13} W/cm^2.

In acoustic system analysis, it is customary to use *decibel* notation to represent *level* of various quantities relative to a chosen reference value. This is a convenient way of handling the wide dynamic range involved in acoustic problems, and also simplifies many system calculations by replacing multiplications with additions of decibel quantities. In the cgs system of units the reference intensity level is the intensity of an acoustic wave with rms pressure equal to 1 μbar. From (2-78) we see that the reference intensity is

$$I_{ref} = 0.667 \times 10^{-12}\text{W/cm}^2 \qquad (2\text{-}80)$$

On the decibel scale, the reference intensity is at a level of 0 dB. The level associated with any other intensity, I, resulting from rms pressure, p, is:

$$\text{intensity level} = L = 10 \log_{10}\left(\frac{I}{I_{ref}}\right) = 10 \log_{10}\left(\frac{p}{p_{ref}}\right)^2$$

$$= 20 \log_{10}\left(\frac{p}{p_{ref}}\right) \qquad (2\text{-}81)$$

As a numerical example, we compute the intensity level corresponding to an intensity of 1 W/cm^2.

$$L = 10 \log\left(\frac{I}{I_{ref}}\right) = 10 \log\left(\frac{1}{0.667 \times 10^{-12}}\right) = 121.76 \text{ dB} \qquad (2\text{-}82)$$

The rms pressure corresponding to this intensity level is

$$p_{rms} = \text{antilog}\left(\frac{121.76}{20}\right) = 1.225 \times 10^6 \text{ } \mu\text{bar} \qquad (2\text{-}83)$$

Thus an intensity of 1 W/cm^2 requires an rms acoustic pressure exceeding 1 standard atmosphere. If this intensity is measured 1 yd from a compact source, the total power of the source is

$$P = 4\pi(1 \times 91.4 \text{ cm/yd})^2(1 \text{ W/cm}^2)$$

$$\approx 105,000 \text{ W}$$

In underwater acoustic systems it is customary to characterize the strength of a source in terms of the intensity produced at a fixed reference range from the source. In the mixed system of units, the reference range is 1 yd. When converted to a decibel level, this source intensity is called *source level*. Thus the source level of a compact source of total power P is defined as

$$\text{source level} = SL = 10 \log \left[\frac{I (1 \text{ yd})}{I_{\text{ref}}} \right]$$

$$= 10 \log \left[\frac{0.952 \times 10^{-5}P}{(1 \text{ yd})^2(0.667 \times 10^{-12})} \right]$$

$$= 71.54 \text{ dB} + 10 \log P \qquad\qquad (2\text{-}84)$$

A compact source of 1 W results in an intensity level at 1 yd that is 71.54 dB above that resulting from an rms pressure of 1 μbar. The source level at any other power is obtained by adding 71.54 dB to the decibel level of the power relative to 1 W. For instance, the source level of a 1000-W source is

$$SL (1000 \text{ W}) = 71.54 + 10 \log(1000) = 71.54 + 30 = 101.54 \text{ dB}$$

To identify the reference levels used, source levels such as this are written in the following alternate forms:

$$SL = 101.54 \text{ dB}//\mu\text{bar @ 1 yd}$$

or

$$SL = 101.54 \text{ dB re 1 } \mu\text{bar @ 1 yd}$$

and indicate that the intensity level, measured at 1 yd, is relative to that obtained if the rms pressure were 1 μbar. One should be careful not to confuse the units following the double slash marks with the units of the parameter whose level is being indicated. All levels in decibel notation actually represent dimensionless ratios.

The intensity level at any range r from a compact source in a homogeneous lossless medium is obtained from the source level as follows:

$$L(r) = 10 \log \left[\frac{I (1 \text{ yd})}{I_{\text{ref}}} \frac{(1 \text{ yd})^2}{(r_{\text{yd}}^2)} \right]$$

$$= SL - 20 \log \left[\frac{r_{\text{yd}}}{1 \text{ yd}} \right] \qquad\qquad (2\text{-}85)$$

The range-dependent term in (2-85) is called the *transmission loss*, *TL*, or *propagation loss*. In the case of spherical waves in a lossless medium, the transmission loss obeys the inverse-square law. In the general case, transmission loss may be a complicated function of range, frequency, and boundary conditions. The transmission loss for other than spherical spreading conditions may be provided in graphical or tabular form. In any case, the intensity at range r for a particular set of conditions may be expressed as

$$L(r) = \text{SL} - \text{TL} \tag{2-86}$$

Transmission loss in the mixed system is simply the source level defined at 1 yd minus the intensity level at range r in yards. For spherical spreading the transmission loss is 20 dB for every decade increase in range. The transmission loss at 10,000 yd is, therefore, 80 dB relative to the level at 1 yd.

2.5.2 MKS System

We shall now consider the system of units based on the meter–kilogram–second (mks) system. For acoustic pressure, this system uses 1 micronewton per square meter as the reference standard for decibel notation. This pressure unit is named the micropascal, abbreviated μ Pa. The reference for range is the meter.

The conversion from microbars to micropascals is easily obtained as follows:

$$1 \text{ newton} = 10^5 \text{ dynes}$$

$$1 \text{ square meter} = 10^4 \text{ square centimeters}$$

Therefore,

$$\frac{1 \text{ N}}{\text{m}^2} = \frac{10 \text{ dyn}}{\text{cm}^2} = 10\mu\text{bar}$$

$$\frac{1\mu\text{N}}{\text{m}^2} = 1 \ \mu\text{Pa} = 10^{-5} \ \mu\text{bar}$$

The micropascal is smaller than 1 μ bar by a factor of 10^5. This means that the reference intensity using the micropascal is smaller by 10^{10}, or 100 dB, compared with the intensity using the microbar as the reference. Hence, given the intensity level with the microbar reference, the level using the micropascal as the reference is

$$L//\mu\text{Pa} = L//\mu\text{bar} + 100 \text{ dB} \tag{2-87}$$

The reference intensity with 1 μ Pa rms pressure level is 10^{10} smaller than with 1 μ bar. Thus, with $p = 1 \ \mu$ Pa,

$$I_{\text{ref}} \doteq 0.667 \times 10^{-22} \text{ W}/\text{cm}^2$$

$$= 0.667 \times 10^{-18} \text{ W}/\text{m}^2$$

The source level of a compact source of P watts in the mks system is obtained from the mixed system source level by adding 100 dB (μbar to μPa) and subtracting 0.77 dB (yard to meter). Thus

$$\text{SL}//\mu\text{Pa} @ 1 \text{ m} = \text{SL}//\mu\text{bar} @ 1 \text{ yd} + 99.23 \text{ dB}$$
$$= 71.54 + 99.23 + 10 \log P$$
$$= 170.77 \text{ dB} + 10 \log P \qquad (2\text{-}88)$$

The source level of a 1-W source in the mks system is, therefore, 170.77 dB//μpa @ 1 m.

2.5.3 Intensity Levels for Broadband Signals

Many acoustic signals, whether originating from specific targets or as ambient noise, are more or less continuous but may not be considered single-frequency sinusoids. The frequency-domain representation of such signals generally occupies a broad band of frequencies rather than the very narrow band associated with a single-frequency signal. The most useful description of such signals is provided by an *intensity spectral density* plot showing the average, or expected, distribution of power density over the frequency band. The ordinate of such a plot is the average intensity that would be observed in a very narrow frequency band at each frequency along the abscissa. In underwater acoustics, the bandwidth used for this measurement is 1 Hz.

Figure 2-12 is a typical intensity spectral density plot of a broadband acoustic signal. The ordinate in this case is plotted as *intensity level* in dB *as measured in a 1-Hz bandwidth*. If this level defines a compact source, it is necessary also to specify the reference range. The level shown in Figure 2-12 is called the *spectrum level* and is identified by the notation:

$$\text{spectrum level, } L_s//\mu\text{Pa}/\text{Hz}$$

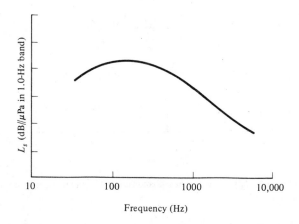

Figure 2-12   Spectral level of typical broadband acoustic signal.

or

$$\text{source spectrum level, } SL_s // \mu\text{Pa}/\text{Hz @ 1 m}$$

Sometimes the notation *dBs* is used to indicate *dB-spectrum,* to avoid the awkward identification of the reference quantities. This is permissible provided that no confusion results concerning the system of units and reference standards.

For the broadband signal the total intensity, or source level, over a frequency band is given by the integral of the intensity spectral density over that band. The integration must be performed using the actual intensity, rather than the spectral level in dB. Because the spectral density is usually available in graphical or tabular form, the integration is often performed by numerical methods. A particularly simple form of numerical integration is indicated in Figure 2-13. The band of interest is divided into narrow rectangles of equal width, Δf. The total intensity level for the signal contained in band β is given by

$$L(\text{in band } \beta) = 10 \log \left[\Delta f \sum_i a_i \right]$$

$$= 10 \log (\Delta f) + 10 \log \left(\sum_i a_i \right) \qquad (2\text{-}89)$$

where

$$a_1 = \text{antilog} \left(\frac{L_{s_i}}{10} \right) = \frac{I_i}{I_{\text{ref}}}$$

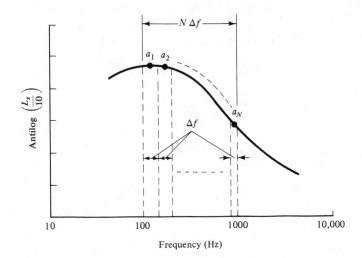

Figure 2-13   Intensity spectrum of broadband acoustic signal to demonstrate numerical integration.

A very simple result is obtained if the spectrum is constant at a level $L_s = 10 \log (a)$ over the band of interest. In that case

$$\sum_{i=1}^{N} a_i = Na$$

and

$$L(\text{in band } \beta) = 10 \log [\Delta f N a]$$
$$= 10 \log (N \Delta f) + 10 \log a \qquad (2\text{-}90)$$
$$= 10 \log (\beta) + L_s$$

2.5.4 Summary of Units and Standards

We conclude this section by listing for comparison the important features of the mixed cgs–English system and the mks system of units.

	Mixed cgs–English	mks
Reference pressure	1 μbar rms	1 μPa rms = 10^{-5} μbar
Reference intensity	0.667×10^{-12} W/cm²	0.667×10^{-18} W/m²
Reference range	1 yd	1 m
Reference bandwidth	1 Hz	1 Hz
$Z_0 = \rho c$	1.5×10^5 g/cm²-sec	1.5×10^6 kg/m²-sec
Source level	SL = 71.54 dB//μbar @ 1 yd	SL = 170.77 dB//μPa @ 1 m
per/watt		= SL//μbar @ 1 yd + 99.23 dB
Transmission loss	TL = $10 \log \left[\dfrac{I(1 \text{ yd})}{I(r \text{ yd})}\right]$	TL = $10 \log \left[\dfrac{I(1 \text{ m})}{I(r \text{ m})}\right]$
		= TL(yd) $-$ 0.77 dB

2.6 EXAMPLE: PULSATING SPHERICAL SOURCE

We now return to the pulsating spherical acoustic source to provide a quantitative example of the concepts introduced in preceding sections.

Assume the following parameters:

$$\text{sphere radius} \qquad a = 0.1 \text{ m}$$

$$\text{operating center frequency, } f_0 = 2388 \text{ Hz}$$

$$\lambda_0 = 0.628 \text{ m}$$

We desire to find the following:

1. Specific acoustic impedance, Z, at the spherical surface at the operating center frequency.

2. Total impedance presented by the water, Z_w, at the spherical surface. Calculate and plot the real and imaginary parts as a function of frequency.

3. The effective mass presented by the water is obtained from $X = \omega M_w$. Derive an expression for M_w and calculate its value at the center frequency.

4. Source level, SL, assuming a total radiated power of 1000 W, in both the mks and mixed system.

5. Calculate and plot the pressure amplitude, particle speed, and displacement amplitude as a function of distance, r, from the spherical surface assuming 1000 W radiated at the center operating frequency.

6. Calculate the peak amplitude of the sinusoidal total force necessary on the spherical surface to achieve a total power of 1000 W at the center frequency.

7. The spherical surface must be driven by a mechanical device of some sort to produce the acoustic output. Assume that we have such a device and it has a simple mechanical impedance that can be written as

$$Z_m = R_m + j\left(\omega M_m - \frac{1}{\omega S_m}\right)$$

where R_m = mechanical resistance to account for power losses caused by friction, etc.
$\quad\quad\quad = 4 \times 10^4 \text{ kg/sec}$
$\quad\quad M_m$ = mass of drive mechanism, including spherical shell
$\quad\quad\quad = 5 \text{ kg}$
$\quad\quad S_m$ = mechanical compliance of drive mechanism; inversely related to Young's modulus of elasticity of the material in the drive mechanism
$\quad\quad\quad = 4 \times 10^{-10} \text{ sec}^2/\text{kg}$

The primary driving force sees a total impedance that is the sum of the mechanical impedance of the drive mechanism and the impedance presented by the water. Thus

$$Z_T = Z_m + Z_w$$

Assuming a sinusoidal driving force, construct an equivalent circuit diagram of the system, including the mechanical drive and the water load. Calculate and plot the real and imaginary parts of the total impedance as a function of frequency.

8. Assuming a constant amplitude sinusoidal force at the input to the total system, calculate and plot the source level, SL, per unit force as a function of frequency.

9. Determine the rms force at the input required for a total output power of 1000 W at the center frequency.

10. Determine the system efficiency at the center frequency.

11. Determine the 3-dB bandwidth of the output power response.

Solution:

1. The specific acoustic impedance at the spherical surface at the center frequency is

$$Z = \rho c \left[1 - j\left(\frac{\lambda_0}{2\pi a}\right)\right]^{-1}$$

In this example, λ_0 and a are chosen so that $\lambda_0/2\pi a = 1.0$. Therefore,

$$Z = \rho c[1 - j1]^{-1} = (1.5 \times 10^6)(0.707\underline{/+45°})$$
$$= (1.06 \times 10^6)\underline{/+45°} \text{ kg/m}^2\text{-s}$$

2. The total impedance Z_w, presented by the water to the sphere is the ratio of the total force at the spherical surface to the resulting surface velocity. Thus

$$Z_w = \frac{f(t, a)}{u(t, a)}$$

For the simple spherical source, the force is equal to the product of pressure and surface area of the sphere. Hence

$$Z_w = (4\pi a^2)\frac{p(t, a)}{u(t, a)}$$

The ratio of pressure and velocity equals the specific acoustic impedance at a. Finally,

$$Z_w = (4\pi a^2)Z$$
$$= \left[\frac{(4\pi a^2)\rho c}{1 + (\lambda/2\pi a)^2}\right]\left[1 + \frac{j\lambda}{2\pi a}\right]$$

The real and imaginary parts are

$$R_w = 4\pi a^2 \rho c\left[1 + \left(\frac{\lambda}{2\pi a}\right)^2\right]^{-1} = (1.88 \times 10^5)(1 + 2.5\lambda^2)^{-1} \text{ kg/sec}$$

$$X_w = \left(\frac{\lambda}{2\pi a}\right)(4\pi a^2 \rho c)\left[1 + \left(\frac{\lambda}{2\pi a}\right)^2\right]^{-1}$$

$$= \left(\frac{\lambda}{0.628}\right)(1.88 \times 10^5)(1 + 2.5\lambda^2)^{-1} \text{ kg/sec}$$

and are shown graphically in Figure 2-14.

3. The effective mass of the water load is

$$M_w = \frac{X_w}{\omega} = \left(\frac{\lambda^2 \rho a}{\pi}\right)\left[1 + \left(\frac{\lambda}{2\pi a}\right)^2\right]^{-1}$$

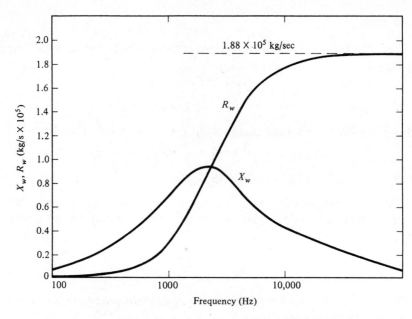

Figure 2-14 Real and imaginary parts of water load impedance for 0.1 meter sphere.

At the center operating frequency we have

$$M_w = \frac{\lambda^2 \rho a}{2\pi} = \frac{(0.394)(10^3)(10^{-1})}{6.28}$$

$$= 6.28 \text{ kg}$$

4. In the mks system the source level is

$$SL = 170.77 + 10 \log P_T$$

For $P_T = 1000$ W this gives

$$SL = 170.77 + 30 = 200.77 \text{ dB}//\mu\text{Pa @ 1 m}$$

In the mixed system we have

$$SL = 71.54 + 10 \log P_T = 101.54 \text{ dB}//\mu\text{bar @ 1 yd}$$

5. From the source level in (4), the rms pressure at 1 m is

$$p_{rms} = \text{antilog}\left(\frac{200.77}{20}\right) \approx 1.1 \times 10^{10} \ \mu\text{Pa}$$

$$= 1.1 \times 10^4 \text{ Pa or N}/\text{m}^2$$

The peak pressure at 1 m is, therefore,

$$p_m = \sqrt{2} \ p_{rms} = 1.55 \times 10^4 \text{ N}/\text{m}^2$$

The pressure as a function of time and range is

$$p(t,r) = \frac{p_m}{r} \exp(j\omega t)$$

$$= \frac{1.55 \times 10^4}{r} \exp(j\omega t) \qquad \text{N/m}^2$$

The particle speed and displacement are obtained directly from the expression for pressure.

$$u(t,r) = \frac{p(t,r)}{Z(r)} = \frac{p_m}{r\rho c}\left(1 - \frac{j\lambda}{2\pi r}\right) \exp(j\omega t)$$

At the center operating frequency this becomes

$$u(t,r) = \left(\frac{1.037 \times 10^{-2}}{r}\right)\left[1 + \left(\frac{0.1}{r}\right)^2\right]^{1/2}$$

$$\exp\left[j \tan^{-1}\left(\frac{-0.1}{r}\right)\right] \exp(j\omega t) \qquad \text{m/sec}$$

and the displacement is

$$\xi(t,r) = \frac{u(t,r)}{j\omega}$$

Figure 2-15 shows the pressure, velocity, and displacement amplitudes normalized relative to their values at 1 m. Also shown is the phase angle of the particle velocity relative to the pressure. The phase of particle displacement lags the velocity term by an additional 90°.

6. The total force at the spherical surface is the product of pressure and area:

$$f(t,a) = (4\pi a^2)p(t,a)$$

For a total power of 1000 W, this is

$$f(t,a) = \frac{4\pi(0.01)(1.55 \times 10^4)}{0.1} \exp(j\omega t)$$

$$= (1.95 \times 10^4) \exp(j\omega t) \qquad \text{newtons}$$

7. The total impedance presented to the primary driving force is the sum of the mechanical impedance of the driving mechanism and the water load.

$$Z_T = Z_m + Z_w$$

$$= R_m + R_w + j\left[\omega(M_m + M_w) - \frac{1}{\omega S_m}\right]$$

The schematic diagram for this system is shown in Figure 2-16. The real part of Z_T is obtained using expressions developed previously.

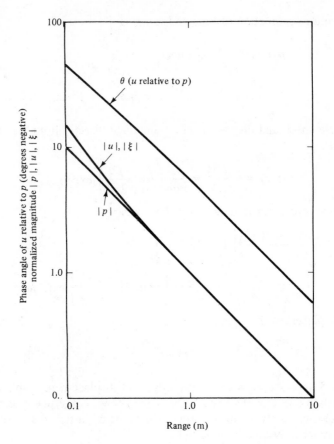

Figure 2-15 Pressure, velocity, and displacement amplitudes versus range for example Section 2.6.

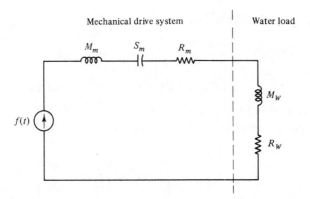

Figure 2-16 Mechanical circuit diagram for simple source and water load.

$$\text{Re}[Z_T] = R_m + R_w = (4 \times 10^4) + (1.88 \times 10^5)(1 + 2.5\lambda^2)^{-1} \quad \text{kg/sec}$$

Similarly,

$$\text{Im}[Z_T] = \omega M_m - \frac{1}{\omega S_m} + X_w$$

$$= 2\pi f(5) - \frac{10}{2\pi f(4)} + \left(\frac{\lambda}{0.628}\right)(1.88 \times 10^5)(1 + 2.5\lambda^2)^{-1} \quad \text{kg/sec}$$

The real and imaginary parts of Z_T are plotted in Figure 2-17 as functions of frequency. Notice that with the parameters assumed in this example, the inertial reactance associated with both the water load and the drive mechanism is exactly balanced by the elastic reactance associated with the stiffness of the drive mechanism at the center frequency.

8. The acoustic power delivered to the water is given by the product of the mean-squared surface velocity and the resistive part of the water load:

$$P_T = \langle u^2 \rangle R_w$$

but

$$\langle u^2 \rangle = \frac{u_m^2}{2} = \frac{f_m^2}{2|Z_T|^2}$$

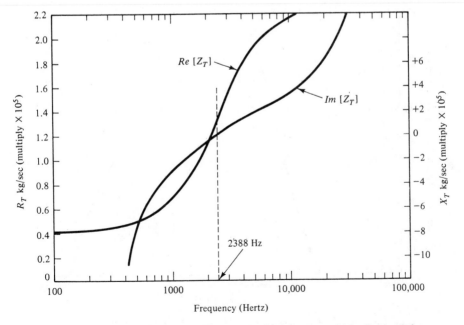

Figure 2-17   Real and imaginary parts of total impedance for example, Section 2-6.

where u_m is the peak value of the velocity and f_m is the peak value of the driving force. Thus

$$P_T = \frac{f_m^2}{2|Z_T|^2}R_w = \frac{f_{\text{rms}}^2}{|Z_T|^2}R_w$$

Let the rms driving force equal 1 N and we obtain the power output per unit force:

$$\frac{P}{\text{newton rms}} = \frac{R_w}{|Z_T|^2}$$

The resulting source level per unit applied force is

$$\text{SL (per newton)} = 170.77 + 10 \log\left(\frac{P}{N_{\text{rms}}}\right)$$

$$= 170.77 + 10 \log\left(\frac{R_w}{|Z_T|^2}\right)$$

The values for R_w and $|Z_T|^2$ are obtained from Figures 2-14 and 2-17. The source level per newton is plotted as a function of frequency in Figure 2-18.

9. At the design center frequency, the source level per newton rms applied force is 117.9 dB$//\mu$pa @ 1 m. For 1000 W output the source level required is 200.77 dB. The required input force to obtain this output level is

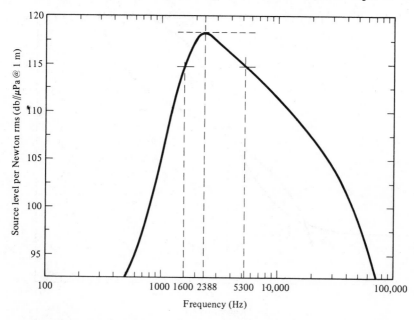

Figure 2-18 Source level per newton as a function of frequency for example, Section 2.6.

$$10 \log \frac{f_{rms}^2}{1} = SL \ (1000 \ W) - SL \ (\text{per newton})$$

$$= 200.77 - 117.9 = 82.87 \ dB$$

or

$$f_{rms}(\text{newtons}) = \text{antilog}\left[\frac{82.87}{20}\right] = 1.39 \times 10^4 \ N$$

10. The efficiency of the system is the ratio of the useful acoustic output to the total power dissipated. This is simply the ratio of the resistive part of the water load to the combined resistance of the mechanical drive and the water load:

$$\text{efficiency} = \frac{R_w}{R_w + R_m} \times 100\%$$

$$= \frac{0.94}{0.94 + 0.4} \times 100 = 70\%$$

This is the efficiency of the mechanical portion of the system only, and does not include any electrical losses that may be incurred in generating the mechanical driving force.

11. Finally, the 3-dB bandwidth for a constant input driving force is read directly from Figure 2-18 as 3700 Hz.

PROBLEMS

2.1. The inductance and capacitance per foot for a lossless electrical transmission line are 0.08 μH and 29.5 pF, respectively. Calculate the propagation speed and the characteristic impedance for an electromagnetic wave on this line.

2.2. A lossless electrical transmission line with characteristic impedance Z_0 is terminated in an impedance $(Z_0/2)(1 + j1)$ at the receiving end. Calculate and plot the magnitude and phase of the line impedance as a function of distance from the receiving end.

2.3. A plane acoustic wave in water is directed normal to an air–water boundary. Assuming that the acoustic signal is a continuous sinusoid, plot the magnitude of pressure at a distance of 3 m from the boundary as a function of frequency over the range 100 to 1000 Hz. Assume that the acoustic propagation speed in water is 1500 m/sec and the characteristic impedance of the air is essentially zero. The effect of all other boundary surfaces is assumed negligible.

2.4. The acoustic pressure and particle speed at one point in the field of a spherical wave are measured in water to be

$$p(t) = 10^9 \sin (2\pi ft) \quad \mu\text{Pa}$$

$$u(t) = 6.7 \times 10^{-4} \sin (2\pi ft - 10°) \quad \text{m/sec}$$

With $\rho = 10^3 \text{kg/m}^3$ and $c = 1500 \text{ m/sec}$, determine for $f = 100$ Hz and 1000 Hz:
 (a) The source level in $dB//\mu$Pa at 1 m
 (b) The total power radiated by the source
 (c) The intensity level at $r = 100$ m

2.5. A broadband source spectrum level in $dB//\mu$Pa at 1 m in a 1-Hz band is tabulated below at several frequencies.

f	L_s
100	130
200	133
300	130
400	127
500	123

Assume that the level at each point is constant over a 100-Hz band centered at each listed frequency. Calculate the intensity level for the broadband signal over the band 50 to 550 Hz. Assuming that the source radiates uniformly in all directions, calculate the total radiated power.

2.6. Assume a plane acoustic wave in water with a measured power flow of 100 W through a rectangular area measuring 0.2×0.1 m. Let the acoustic signal be sinusoidal at a frequency of 10 kHz and $Z_0 = 1.5 \times 10^6 \text{ kg/m}^2\text{-sec}$.
 (a) Find the acoustic intensity in W/m^2.
 (b) Find the intensity level in $dB//\mu$Pa and $dB//\mu$bar.
 (c) Assuming that the plane wave is an approximation to a spherical wave originating at a distance of 10 yd from the point of measurement, determine the source level in $dB//\mu$Pa at 1 m and in $dB//\mu$bar at 1 yd, and the total source power in watts.
 (d) At the point of measurement, determine the peak and rms pressure in μPa.
 (e) At the point of measurement, determine the peak particle speed and displacement.

SUGGESTED READING

1. *Physics of Sound in the Sea,* Part I: *Transmission;* originally issued as Division 6, Vol. 8, NDRC Summary Technical Reports, Wakefield Printing Co., Wakefield, Mass.; reprinted in 1969 by the U.S. Government Printing Office, Washington, D.C.

2. Officer, C. B., *Introduction to the Theory of Sound Transmission.* New York: McGraw-Hill Book Company, 1958.

3. Kinsler, L. E., and A. R. Frey, *Fundamentals of Acoustics,* 2nd ed. New York: John Wiley & Sons, Inc., 1962.

4. Beyer, R. T., *Nonlinear Acoustics.* Washington, D.C.: U.S. Government Printing Office, 1975, Stock No. 0-596-215.

3   Acoustic Transducers

3.0 INTRODUCTION

A transducer is a device that converts energy from one form into another form. In underwater acoustics this definition includes such diverse devices as explosive charges that convert chemical energy into acoustic energy, and hydrodynamic oscillators that mechanically convert a static hydraulic pressure into an oscillating acoustic pressure wave [1, 2]. For our purposes we confine our attention to a special class of acoustic transducers capable of providing an electrical output signal in response to an acoustic input, or an acoustic output in response to an electrical input. These devices generally can be modeled as linear devices, and therefore are amenable to straightforward analysis.

In underwater acoustics, a transducer designed primarily to produce an electrical output in response to an acoustic input is called a *hydrophone*. Hydrophones constitute the basic acoustic receiving sensor element in underwater acoustic systems. In the audio acoustic range, hydrophones are typically designed to operate over a broad frequency band and are small relative to the wavelength of the highest intended frequency. In the ultrasonic region, hydrophones may of necessity be appreciable in size relative to the operating wavelength.

A transducer intended primarily for the generation of an acoustic output signal in response to an electrical input is called a *projector*. In the interest of efficiency, projector dimensions are often of the same order of magnitude as the operating wavelength and may be relatively narrowband devices compared with hydrophones.

Most linear acoustic transducers take advantage of either piezoelectric or magnetostrictive phenomena. For the purposes of system analysis these phenomena result in equations of performance that are similar in form since they rely on the reciprocal interaction between material strains and concurrent electric or magnetic fields. We consider here only the principles of operation of a simple class of piezoelectric devices used either as hydrophones or as projectors.

The piezoelectric effect is normally associated with the behavior of certain crystalline substances. When these materials are subjected to a mechanical stress, electrical charges of opposite polarity appear on portions of the crystal surfaces. The magnitude of these charges is in proportion to the magnitude of the stress, and the charges change in polarity as the stress changes from a compression to a tension. Conversely, an applied electric field results in a mechanical strain proportional to the electric field strength. The sense of the strain reverses with a reversal of the electric field polarity. Early underwater transducers used piezoelectric quartz and Rochelle-salt crystals in both hydrophone and projector applications.

Ceramic materials exhibit behavior similar to the piezoelectric effect provided that the material is properly conditioned. In their natural state, ceramic materials are polycrystalline in structure and exhibit no piezoelectric properties. This condition is altered by impressing a static electric field across the ceramic element in the presence of high heat. The element experiences a permanent increase in dimension in the direction of the polarizing field and a decrease in dimension at right angles to the field. This bias strain persists after removal of the "poling" field. Subsequent application of an electric field of the same polarity in the poling direction causes a further, but temporary, expansion in that direction. A reversal of the polarity of the field results in a contraction (or a reduced expansion) in the poling direction. This behavior is linear in form provided that the magnitude of the applied field is less than that of the original polarizing field. The application of a mechanical compression in the poling direction results in a voltage with the same polarity as the original polarizing field, and a tensile force causes a polarity reversal. Again, for linear operation, these applied stresses must not exceed the bias strain created in the original poling operation. Because of the ease with which ceramic materials can be molded into various shapes, they have largely replaced crystalline materials in underwater transducers in the audio and low ultrasonic frequency range.

In this chapter we develop electroacoustic equivalent circuits for simple hydrophone and projector configurations using piezoelectric ceramic materials. The equivalent circuits provide a convenient means for assessing the performance of these devices as elements of the overall underwater system.

3.1 LONGITUDINAL VIBRATION OF A SIMPLE BAR

An understanding of the principles of transducers using piezoelectric ceramic materials is facilitated by a study of the electroacoustic properties of a simple bar of the material. Although practical transducers may use more complicated

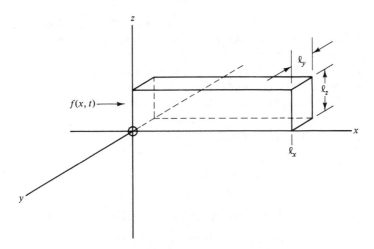

Figure 3-1   Rectangular ceramic transducer element.

shapes, the behavior of a rectangular bar of the ceramic material demonstrates the essential elements of transducer operation.

Consider a bar of the ceramic material with rectangular cross section as shown in Figure 3-1. The bar is rigidly clamped at $x = \ell_x$ and a time-varying force applied in the x-direction at $x = 0$. The time-varying applied force results in particle velocities and displacements within the bar that are described by the same equations as developed in Chapter 2 for plane waves in the water. Thus let

$$p(x, t) = \frac{f(x, t)}{\ell_y \ell_z} = \text{acoustic pressure}$$

$$\rho_m = \text{density of ceramic material}$$

$$s_m = \text{compliance of material, equal to the inverse of Young's modulus in the } x\text{-direction}$$

$$u(x, t) = \frac{\partial \xi}{\partial t} = \text{particle speed in } x\text{-direction}$$

and write

$$\frac{\partial p}{\partial x} = -\rho_m \frac{\partial u}{\partial t}$$

$$\frac{\partial u}{\partial x} = -s_m \frac{\partial p}{\partial t} \tag{3-1}$$

$$\frac{\partial^2 p}{\partial t^2} = \frac{1}{\rho_m s_m} \frac{\partial^2 p}{\partial x^2}$$

The *longitudinal waves* travel in the bar with a propagation speed $c_m = 1/\sqrt{\rho_m s_m}$. The characteristic impedance of the bar is $Z_{0m} = \sqrt{\rho_m / s_m}$.

With the bar rigidly clamped at $x = \ell_x$, the particle speed at that point is constrained to be zero. Recalling the results from Chapter 2, this is analogous to an open-circuit termination for an electrical transmission line, or a rigid boundary termination for an acoustic plane wave. In the general case, the input impedance presented to the applied force at the origin (assuming a sinusoidal driving force) is given by direct analogy to (2-23) and (2-44) as

$$Z(\text{input}) = \ell_z \ell_y Z_{0m} \left[\frac{Z_\ell \cos \alpha + jZ_{0m} \sin \alpha}{Z_{0m} \cos \alpha + jZ_\ell \sin \alpha} \right] \tag{3-2}$$

$$\alpha = \frac{2\pi \ell_x}{\lambda_m}$$

where λ_m = acoustic wavelength in the material = c_m / f_o

Z_{0m} = characteristic impedance of the material

Z_ℓ = termination impedance at $x = \ell_x$

$\ell_z \ell_y$ = area of the end of the bar

With the bar rigidly clamped, Z_ℓ approaches infinity with the result that

$$Z(\text{input}) = -j\ell_z \ell_y Z_{0m} \cot \alpha \qquad (Z_\ell \to \infty) \tag{3-3}$$

The input impedance at $x = 0$ is purely reactive, with a sign determined by the sign of $\cot \alpha$. The normalized input impedance as a function of ℓ_x / λ_m is plotted in Figure 3-2. For $\ell_x < \lambda_m / 4$ the reactance is negative, indicating that the behavior is dominated by the elastic properties of the material. For ℓ_x greater than $\lambda_m / 4$ but less than $\lambda_m / 2$, the reactance is positive indicating a net inertial reaction.

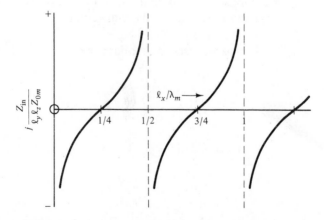

Figure 3-2 Input mechanical impedance for a clamped bar.

It is tempting at this point to attempt to derive a mechanical "equivalent circuit" containing lumped inertial and elastic elements that will duplicate the impedance in (3-3) over the frequency band. Because of the nature of the cotangent function, this is not possible with constant value elements. However, we can derive equivalent circuits containing fixed elements that are satisfactory approximations over limited frequency ranges. Two frequency ranges that are of particular interest are the low-frequency range where $\ell_x \ll \lambda_m/4$, and the range in the vicinity of the frequency that results in zero input impedance.

First we consider the low-frequency range. As α approaches zero, we have

$$Z_{in} = -j\ell_z\ell_y Z_{0m} \cot \alpha \simeq \frac{-j\ell_z\ell_y Z_{0m}}{\alpha}$$

By substitution of the definitions for Z_{0m} and α, this can be put in the form

$$Z_{in} = -j\left(\frac{\ell_z\ell_y}{\ell_x s_m}\right)\frac{1}{2\pi f} \tag{3-4}$$

Define a total mechanical compliance for the bar as

$$S_m = \frac{\ell_x s_m}{\ell_z\ell_y} \tag{3-5}$$

and write

$$Z_{in} = -j\left(\frac{1}{2\pi f S_m}\right) \qquad \left(\text{for } \ell_x \ll \frac{\lambda_m}{4}\right) \tag{3-6}$$

The mechanical impedance in the low-frequency region is the mechanical analog of a capacitive reactance in electrical circuits.

Now consider the frequency region near the value where $Z_{in} = 0$. In this region the impedance is nearly a linear function of frequency, changing from a negative value to a positive value as the frequency increases. The equivalent circuit in this region may be approximated by a series combination of an inertial element and a compliant element. The lowest frequency giving a zero input impedance is defined by

$$\alpha_r = \frac{\pi}{2} = 2\pi f_r \ell_x \sqrt{\rho_m s_m}$$

or

$$f_r = (4\ell_x \sqrt{\rho_m s_m})^{-1} = \frac{c_m}{4\ell_x} \tag{3-7}$$

In the vicinity of f_r let

$$Z_{in} = -j\ell_z\ell_y Z_{0m} \cot \alpha \simeq -j\left(\frac{1}{2\pi f S_m} - 2\pi f M_m\right) \tag{3-8}$$

where f = frequency near f_r

S_m = effective total compliance near f_r

M_m = effective total mass near f_r

Now substitute (3-7) for f in (3-8), obtaining zero input impedance by definition, and obtain

$$M_m S_m = \frac{1}{4\pi^2 f_r^2} = \frac{4\ell_x^2 \rho_m S_m}{\pi^2} \qquad (3\text{-}9)$$

The slope of Z_{in} at $f = f_r$ from (3-8) is

$$\left. \frac{dZ_{in}}{df} \right|_{f=f_r} = +j\ell_z\ell_y \sqrt{\frac{\rho_m}{S_m}} \left(\frac{\pi}{2f_r}\right) = +j\left(2\pi M_m + \frac{1}{2\pi f_r^2 S_m}\right) \qquad (3\text{-}10)$$

Multiply both sides of (3-10) by S_m, substitute (3-9) for $M_m S_m$ and (3-7) for f_r, and finally obtain

$$S_m = \frac{8}{\pi^2}\left(\frac{\ell_x S_m}{\ell_z \ell_y}\right) \qquad (3\text{-}11)$$

Notice that the compliance defind by (3-11) is slightly smaller than the low-frequency value defined by (3-5).

Substitution of (3-11) in (3-9) provides the value of the effective mass in the vicinity of f_r. Thus

$$M_m = \frac{4\ell_x^2 \rho_m S_m}{\pi^2 S_m} = \frac{\ell_z \ell_y \ell_x \rho_m}{2} \qquad (3\text{-}12)$$

The effective mass given by (3-12) is one-half the actual mass of the bar. This is reasonable because one end of the bar is clamped, with the result that all parts of the bar are not moving to the same degree.

To this point we have neglected mechanical losses in the material or in the mounting structure that holds the bar. These losses are generally small and in the low-frequency case may be neglected. In the vicinity of f_r, the reactive impedance is near zero so that the loss term, represented by a mechanical resistance, may not be insignificant. The input impedance near f_r therefore takes the following form:

$$Z_{in} = R_m - j\left(\frac{1}{\omega S_m} - \omega M_m\right) \qquad (3\text{-}13)$$

where R_m is the mechanical resistance caused by losses in material and mount.

We may now construct mechanical equivalent circuits as shown in Figure 3-3. The mechanical circuit in Fig. 3-3(a) is useful in analyzing hydrophone operation because hydrophone dimensions are typically small compared with a

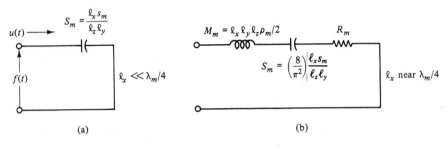

Figure 3-3 Mechanical equivalent circuits for ceramic bar: (a) low-frequency equivalent circuit; (b) equivalent circuit near resonance.

wavelength. Projector design calls for operation near the natural resonant frequency of the device for the efficient generation of acoustic output power. The equivalent circuit in Figure 3-3(b) is appropriate for the projector application.

3.2 HYDROPHONE OPERATION

For hydrophone operation conducting surfaces are provided on opposite faces of the ceramic bar as shown in Figure 3-4. The ceramic material is polarized by application of a steady electric field parallel to the z-axis. With voltage applied between the conducting electrodes, dimensional changes occur in the material and for this discussion we shall be concerned primarily with changes in the x-direction. Assume that the ceramic bar is rigidly clamped at $x = \ell_x$, and that the bar is perfectly shielded from an external acoustic pressure field except on the exposed face at the origin with area $\ell_z\ell_y$.

With the electrical terminals open-circuited, we now apply a static force, f_x, to the exposed face. This results in an electric charge accumulating on the conducting surfaces with a magnitude given by

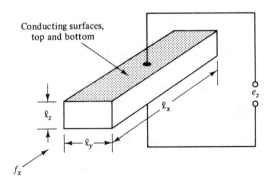

Figure 3-4 Elementary hydrophone configuration.

$$Q = -d\left(\frac{f_x}{\ell_y \ell_z}\right) \ell_x \ell_y$$

$$= -dp_x \ell_x \ell_y \tag{3-14}$$

where Q = electrical charge in coulombs

p_x = pressure on the y-z face (N/m^2)

d = piezoelectric strain coefficient of the ceramic material in coulombs per newton or meters per volt, appropriate for the direction of applied force and polarization

An electric charge also accumulates on the conducting surfaces if a static voltage, e_z, is applied to the electrical terminals. This charge is given by

$$Q = \left[\epsilon \epsilon_0 \left(\frac{\ell_x \ell_y}{\ell_z}\right)\right] e_z \tag{3-15}$$

where ϵ = dielectric constant of the material

ϵ_0 = permittivity of free space, 8.85×10^{-12} farads/meter

The bracketed term in (3-15) will be recognized as the ordinary capacitance, C, of a parallel-plate capacitor of area $\ell_x \ell_y$, thickness ℓ_z, and dielectric constant ϵ.

In the presence of both an applied force and an applied voltage, (3-14) and (3-15) are combined to give the total charge.

$$Q = \epsilon \epsilon_0 \left(\frac{\ell_x \ell_y}{\ell_z}\right) e_z - dp_x \ell_x \ell_y \tag{3-16}$$

The inverse piezoelectric effect requires a mechanical strain in the material resulting from an impressed voltage. The strain resulting from both an applied force and an impressed voltage is

$$\frac{\partial \xi}{\partial x} = \frac{e_z d}{\ell_z} - s_m p_x \tag{3-17}$$

where $\dfrac{\partial \xi}{\partial x}$ = longitudinal strain in the x-direction

d = piezoelectric strain coefficient

s_m = material compliance

3.2.1 Hydrophone Equivalent Circuit

Equations (3-16) and (3-17) define the interactions between the acoustic (mechanical) and the electrical portions of the hydrophone element. These relationships are more conveniently displayed in the form of an electroacoustic

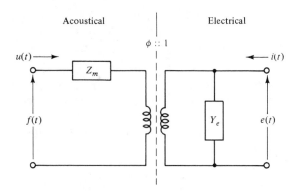

Figure 3-5   Transducer electroacoustic equivalent circuit.

equivalent circuit, one possible form of which is shown in Figure 3-5. The acoustic side consists of a force and velocity at the terminals with a series mechanical impedance Z_m. The electrical and mechanical sections are separated by a perfect transformation device with a transformation ratio, ϕ, that transforms electrical quantities into mechanical, and vice versa. The electrical side consists of a parallel admittance, Y_e, with a voltage across the terminals and an input current.

This equivalent circuit is similar to that used to describe a simple electrical transformer. For an electrical transformer, the parameters, Y, ϕ, and Z are determined by a set of open-circuit and short-circuit measurements. With a voltage applied to the side containing the admittance, the transformation ratio is obtained as the ratio of open-circuited output voltage to applied voltage. The admittance, Y, is the ratio of input current to applied voltage with the output open-circuited. The impedance, Z, is determined by short-circuiting the side containing Y and measuring the ratio of applied voltage to current on the side containing Z.

This same procedure may be used to define the parameters of the equivalent circuit in Figure 3-5. The mechanical equivalent of an "open circuit" is obtained by setting the velocity, $u(t)$, equal to zero. A zero velocity implies that the mechanical side is constrained, or clamped, so that it is not free to move. If the element is constrained at the equilibrium condition, the displacement, and therefore the strain in the material, is zero. This will remain true for a static applied voltage or for a low-frequency applied sinusoidal voltage (such that $\ell_x \ll \lambda_m/4$). Setting the strain in (3-17) to zero, we obtain the transformation factor, ϕ, by solving for the ratio of output force to applied voltage. Thus, for $\ell_x \ll \lambda_m/4$, let

$$\frac{\partial \xi}{\partial x} = 0 = \frac{e_z(t)d}{\ell_z} - s_m p_x(t)$$

from which

$$\phi = \frac{p_x(t)\ell_y\ell_z}{e_z(t)} = \frac{f_x(t)}{e_z(t)} = \frac{d\ell_y}{s_m} \tag{3-18}$$

The transformation from mechanical force to electrical voltage is proportional to the piezoelectric strain coefficient and inversely proportional to the material compliance.

    With the mechanical side still clamped, solve (3-18) for $p_x(t)$ in terms of $e_z(t)$ and substitute in (3-16).

$$Q(t) = \epsilon\epsilon_0\left(\frac{\ell_x\ell_y}{\ell_z}\right)e_z(t) - \frac{d^2}{s_m}\left(\frac{\ell_x\ell_y}{\ell_z}\right)e_z(t)$$

$$= \epsilon\epsilon_0\left(1 - \frac{d^2}{\epsilon\epsilon_0 s_m}\right)\left(\frac{\ell_x\ell_y}{\ell_z}\right)e_z(t) \tag{3-19}$$

Now define a new dielectric constant, ϵ' as

$$\epsilon' = \epsilon\left(1 - \frac{d^2}{\epsilon\epsilon_0 s_m}\right) \tag{3-20}$$

and write

$$Q(t) = \epsilon'\epsilon_0\left(\frac{\ell_x\ell_y}{\ell_z}\right)e_z(t) \tag{3-21}$$

The constant ϵ' is the effective dielectric constant of the material under the clamped condition and is called the *clamped dielectric constant*.

    Current is defined as the time rate of change of charge. Thus, with the mechanical side clamped,

$$i(t) = \frac{\partial Q(t)}{\partial t} = \epsilon'\epsilon_0\left(\frac{\ell_x\ell_y}{\ell_z}\right)\frac{\partial e_z(t)}{\partial t} \tag{3-22}$$

$$= C_0\frac{\partial e_z(t)}{\partial t} \tag{3-23}$$

where C_0 is the *clamped capacitance* of the element. The admittance of the electrical side is thus the admittance of a capacitor, C_0, in parallel with the input terminals with a value

$$Y_e = j\omega C_0 \tag{3-24}$$

In this discussion we have assumed that the frequency is sufficiently low that inductive effects on the electrical side may be neglected. We have also neglected dielectric losses that occur in the ceramic material in the presence of an alternating electric field. These losses are generally small at audio and low ultrasonic frequencies and have little effect in the hydrophone application.

To determine the impedance of the mechanical portion of the circuit, the ratio of applied force to input velocity at the face of the element is evaluated with the other end rigidly clamped and the electrical side short circuited. The input impedance so obtained is identical to that discussed in Section 3.2 and, for the lossless bar clamped on one end, gives, in accordance with (3-3),

$$Z_m = -j\ell_z\ell_yZ_{0m}\cot\alpha$$

For the hydrophone application, we assume that $\ell_x \ll \lambda_m/4$, with the result that

$$Z_m \simeq -j\left(\frac{1}{2\pi f S_m}\right) \tag{3-25}$$

where

$$S_m = \frac{\ell_x S_m}{\ell_z\ell_y}$$

The equivalent circuit of Figure 3-5 may now be redrawn as in Figure 3-6.

3.2.2 Hydrophone Sensitivity

To define hydrophone performance, it is convenient to transform the terms on the acoustic side to the electrical side to obtain an all-electric equivalent circuit. From (3-18) the voltage equivalent of force is obtained by dividing the force by the transformation factor. Remembering that impedances transform as the square of the transformation factor, the mechanical compliance, S_m is multi-

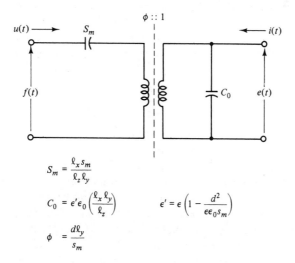

$$S_m = \frac{\ell_x S_m}{\ell_z\ell_y}$$

$$C_0 = \epsilon'\epsilon_0\left(\frac{\ell_x\ell_y}{\ell_z}\right) \qquad \epsilon' = \epsilon\left(1 - \frac{d^2}{\epsilon\epsilon_0 S_m}\right)$$

$$\phi = \frac{d\ell_y}{S_m}$$

Figure 3-6   Hydrophone electroacoustic equivalent circuit.

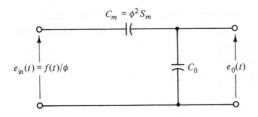

Figure 3-7 All-electric hydrophone equivalent circuit.

plied by ϕ-squared to obtain the equivalent electrical capacitance, C_m. With these changes, the all-electric equivalent circuit is shown in Figure 3-7.

From the equivalent circuit, the open-circuit output voltage for a given input force is

$$e_0(t) = \left(\frac{f(t)}{\phi}\right)\frac{C_m/C_0}{1 + C_m/C_0} \tag{3-26}$$

Converting the force to the equivalent plane wave pressure acting on the $\ell_y\ell_z$ face, we write

$$\frac{e_0(t)}{p(t)} = \left(\frac{\ell_y\ell_z}{\phi}\right)\frac{C_m/C_0}{1 + C_m/C_0} \tag{3-27}$$

Equation (3-27) defines the open-circuit hydrophone sensitivity in terms of output voltage per unit input pressure. From the defining equations for C_m, C_0, and ϕ, (3-27) can be put in the alternative forms

$$\frac{e_0}{p} = \frac{\ell_z d}{\epsilon\epsilon_0} \tag{3-28}$$

$$= (\ell_x\ell_y)\frac{d}{C} \tag{3-29}$$

where

$$C = \frac{(\ell_x\ell_y)\epsilon\epsilon_0}{\ell_z}$$

The capacitance, C, in (3-29) is the ordinary *unclamped* capacitance of the element. The hydrophone sensitivity is related to the piezoelectric strain coefficient, dielectric constant, and element dimensions.

Equations (3-28) and (3-29) indicate that hydrophone performance is improved by increasing the separation between the conducting electrodes, thereby reducing the electrical capacitance of the element. It must be remembered that these equations give the *open-circuit* sensitivity of the device. Ultimately, the hydrophone element must be connected to other electronic circuit elements that amplify and transmit the signal to its final destination. This additional circuitry connected to the electrical hydrophone terminals presents a parallel load impedance that must be included in evaluating the actual achieved hydrophone sensitivity. Increasing the hydrophone open-circuit sensitivity by reducing the element capacitance also results in an increased internal impedance for the hydrophone as a signal source. The performance of the hydrophone in combination with follow-on circuits is considered further in Section 3.3.4.

3.2.3 Cylindrical Hydrophone Elements

The rectangular bar configuration considered so far provides a simple model for demonstrating the fundamental relationships among various electrical and mechanical properties of materials, but for several reasons other hydrophone geometric configurations are more practical, at least in the audio-frequency region. In particular, a ceramic cylindrical shell is the basic shape used in many hydrophone applications in the audio and low ultrasonic frequency regions. The use of the cylindrical shell in three possible modes of operation is shown in Figure 3-8.

Figure 3-8(a) shows a cylindrical shell operating in the *radial mode*. The inner and outer cylindrical surfaces are plated with conducting material and the poling field applied between these surfaces. In this mode, as well as in the other modes to be considered, the inner surface of the cylinder is acoustically shielded from the applied acoustical field. This configuration is characterized by moderate sensitivity and high output capacitance. Because of the large capacitance, the design is capable of driving long cables directly without suffering much loss in sensitivity.

Figure 3-8(b) depicts the *tangential mode* of operation. Conducting stripes formed lengthwise on the ceramic cylindrical surface divide the cylinder into an even number of curved segments. Alternate stripes are electrically connected and a poling field applied in the directions indicated by the arrows. The cylindrical segments are thus electrically in parallel but mechanically in series. This geometry is characterized by high sensitivity and low capacitance. Because of the low capacitance, the output impedance is high, making it unsuitable for directly driving a long cable. It is an excellent design where high sensitivity is required and suitable electronic isolation from the cable can be provided close to the hydrophone element.

The configuration shown in Figure 3-8(c) operates in the *longitudinal* mode. The ends of the ceramic cylinder are made electrically conducting and the poling field applied in the direction parallel to the cylindrical axis. This configuration has somewhat higher sensitivity than the radial mode and, as with the tangential mode, the electrical impedance of the device is high.

In the discussion of the simple rectangular bar hydrophone, we defined a piezoelectric strain coefficient, d, relating strain along the x-direction to applied field in the z-direction. Actually, an applied field in the z-direction results in strain in the z-direction as well as in directions perpendicular to the z-axis. In the simple bar hydrophone we neglected strain in any direction other than the x-direction. This is a reasonable approximation assuming that the bar length, ℓ_x, is long compared to either ℓ_y or ℓ_z.

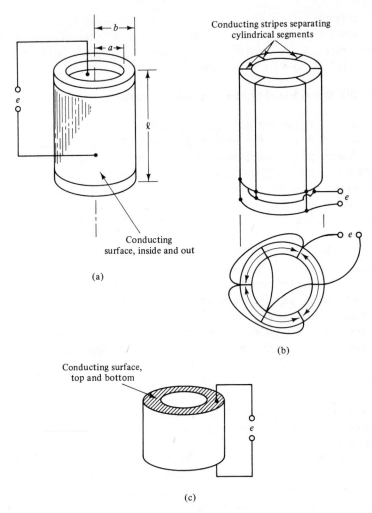

Figure 3-8 Ceramic cylindrical shell hydrophones: (a) radial mode; (b) tangential mode; (c) longitudinal mode.

With the cylindrical shell hydrophones, it is necessary to consider strains in both the direction of the applied electric field and in a direction perpendicular to the field. The piezoelectric strain coefficients are in general different in these two directions, and in fact have opposite signs. Thus, considering the rectangular bar, if the poled direction coincides with the z-axis, an applied voltage of the same polarity as the original polarization results in an *expansion* of the material in the z-direction and a *contraction* in the x and y directions. The piezoelectric strain coefficients for these orthogonal directions are identified as follows:

d_{33}: piezoelectric strain coefficient in a direction aligned with the polarizing field

d_{31}: piezoelectric strain coefficient in a direction orthogonal to the polarizing field

In the analysis of the simple bar hydrophone, the applied field and the strain axis are orthogonal, so that the coefficient d_{31} is the one appropriate in (3-28) and (3-29).

The hydrophone sensitivities for ceramic tubes operated in the three modes described above have been derived by Langevin [3]. His analysis assumes a frequency of operation well below any mechanical resonances in the cylindrical tube. As examples of hydrophone performance, we shall calculate sensitivity using Langevin's results for a cylindrical hydrophone in the radial mode and in the tangential mode. In both cases, we assume that the interior surfaces and the cylindrical ends are completely shielded from the acoustic field.

The voltage sensitivity for the radially polarized ceramic cylinder is

$$\frac{e}{p} = b \left[\frac{d_{33}}{\epsilon \epsilon_0} \left(\frac{b - a}{b + a} \right) + \frac{d_{31}}{\epsilon \epsilon_0} \right] \qquad (3\text{-}30)$$

where a and b are the inner and outer radii of the cylindrical shell, respectively. Comparing (3-30) with (3-28) we see that in both cases the sensitivity is determined by the piezoelectric strain coefficients, the dielectric constant, and a linear dimension.

Ceramic materials commonly used in hydrophone and projector applications include barium titanate and a composition of lead zirconate and lead titanate commonly refered to as PZT.[1] To provide a numerical example of hydrophone sensitivity, the material constants associated with the composition PZT-5A are as follows:

$$d_{33} = 374 \times 10^{-12} \text{ C/N}$$

$$d_{31} = -171 \times 10^{-12} \text{ C/N}$$

$$\epsilon = 1700$$

Assume a radially polarized ceramic cylindrical shell with outer diameter of 1 in., length of 1 in., and wall thickness of 0.0625 in. Converting these units to meters and substituting in (3-30), we obtain

$$\frac{e}{p} = -1.23 \times 10^{-4} \text{ V/Pa} = -1.23 \times 10^{-10} \text{ V/}\mu\text{Pa}$$

It is customary to express hydrophone sensitivity as a decibel level relative to 1 V rms assuming an acoustic pressure field of 1 μPa rms. This level is

---

[1] PZT is a registered trademark used by the Clevite Corporation to identify a family of lead zirconate–lead titanate compositions.

obtained by converting the square of the voltage-to-pressure ratio to decibels. Thus, in the example above, the hydrophone sensitivity, H, in decibels is

$$H = 10 \log \left(\frac{e}{p}\right)^2 = 20 \log \left(\frac{e}{p}\right)$$

$$= -198.2 \text{ dB}//V/\mu\text{Pa} \tag{3-31}$$

The electrical output impedance of the ceramic cylinder is determined by the unclamped capacitance at the electrical terminals. With the inner and outer surfaces fully covered by conducting electrodes, this capacitance is given by

$$C = \frac{2\pi\epsilon\epsilon_0\ell}{\ln(b/a)} = \frac{(2\pi)(1700)(8.85 \times 10^{-12})(0.0254)}{\ln(0.0127/0.0111)}$$

$$= 0.0176 \ \mu\text{F} \tag{3-32}$$

For the tangential mode, assume the same ceramic cylindrical shell as above, except that it is divided into six segments as described in Figure 3-8(b). The sensitivity given by Langevin for this configuration for N segments is

$$\frac{e}{p} = \frac{1}{N} \frac{2\pi b}{\ln(b/a)} \left[\frac{d_{31}}{\epsilon\epsilon_0}\left(\frac{b-a}{b+a}\right) + \frac{d_{33}}{\epsilon\epsilon_0}\right] \tag{3-33}$$

Substitution of the previously listed numerical values gives

$$\frac{e}{p} = 2.39 \times 10^{-9} \text{ V}/\mu\text{Pa}$$

Converting to a decibel level, this becomes

$$H = 20 \log(2.39 \times 10^{-9}) = -172.4 \text{ dB}//V/\mu\text{Pa}$$

This result indicates an improvement of almost 26 dB in sensitivity as compared with the radial mode configuration. However, for a fair comparison, we must also consider the output impedance of the two devices. For the tangentially polarized and segmented cylinder, the output capacitance is the parallel combination of the capacitance of the six cylindrical segments. The resulting capacitance is approximately

$$C = \frac{N^2(b-a)\ell \ \epsilon\epsilon_0}{2\pi b}$$

$$= \frac{(36)(1.59 \times 10^{-3})(0.0254)(1700)(8.85 \times 10^{-12})}{(6.28)(0.0127)}$$

$$= 0.000274 \ \mu\text{F} \tag{3-34}$$

Comparing (3-34) with (3-32), the effective source capacitance with the tangentially polarized device is smaller by a factor of 64 than the source capacitance of the radially polarized hydrophone. To achieve the sensitivity predicted by (3-33), it is necessary to ensure that the following electronic circuit is high impedance compared with the reactance of the capacitance obtained in (3-34).

3.2.4 Hydrophone as a System Element

The equivalent circuit of a hydrophone driving an electrical cable and an electronic amplifier is shown in Figure 3-9. The hydrophone is characterized by an open-circuit voltage e_{in} and an internal impedance determined by the un-clamped electrical capacitance C. In the audio-frequency range, the electrical cable is short compared with the electrical wavelength so that the cable may be represented by a lumped shunt capacitance C_ℓ. The input impedance of the amplifier is represented by a shunt capacitor C_a and a shunt resistor R_a. The voltage at the amplifier input resulting from the hydrophone signal is e_a. An additional amplifier input voltage, e_n, is shown that represents unwanted electronic noise developed in the amplifier circuit and referred to the input terminals. The purpose of the amplifier is to provide voltage gain as necessary and to provide a low-impedance drive for any subsequent circuitry.

Assuming a sinusoidal input signal, the signal at the amplifier terminals is

$$e_a = e_{in} \left(\frac{1}{1 + ZY} \right) \tag{3-35}$$

where $Z = 1/j\omega C$

$\quad\quad\quad Y = j\omega(C_\ell + C_a) + 1/R_a$

Consider first the case where R_a is infinite. The response is then independent of frequency (for frequencies well below any natural mechanical resonance in the ceramic element) and is given by

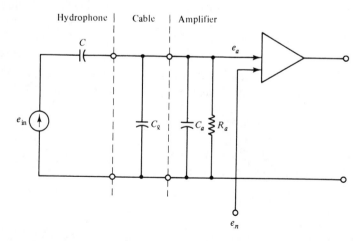

Figure 3-9   Equivalent circuit of hydrophone, cable, and amplifier.

$$e_a = e_{in}\left(\frac{C}{C + C_\ell + C_a}\right) \qquad \text{(for } R_a = \infty\text{)} \qquad (3\text{-}36)$$

To preserve the open-circuit sensitivity given by (3-30) or (3-33), the hydrophone capacitance, C, must be large compared with the sum of the cable capacitance and the amplifier input capacitance.

The cables used with hydrophone elements may be either coaxial or twisted shielded pairs. The capacitance per unit length for these cables ranges from 15×10^{-12} F/ft for some coaxial cables to 100×10^{-12} F/ft or more for twisted shielded pairs. The amplifier input capacitance is generally on the order of 20×10^{-12} F for high-quality, high-input impedance amplifiers.

Consider now the radially polarized ceramic element in the example in Section 3.2.3. The open-circuit sensitivity is -198.2 dB//V/μPa and the capacitance is 0.0176 μF. Assume a cable with 100×10^{-12} F/ft capacitance is used to connect this hydrophone to an amplifier with 20×10^{-12} F input capacitance. The attenuation in voltage caused by the cable and amplifier input capacitance is given by

$$\frac{e_a}{e_{in}} = \frac{C}{C + C_\ell + C_a} = \frac{0.0176}{0.0176 + 0.0001\ell} \qquad (3\text{-}37)$$

where ℓ is the cable length. For example, a cable length of 21 ft results in a sensitivity loss of 1 dB, and the loss is 3 dB with a 73-foot cable. The system designer may increase the allowed length of cable by selecting a cable with lower capacitance or by increasing the hydrophone capacitance (doubling the length of the ceramic cylinder doubles the value of C).

For the tangentially polarized element described in Section 3.2.3, the element capacitance is 0.000274 and the sensitivity is -172.4 dB//V/μPa. For this unit we assume a coaxial cable with 15×10^{-12} F/ft and an amplifier input capacitance of 20×10^{-12} F. The attenuation, using (3-37) is, therefore,

$$\frac{e_a}{e_{in}} = \frac{0.000274}{0.000294 + 0.000015\ell}$$

With zero cable length, the sensitivity loss is 0.6 dB compared with the open-circuit value. A cable length of approximately 6 ft results in a 3 dB loss. Obviously, with this type of hydrophone the amplifier should be located as close as possible to the hydrophone element to realize the inherent sensitivity of the device.

We now return to (3-35) and assume a finite value for R_a. The frequency response of the output relative to the input is

$$\frac{e_a}{e_{in}} = \frac{j\omega R_a C}{1 + j\omega R_a(C + C_\ell + C_a)} \qquad (3\text{-}38)$$

This response is sketched as a function of frequency in Figure 3-10. The shunt resistance causes the response to fall off at low frequencies, with a corner

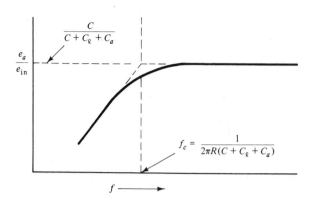

Figure 3-10 Frequency response of hydrophone in combination with cable and amplifier.

frequency defined by

$$f_c = \frac{1}{2R_a\pi(C + C_\ell + C_a)} \tag{3-39}$$

The shunt resistance in this case is associated with the amplifier input impedance. Any additional sources of leakage resistance in the cable or hydrophone assembly should be added to this term to determine the combined effect on low-frequency characteristics.

A typical high-quality, high-input impedance amplifier may have a value of R_a on the order of 100 MΩ or more. For the tangential mode hydrophone of the previous example connected directly to the amplifier input terminals, a corner frequency of 5.4 Hz is obtained if the input impedance is 100 MΩ. Obviously, care must be exercised to eliminate other leakage paths to achieve this result.

3.2.5 Effect of Amplifier Noise

The amplifier output signal is the result of the linear combination of the input signal from the hydrophone and the undesired electronic noise signal referred to the amplifier input. To be a useful device, the desired output resulting from the hydrophone signal should exceed the undesired output caused by the electronic noise. The minimum acceptable ratio of desired to undesired signal is somewhat arbitrary and may depend on the application. Generally, a 10 dB or better advantage is sought when selecting hydrophone and amplifier characteristics.

The manufacturers of low-noise amplifiers (or preamplifiers) suitable for use with ceramic hydrophone elements provide the noise characteristics of the device in the format shown in Figure 3-11. The ordinate is the rms noise level as measured in a 1-Hz band, expressed in decibels relative to 1 V rms. To obtain

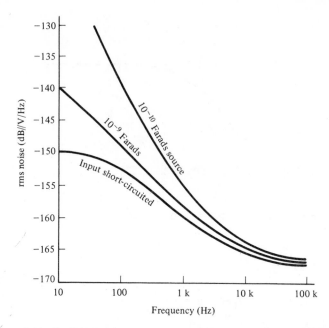

Figure 3-11   Equivalent electronic input noise for a hydrophone preamplifier.

the actual noise voltage, divide the ordinate by 20 and take the antilog. The noise level increases rapidly as the frequency or source capacitance decreases.

The effect of source impedance on amplifier electronic noise must be considered when comparing the system performance of the radial mode hydrophone with the tangential mode design. For a fixed amplitude sinusoidal pressure input signal of 1 μPa rms, the ratio of desired signal power to electronic noise power at the amplifier input terminals (expressed in dB) is given by the difference between hydrophone sensitivity, H, measured at the amplifier terminals, and the amplifier noise level as shown in Figure 3-11. Thus for electronic noise

$$(SNR)dB/\mu Pa = H - N_a \qquad (3\text{-}40)$$

To obtain the SNR at any other input signal pressure level, simply add the input intensity level relative to the reference level of 1 μPa. The value of SNR from (3-40) is plotted in Figure 3-12 for both the radial mode and the tangential mode designs from the previous examples. Both hydrophone elements are assumed connected directly to the amplifier terminals, and the amplifier noise is taken from Figure 3-11. Noise curves for 0.000274 and 0.0176 μF are roughly estimated from Figure 3-11. Notice that below 10 Hz, the advantage of high hydrophone sensitivity for the tangential mode is offset by the increased electronic noise.

The lower bound on the expected input acoustic signal is normally established by the minimum value of *ambient acoustic noise* present in the ocean. The subject of ambient noise in the ocean is discussed in detail in Chapter 10. For

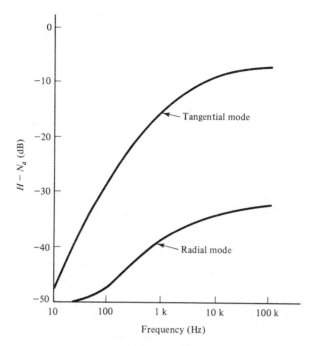

Figure 3-12   Signal-to-electronic-noise ratio in a 1-Hz band for an acoustic input of 1 μPa rms.

now it is sufficient to recognize that random pressure fluctuations exist that cover a broad frequency band and that arise from a number of different causes. It is convenient to describe the characteristics of ambient noise by means of an intensity spectral density plot as discussed in Section 2.5.3.

Assume that the minimum expected level of ambient noise over the frequency band is that shown in Figure 3-13. Our system design goal is to select a hydrophone–preamplifier combination such that at any frequency over the band, the preamplifier output signal, caused by ambient noise, will exceed the output level caused by electronic noise at that frequency by an acceptable amount. The output acoustic signal-to-electronic-noise ratio is obtained by adding N_s from Figure 3-13 to (3-40):

$$\text{output (SNR) dB} = N_s + H - N_a \tag{3-41}$$

In Figure 3-14, the resulting SNR is shown for both the radial and tangential mode hydrophone designs. For the assumed conditions, the radial mode design is only marginally acceptable below 1 kHz, and unsatisfactory above 1 kHz. The tangential design provides an adequate margin over the whole range. The performance of the radial mode hydrophone in this application may be improved by increasing its diameter. For instance, if the diameter is doubled, the output SNR will increase approximately 6 dB, thereby increasing the usable frequency band considerably.

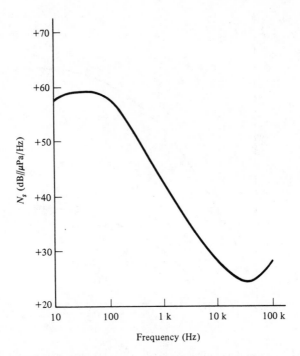

Figure 3-13 Minimum expected acoustic noise level.

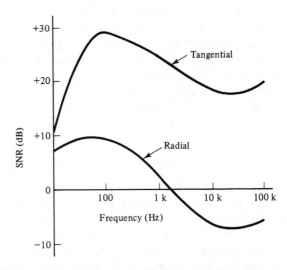

Figure 3-14 SNR: minimum ambient noise to amplifier electronic noise.

3.2.6 Upper-Frequency Limit for Cylindrical Hydrophones

The response of the cylindrical hydrophone element is essentially constant at frequencies below the lowest natural resonant frequency of the cylinder. At the resonant frequency, there is an increase in hydrophone output followed by a rapid decrease at higher frequencies.

The ceramic cylindrical shell has three possible modes of vibration. These are the *radial mode,* in which the mean cylindrical radius changes; the *length mode,* causing the length to change; and the *thickness mode,* resulting in a variation in cylindrical wall thickness.

The radial mode resonance occurs when the mean circumference of the cylinder is approximately one wavelength at the acoustic velocity of the material. Thus if D is the mean diameter and c is the speed of longitudinal waves in the material, the natural resonant frequency of this mode is

$$f_r \simeq \frac{c}{\pi D} \qquad (3\text{-}42)$$

In a similar manner, the length mode and the thickness mode require that the length and thickness, respectively, be about one-half wavelength with the appropriate material sound speed. Typically, the lowest natural frequency is obtained with the radial vibration mode.

For the radially poled ceramic tube of the previous examples, assume the following parameters:

ρ = density = 7.75×10^3 kg/m^3

s_{11} = material compliance in a direction perpendicular to the polarizing field with the electrical terminals open-circuited = 14.4×10^{-12} m^2/N

$D = 0.0238$ m

The sound speed in the material is

$$c = \sqrt{\frac{1}{\rho s_{11}}} \simeq 3000 \text{ m/sec}$$

Therefore,

$$f_r = \frac{3000}{0.0238 \pi} \simeq 40{,}000 \text{ Hz}$$

At this resonant frequency the wavelength of sound in water is approximately 0.04 m, so that the hydrophone element can no longer be considered small compared with the in-water wavelength.

3.3 PROJECTOR OPERATION

When a transducer is used as a projector, the simple equivalent circuits of
Figures 3-6 and 3-7 are no longer adequate. Acoustic power must be delivered
to the medium, requiring consideration of the impedance of the medium reacting
on the active surface of the projector. For a reasonable output power, the ceramic
element will not be small compared to a wavelength, requiring both inertial and
stiffness effects to be included in the equivalent circuit. Finally, mechanical and
electrical losses in the transducer material and support structure must be included
to determine overall efficiency.

   As an example we consider a simple rectangular bar, electrically and
mechanically arranged similar to the hydrophone configuration shown in Figure
3-4. One end of the bar is rigidly fixed and the other end is exposed to the water.
All other surfaces are isolated from the acoustic field in the water. An alternating
voltage applied between the electrode faces of the bar causes a longitudinal
vibration that couples acoustic energy into the water.

   In the vicinity of the natural resonant frequency of the system, the electroacoustic equivalent circuit of this simple projector is shown in Figure 3-15. On
the electrical side, the input admittance is determined by the parallel combination of the clamped capacitance, C_0, and a resistance, R_0, representing the
electrical losses. The value of C_0 is determined from the equations listed on
Figure 3-6. The resistance, R_0, is the result primarily of losses in the dielectric
material in the presence of the applied alternating electric field. The losses
associated with a dielectric material are defined in terms of the *loss tangent*. The
loss tangent is defined at a particular frequency, usually 1 kHz, and for the
parallel equivalent electric circuit of Figure 3-15 is approximately

$$\text{loss tangent} = \tan \delta = \frac{1}{2\pi f R_0 C} \tag{3-43}$$

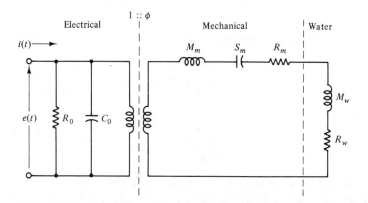

Figure 3-15 Projector equivalent circuit.

where $f = 1000$ Hz

C = unclamped capacitance

R_0 = parallel resistor representing dielectric losses

For ceramic materials used in projector applications the loss tangent is typically in the range 0.004 to 0.04. As an example, assume a capacitance of 0.002 μF and a loss tangent of 0.01. The value of R_0 at 1000 Hz is then

$$R_0 = \frac{10^6}{(2000\pi)(0.002)(0.01)} = \frac{10^8}{4\pi}$$

$$\simeq 8 \text{ M}\Omega$$

The loss associated with R_0 in this case would be small.

For operation in the vicinity of the natural resonance of the bar, the values of S_m, M_m and ϕ are obtained from (3-11), (3-12), and (3-18) respectively. These equations are repeated here for convenience:

$$S_m = \frac{8}{\pi^2}\left(\frac{\ell_x S_m}{\ell_z \ell_y}\right)$$

$$M_m = \frac{\ell_z \ell_y \ell_x \rho_m}{2}$$

$$\phi = \frac{d\ell_y}{S_m}$$

The mechanical resistance term, R_m, represents the mechanical losses in the ceramic element and associated structure resulting from the vibrating motion of the system. The ceramic mechanical losses are characterized by a mechanical quality factor, Q_m, defined as

$$Q_m = \frac{2\pi f_r M_m}{R_m} \tag{3-44}$$

where f_r is the bar resonant frequency. The value of Q_m is typically in the range 100 to 500 for ceramic materials used in projectors. Substitution for f_r and M_m, using (3-7) and (3-12), gives

$$R_m = \frac{\pi}{4}\left(\frac{A}{Q_m}\right) Z_{0m} \tag{3-45}$$

where A = cross-sectional area of bar

$Z_{0m} = \sqrt{\rho_m/s_m}$ = characteristic impedance of the ceramic

The mechanical losses in the supporting structure and in such things as ceramic bonding joints may be appreciable compared with losses in the ceramic material.

Accurate accounting of all losses generally requires experimental measurement for a given mechanical transducer configuration.

The impedance presented by the water to the active projector surface depends in detail on the transducer geometry. For the simple pulsating sphere discussed in Sections 2.4 and 2.6, this impedance is easily obtained as the product of the spherical specific acoustic impedance at the surface of the sphere and the spherical area. The form of the real and imaginary parts of this impedance is shown in Figure 2-9 and again in Figure 2-14. For a nonspherical transducer surface, the motion of each surface element affects the pressure at all points on the vibrating surface. The reaction force must be obtained by integrating the resulting pressure distribution over the entire surface. This is typically a difficult mathematical problem except for simple geometric shapes. A flat circular disk, or piston shape, is one shape of practical interest for which the impedance presented by the water, or *radiation impedance,* is readily available. Even for this case the resistive and reactive components of the radiation impedance must be expressed mathematically in the form of infinite series. For a circular piston transducer face of diameter D, the resistive and reactive terms of the piston impedance functions take the form

$$R_w = \frac{\rho c \pi D^2}{4} \left(\frac{\nu^2}{2 \cdot 4} - \frac{\nu^4}{2 \cdot 4^2 \cdot 6} + \frac{\nu^6}{2 \cdot 4^2 \cdot 6^2 \cdot 8} - \cdots \right)$$

$$X_w = \frac{\rho c \pi D^2}{4} \left(\frac{4}{\pi} \right) \left(\frac{\nu}{3} - \frac{\nu^3}{3^2 \cdot 5} + \frac{\nu^5}{3^2 \cdot 5^2 \cdot 7} - \cdots \right) \tag{3-46}$$

where $\nu = 2\pi D / \lambda$. These functions are plotted in normalized form in Figure 3-16. Notice that the piston impedance functions are actually very similar to the spherical impedance functions of Figure 2-9. For large values of $2\pi D / \lambda$, the resistive component approaches a constant equal to the product of the piston area and the characteristic impedance of water. In this same region, the reactive component approaches zero. For $2\pi D / \lambda \ll 1$ the resistive component is proportional to the square of the argument and the reactive component is a linear function of the argument.

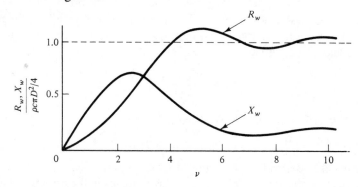

Figure 3-16   Radiation impedance functions for circular piston.

3.3.1 Example

Assume a longitudinal bar projector with the following characteristics:

$$\ell_y = \ell_z = 0.02 \text{ m}$$

$$\ell_x = 0.1 \text{ m}$$

$$d = -123 \times 10^{-12} \text{ m/V}$$

$$s_m = 10.9 \times 10^{-12} \text{ m}^2/\text{N}$$

$$\epsilon = 1300$$

$$\rho_m = 7.5 \times 10^3 \text{ kg/m}^3$$

$$\tan \delta = 0.01$$

$$Q_m = 100$$

In addition, we assume that the active projector face is formed in the shape of a circular piston with a diameter, $D = 0.028$ m. We desire to determine the electroacoustic equivalent circuit in the vicinity of the natural resonant frequency of the system, and the all-electric equivalent circuit. We shall determine and plot the input admittance in the vicinity of resonance, the bandwidth, the efficiency, and the acoustic output power as a function of frequency.

We first calculate the velocity of propagation and characteristic impedance for the ceramic material.

$$c_m = \sqrt{\frac{1}{\rho_m s_m}} = 3497 \text{ m/sec}$$

$$Z_{0m} = \rho_m c_m = (7500)(3497) = 2.62 \times 10^7 \text{ kg/m}^2\text{-sec}$$

From (3-7), the natural resonant frequency of the bar (clamped at one end) is

$$f_r = \frac{1}{4\ell_x \sqrt{\rho_m s_m}} = \frac{c_m}{4\ell_x} = 8742 \text{ Hz}$$

We now calculate the value of the various terms in the equivalent circuit.

Unclamped electrical capacitance:

$$C = \frac{\epsilon \epsilon_0 \ell_y \ell_x}{\ell_z}$$

$$= 1150 \times 10^{-12} \text{ F}$$

Dielectric electrical loss resistance (at $f = 8742$ Hz):

$$R_0 = \frac{1}{2\pi f C \tan \delta}$$

$$= 1.58 \times 10^6 \ \Omega$$

Clamped capacity:

$$C_0 = \epsilon\epsilon_0\left(1 - \frac{d^2}{\epsilon\epsilon_0 S_m}\right)\frac{\ell_x\ell_y}{\ell_z} = 1011 \times 10^{-12} \text{ F}$$

Mechanical compliance (3-11):

$$S_m = \left(\frac{8}{\pi^2}\right)\frac{\ell_x S_m}{\ell_z\ell_y} = 2209 \times 10^{-12} \text{ m/N}$$

Effective mass:

$$M_m = \frac{\ell_x\ell_y\ell_z\rho_m}{2} = 0.15 \text{ kg}$$

Ceramic mechanical loss resistance:

$$R_m = \frac{2\pi f_r M_m}{Q_M} = 82.3 \text{ kg/s}$$

The radiation impedance is calculated from (3-46) at the mechanical resonant frequency of the ceramic element. The wavelength in water at the resonant frequency is

$$\lambda_r = \frac{c}{f_r} = \frac{1500}{8742} = 0.17 \text{ m}$$

from which

$$\nu = \frac{2\pi D}{\lambda_r} = 1.03$$

and

$$R_w = \frac{\rho c \pi D^2}{4}\left(\frac{\nu^2}{8} - \cdots\right) = 122 \text{ kg/s}$$

$$X_w = \frac{\rho c \pi D^2}{4}\left(\frac{4\nu}{3\pi} - \cdots\right) = 403 \text{ kg/s}$$

$$M_w = \frac{X_w}{2\pi(8742)} = 0.007 \text{ kg}$$

This completes the determination of the values of the elements for the equivalent circuit of Figure 3-15. Using the transformation factor, ϕ, the values of elements for the all-electrical circuit shown in Figure 3-17 are obtained:

$$\text{transformation factor } \phi = \frac{d\ell_y}{S_m} = 0.226$$

$$L_M = \frac{M_m}{\phi^2} = 2.94 \text{ H}$$

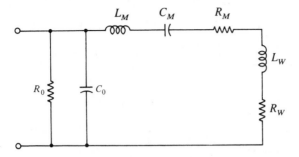

Figure 3-17 All-electrical projector equivalent circuit.

$$C_M = S_m \phi^2 = 112.8 \times 10^{-12} \text{ F}$$

$$R_M = \frac{R_m}{\phi^2} = 1611 \ \Omega$$

$$L_W = \frac{M_w}{\phi^2} = 0.12 \text{ H}$$

$$R_W = \frac{R_w}{\phi^2} = 2389 \ \Omega$$

From the all-electric equivalent circuit, the input admittance as seen at the electrical terminals is readily obtained as

$$Y_{in} = \underbrace{\left(\frac{1}{R_0} + j\omega C_0\right)}_{\substack{\text{electrical} \\ \text{branch}}} + \underbrace{\frac{1}{R_M + R_W + j[\omega(L_M + L_W) - 1/\omega C_M]}}_{\text{mechanical branch}} \qquad (3\text{-}47)$$

Substitution of numerical values determined for each of the terms in (3-47) permits the calculation and plotting of the conductance and susceptance elements of the input admittance as shown in Figure 3-18. Notice the sharp peak in the conductance term at the system resonant frequency at 8569 Hz. The susceptance in the vicinity of resonance is dominated by the behavior of the branch representing the mechanical portion of the system. The series resonance of the mechanical branch results in a rapid fluctuation in the net susceptance in the vicinity of resonance. For this particular design, the susceptance actually reverses sign and behaves inductively over a small frequency range above resonance. For frequencies far removed from resonance, the conductance is near zero and the susceptance is controlled by the value of C_0.

The acoustic power delivered to the water is the product of the radiation resistance term and the mean-squared current in the mechanical branch. Thus

$$P_0 = \overline{i_M^2} R_W = \overline{e^2} \, |Y_M|^2 R_W \qquad (3\text{-}48)$$

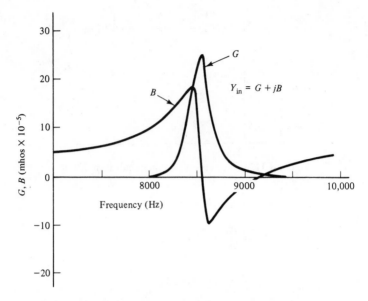

Figure 3-18 Input admittance for a longitudinal bar projector.

(where Y_M is the admittance of the mechanical branch), or

$$\frac{P_0}{e^2} = |Y_M|^2 R_W \tag{3-49}$$

The power output for a 1-V rms input is shown in Figure 3-19 as a function of frequency. From this figure the half-power bandwidth of the device is seen to be about 200 Hz. This bandwidth is also the ratio of the resonant frequency and the

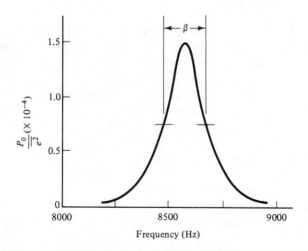

Figure 3-19 Normalized acoustic output power for a simple projector.

Q of the series resonant branch. That is,

$$Q = \frac{1}{2\pi f_r C_M(R_M + R_W)} \tag{3-50}$$

$$\beta = \frac{f_r}{Q} = \frac{8569}{41.4} = 206 \text{ Hz}$$

The efficiency of the device is the ratio of the acoustic power output to the total power input. Neglecting the very small electrical dielectric loss, the total input power, normalized to 1 V input, is

$$\frac{P_T}{e^2} = |Y_M|^2(R_W + R_M) \tag{3-51}$$

From which the efficiency is simply

$$\text{efficiency} = \frac{R_W}{R_W + R_M}$$

$$= \frac{2389}{4000} = 0.6 \text{ or } 60\%$$

3.3.2 Practical Projector Configurations

The longitudinal bar projector configuration is of limited usefulness, at least in the audio-frequency region. As with the hydrophones, the cylindrical shell forms the basis for several practical projector designs. For instance, the ceramic ring projector uses a segmented cylindrical configuration similar to that shown in Figure 3-8(b). For projector operation, the ceramic segments are actually physically separated by a conducting surface. The segments are tangentially polarized and the cylinder tightly wrapped to place the segments in compression.

A very common configuration makes use of a stacked set of longitudinally polarized elements of the type shown in Figure 3-8(c). A projector design using this configuration is called a Tonpilz projector and is shown in Figure 3-20. The ceramic elements are stacked with their polarization directions alternately reversed so that they may be connected electrically in parallel, while remaining mechanically in series. This arrangement allows the designer some flexibility with respect to the electrical input impedance of the device.

A metal rod through the center of the ceramic stack connects the baseplate, or tail mass, to the projector face, and provides a means for prestressing the ceramic stack. The active face, or head mass, is a solid metal piece with a flat circular, or piston-like, face. The mass of the active face must be considered, together with the effective mass of the ceramic stack and water load, when calculating the resonant frequency of the system.

Detailed electroacoustic equivalent circuits for projectors using stacks of longitudinally poled ceramic disks have been developed by Martin [4–6]. The

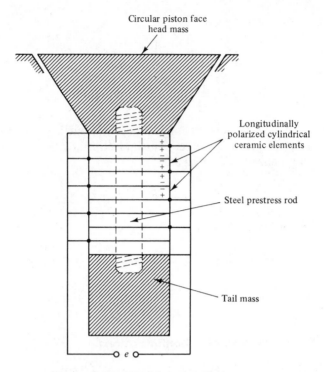

Figure 3-20 Tonpilz projector.

mechanical properties of each disk are modeled by a T-network, with the stack represented by a cascade of the individual T-networks. The electrical drive is applied to the common base connection of the cascaded mechanical elements. This is obviously much more complex than the simple equivalent circuits previously described. However, in the vicinity of resonance, the behavior of the device is often adequately described using an equivalent circuit of the form shown in Figure 3-15, provided that the elements in the equivalent circuit are appropriately redefined or measured experimentally.

Another projector configuration using a stack of longitudinally polarized ceramic elements is the *flextensional* transducer [7] shown in Figure 3-21. In this configuration the ceramic stack is mounted on the major axis of a metallic elliptical cylinder. Prestress is applied by compressing the ellipse along its minor axis, thereby extending the major axis dimension. The slightly oversized stack is then put in place on the major axis and the compressive force released on the minor axis. This places the ceramic stack in compression. Shims may be used at the ends of the ceramic stack to achieve the desired prestress level.

With the flextensional configuration, the elliptical shell acts as a mechanical impedance transformer. A small velocity imparted at the ends of the ceramic stack is converted to a larger velocity at the major faces of the ellipse. Thus both the radiation resistance and the effective mass of the water load transform to

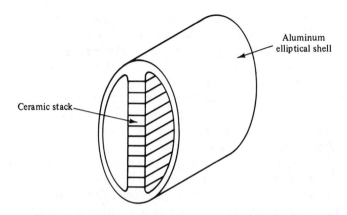

Figure 3-21 Flextensional transducer.

larger values as seen by the ceramic stack. This provides the designer with a means for modifying the impedance match between the source and the water load, thereby improving efficiency and increasing bandwidth.

3.3.3 Cavitation

The maximum acoustic output power obtainable from a projector is limited by the usual electrical, mechanical, and thermal limitations. The peak applied voltage is limited by the dielectric strength of the ceramic and of the electrical connections. Internal mechanical and dielectric losses will raise the temperature to unacceptable levels if the projector is driven too hard. An additional limitation is imposed by the phenomenon of *cavitation* of the water medium at the vibrating surface of the projector. At shallow depths the output power may be limited by cavitation at a level far below that imposed by other factors.

The alternating acoustic pressure produced at the face of a projector is superimposed on the ambient static pressure of the medium. On the rarefaction half of the acoustic pressure cycle, the absolute pressure is reduced below the ambient pressure. As the acoustic intensity is increased, the absolute pressure may be reduced to zero or become negative. At this point the medium literally ruptures, causing the formation of large numbers of bubbles from the release of dissolved gases in the water. Cavitation is the name applied to the process of bubble formation and collapse.

Several undesirable effects occur in the presence of cavitation. The presence of large numbers of bubbles near the projector face act like a pressure release surface and, therefore, drastically reduce the radiation impedance seen by the projector. This "short-circuit" effect limits the radiated acoustic power at the level where cavitation begins. If the projector input power is increased further, the acoustic output power may actually decrease while the internal projector losses increase rapidly.

Near the cavitation threshold the bubble formation on the negative half pressure cycle causes a second harmonic distortion of the acoustic pressure wave. This harmonic distortion becomes noticeable before the cavitation bubbles become visible to the naked eyes and, therefore, may be used as a sensitive indicator of the onset of cavitation.

The violent action of the oscillating and collapsing bubbles formed during cavitation may actually cause erosion of the projector face. Finally, the collapsing bubbles generate significant amounts of acoustic noise that may interfere with the intended purpose of the projector.

The acoustic intensity at cavitation threshold is primarily a function of depth, although many other factors must also be considered. At sea level, the absolute pressure is reduced to zero if the peak acoustic pressure is 1 atm. Define this condition as the cavitation threshold, with a threshold intensity T. With the peak acoustic pressure equal to 10^5 N/m^2, or 1 atm, T is given by

$$T = \frac{p_m^2}{2\rho c} = \frac{(10^5)^2}{(2)(1.5 \times 10^6)} = 0.33 \times 10^4 \text{ W/m}^2 = 0.33 \text{ W/cm}^2$$

As an example, a piston projector with a 6-in.-diameter face is limited by cavitation to a maximum power of about 60 W when operated near the water surface. Since the cavitation threshold is proportional to the square of the ambient pressure, the power that can be radiated increases rapidly with depth. At a depth of 100 ft, the pressure is about 4 atm. At this depth, the 6-in.-diameter projector is capable of about 16 times the maximum power at the surface, or 960 W.

Assume a total radiated acoustic power, P, at the onset of cavitation, distributed uniformly over an effective projector area, A. With a threshold T at sea level, the relationship between maximum power and depth is given by

$$P = AT\left(1 + \frac{h}{33}\right)^2 \tag{3-52}$$

where h = depth, ft.

The cavitation threshold is generally lowered by the presence of excessive amounts of dissolved gases or microscopic bubbles in the water. This makes the accurate prediction of cavitation threshold near the ocean surface quite difficult. The problem is further complicated by the fact that the acoustic intensity may not be uniform over the active face of the projector. Local "hot spots" on the projector face may initiate cavitation even though the average intensity over the surface may be below the cavitation threshold.

The process of bubble formation during the cavitation process requires a finite time. As the acoustic frequency increases, the time available on each negative half-cycle decreases with the result that a higher peak acoustic intensity is required to initiate cavitation. Figure 3-22 presents the cavitation threshold at sea level for fresh water as a function of frequency. The frequency dependence

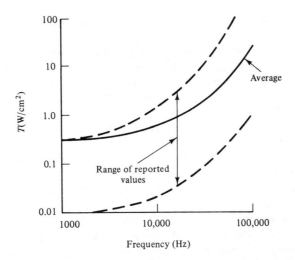

Figure 3-22   Cavitation threshold versus frequency: CW data in fresh water at 1 atm. (Data from Flynn [12] and Esche [13].)

is very slight below 10 kHz, but thereafter the cavitation threshold increases rapidly as frequency increases. At 30 kHz, the average threshold is approximately 1 W/cm². Note, however, the wide variation in cavitation threshold about the average value. This variation results from the wide variation in the condition of the water with respect to dissolved gases and bubbles and from differences in measurement criteria and technique of various investigators.

For a finite duration pulse of acoustic energy, the cavitation threshold increases for pulse lengths below about 5 msec. Again the finite time required for the cavitation process results in a higher threshold intensity as the pulse duration decreases.

PROBLEMS

3.1. The ceramic barium titanate is often used in transducer applications. The density of this material is 5.55×10^3 kg/m³ and the elastic compliance is 8.3×10^{-12} m²/N. Calculate the characteristic impedance and the sound speed of longitudinal compression waves for this material, and compare with the corresponding parameters for water.

3.2. A slender bar of barium titanate is rigidly clamped at one end. A longitudinal sinusoidal force is applied at the other end and it is determined that the lowest frequency resulting in a zero input impedance is 30 kHz. Determine the length of the bar.

3.3. The dielectric constant of barium titanate is 1200 and the piezoelectric strain coefficient is -54×10^{-12} C/N. Using the characteristics listed in Section 3.2.3 for the material PZT-5, compare the open-circuit hydrophone sensitivities for these two materials, assuming a simple longitudinal bar hydrophone configuration.

3.4. A radially poled ceramic cylindrical shell hydrophone has an unclamped electrical capacitance of 0.02 μF and a natural radial mode resonant frequency of 25 kHz. Assuming that the ceramic is PZT-5A and the cylindrical length is 2 in., determine the diameter and wall thickness of the cylinder and the open-circuit sensitivity of the hydrophone.

3.5. The hydrophone in Problem 3.4 is connected to a preamplifier through a cable having 50×10^{-12} F/ft. Assuming the preamplifier has an input capacity of 20×10^{-12} F, determine the length of cable that will result in a loss of 1 dB in effective hydrophone sensitivity.

3.6. A radial-mode cylindrical hydrophone using PZT-5A has a natural resonant frequency of 10 kHz. Assuming a wall thickness small compared with the diameter, determine the maximum possible hydrophone sensitivity for this design.

3.7. Using (3-33) it can be shown that the sensitivity of a tangential mode cylindrical hydrophone increases without bound as b/a approaches unity. However, the capacitance of the device approaches zero with this same limiting condition, resulting in an infinite internal impedance. A more useful description of the limiting performance of the tangential design is given by the product of the sensitivity from (3-33) and the capacitance from (3-34). Determine the form of this product as b/a approaches unity using the approximation $\ln(b/a) = (b/a) - 1$.

3.8. A tangential mode cylindrical hydrophone has a natural resonant frequency of 30 kHz. The capacitance of the device is designed to be 10 times the input capacitance of the preamplifier, which is 20×10^{-12} F. Using the result from Problem 3.7, determine the limiting value of the hydrophone sensitivity assuming the following parameters:

$$\text{material} = \text{PZT-5A}$$

$$N = 6$$

$$\ell = 2b$$

3.9. With reference to Figure 3-9, assume that

$$C = 0.015 \ \mu\text{F}$$

$$C_\ell + C_a = 0.005 \ \mu\text{F}$$

$$e_n = 3 \times 10^{-8} \text{ V rms (in 1-Hz band)}$$

With an input acoustic noise spectral level of $+60$ dB//μPa/Hz, determine the required *open-circuit* hydrophone sensitivity to obtain a $+10$-dB ratio of acoustic signal-to-electronic noise power ratio at the input to the preamplifier. With a 20-dB preamplifier voltage gain, determine the rms voltage in a 1-Hz band at the preamplifier output resulting from the acoustic signal input.

3.10. A high-frequency omnidirectional projector operating at 30 kHz has an effective radiating area of 30 cm^2. With a cavitation threshold of 1 W/cm^2, determine the minimum noncavitating depth for a source level of 195 dB//μPa/m. Determine the required radiating area that would result in a cavitation threshold depth of 20 ft.

SUGGESTED READING

1. Weston, D. E., "Underwater Explosions as Acoustic Sources," *Proc. Phys. Soc. Lond.*, Vol. 76 (Pt. 2), p. 233 (1960).

2. Bouyoucos, J. V., "Self-Excited Hydrodynamic Oscillators," Acoustic Research Laboratory, Harvard University, TM No. 36, July 31, 1955.

3. Langevin, R. A., "The Electro-acoustic Sensitivity of Cylindrical Ceramic Tubes," *J. Acoust. Soc. Am.*, Vol. 26, No. 1, p. 421 (May 1954).

4. Martin, G. E., "Vibrations of Longitudinally Polarized Ferroelectric Cylindrical Tubes," *J. Acoust. Soc. Am.*, Vol. 35, No. 4, pp. 510–521 (Apr. 1963).

5. Martin, G. E., "On the Theory of Segmented Electromechanical Systems," *J. Acoust. Soc. Am.*, Vol. 36, No. 7, pp. 1366–1370 (July 1964).

6. Martin, G. E., "Vibrations of Coaxially Segmented Electromechanical Systems," *J. Acoust. Soc. Am.*, Vol. 36, No. 8 (Aug. 1964).

7. Toulis, W. J., "Design Problems for High Power Flextensional Transducers," *U.S. Navy J. Underwater Acoust.*, Vol. 15, No. 2, Pt. III, Appendix D (Apr. 1965).

8. Urick, R. J., *Principles of Underwater Sound for Engineers*, 2nd ed. New York: McGraw-Hill Book Company, 1975, Chap. 4.

9. Albers, V. M., *Underwater Acoustics Handbook—II*, University Park, Pa.: The Pennsylvania State University Press, 1965, Chaps. 10–12.

10. Woollett, R. S., "Ultrasonic Transducers: Part 2, Underwater Sound Transducers," *Ultrasonics*, Vol. 3, pp. 243–253 (Oct. 1970).

11. Hueter, T. F., "Twenty Years in Underwater Acoustics: Generation and Reception," *J. Acoust. Soc. Am.*, Vol. 51, No. 3 (Pt. 2), pp. 1025–1040 (1972).

12. Flynn, H. G., "Physics of Acoustic Cavitation in Liquids," in *Physical Acoustics*, Vol. I, Part B, W. P. Mason, Ed. Academic Press, Inc., New York: 1964, pp. 57–172.

13. Esche, R., "Schwingungskavitation in Flussigkeiten," *Akust. Beih. (Acustica)*, Vol. 4 (1952).

4

Reflection, Transmission, and Refraction

~~~~~~

4.0 INTRODUCTION

In Chapter 2 equations were developed describing the propagation of both plane and spherical acoustic waves in an idealized homogeneous medium. Actually, factors affecting acoustic transmission in the ocean are variable in all three spatial dimensions, and may vary with time as well. By far the most important variation in acoustic parameters occurs with depth. The ocean tends to be structured in horizontal layers, starting with the air–water interface at the surface and ending in the various sedimentary layers at the bottom.

In this chapter we consider the modifications imposed on an acoustic plane wave at the interface between fluid layers with differing characteristics. In general, a portion of the plane wave energy is reflected from the interface, and a portion is transmitted through the interface, possibly with a change in direction (refraction). Of particular importance to the long-range transmission of acoustic energy are the losses that may occur on reflection from the surface and bottom interfaces. As an introduction to determining propagation paths in the ocean, the refraction resulting from a sound speed that is a linear function of depth is considered. The concept of acoustic ray diagrams is introduced as an aid in determining the variation in acoustic intensity with range and depth.

4.1 NORMAL INCIDENCE

Assume a medium consisting of two stratified fluids as shown in Figure 4-1. The acoustic properties of interest are the densities and sound speeds for the two fluids, which determine the plane wave characteristic impedances, Z_1 and Z_2.

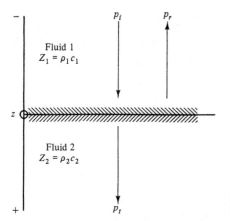

Figure 4-1 Plane wave normal to a boundary.

A plane acoustic wave with pressure p_i travels in the positive z-direction (downward) normal to the fluid interface. Assuming that $Z_1 \neq Z_2$, a *reflected wave* with pressure p_r originates at the fluid interface and travels in the negative z-direction. A portion of the energy in the *incident wave* proceeds into fluid 2, resulting in a *transmitted wave* with pressure p_t traveling in the positive z-direction. We shall derive the relationships among p_i, p_r, and p_t in terms of the characteristic impedances, Z_1 and Z_2.

Let the pressure waves be represented by complex sinusoidal functions of time and distance as follows:

$$p_i(t, z) = A_1 \exp\left[j\omega\left(t - \frac{z}{c_1} \right) \right]$$

$$p_r(t, z) = B_1 \exp\left[j\omega\left(t + \frac{z}{c_1} \right) \right] \qquad (4\text{-}1)$$

$$p_t(t, z) = A_2 \exp\left[j\omega\left(t - \frac{z}{c_2} \right) \right]$$

Remember that the actual pressures are given by the real parts of (4-1).

The relationships among A_1, A_2, and B_1 are found by introducing the physical conditions that must be satisfied at the fluid interfaces. For the two fluids to remain in contact, it is necessary that neither the acoustic pressure nor the particle speed normal to the interface be discontinuous at the interface. Thus, letting $z = 0$, we may write

$$p_i(t, 0) + p_r(t, 0) = p_t(t, 0) \qquad (4\text{-}2)$$

Similarly, the resultant particle speed in fluid 1 in the z-direction must equal that in fluid 2 at $z = 0$.

$$u_i(t, 0) + u_r(t, 0) = u_t(t, 0) \qquad (4\text{-}3)$$

From the requirement (4-2) and the functions in (4-1), with $z = 0$, we immediately obtain

$$A_1 + B_1 = A_2 \tag{4-4}$$

In Chapter 2 we learned that plane wave acoustic pressure and particle speed are related through the characteristic impedance [see (2-38) and (2-39)]. Hence, with $z = 0$ we write

$$u_i(t) = \frac{p_i(t)}{Z_1}$$

$$u_r(t) = -\frac{p_r(t)}{Z_1} \tag{4-5}$$

$$u_t(t) = \frac{p_t(t)}{Z_2}$$

Substitution of the relationships in (4-5) into (4-3) gives

$$\frac{1}{Z_1}[p_i(t) - p_r(t)] = \frac{1}{Z_2}p_t(t)$$

or

$$\frac{1}{Z_1}(A_1 - B_1) = \frac{A_2}{Z_2} \tag{4-6}$$

Now use (4-4) to eliminate A_2 in (4-6) and obtain the relationship between A_1 and B_1.

$$\frac{B_1}{A_1} = \frac{Z_2 - Z_1}{Z_2 + Z_1} = \frac{\rho_2 c_2 - \rho_1 c_1}{\rho_2 c_2 + \rho_1 c_1} \tag{4-7}$$

Notice that the ratio of the reflected presssure amplitude to incident pressure amplitude is identical in form to (2-13), which gives the ratio of reflected to incident voltage amplitudes for an electrical transmission line.

Equation (4-4) may also be used to eliminate B_1 from (4-6), thereby giving the relationship between the incident and the transmitted pressure amplitudes. Thus

$$\frac{A_2}{A_1} = \frac{2Z_2}{Z_2 + Z_1} \tag{4-8}$$

The reflection and transmission coefficients defined by (4-7) and (4-8) are both real because the plane wave characteristic impedances are real. Notice, however, that (4-7) can be either positive or negative, depending on the relationship between Z_1 and Z_2.

If Z_2 exceeds Z_1, the reflection coefficient is positive and the incident and reflected pressures add at the interface. In the extreme, as Z_2 approaches infinity the reflection coefficient is unity, with the result that the pressure at the interface is double the pressure in the incident wave. This result is analogous to the open-circuit termination for an electrical transmission line.

If Z_2 is less than Z_1, the reflection coefficient is negative, indicating a 180° relationship between the incident and reflected pressures at the interface. As Z_2

approaches zero, the reflection coefficient approaches -1, resulting in perfect cancellation of the incident and reflected pressures at the interface. This corresponds to the short-circuited electrical transmission line with zero voltage at the short-circuit termination.

The total pressure at any point above the interface in Figure 4-1 is obtained as the sum of the incident and reflected waves. From (4-1) and (4-7) for $z < 0$ we write

$$p(t,z) = p_i + p_r$$

$$= A_1 \left\{ \exp\left[j\omega\left(t - \frac{z}{c_1} \right) \right] + R \exp\left[j\omega\left(t + \frac{z}{c_1} \right) \right] \right\}$$

where

$$R = \frac{B_1}{A_1} = \frac{Z_2 - Z_1}{Z_2 + Z_1}$$

The real part of this expression is

$$\mathrm{Re}[p(t,z)] = \pm A_1 \left[(1 + R)^2 \cos^2\left(\frac{\omega z}{c_1} \right) + (1 - R)^2 \sin^2\left(\frac{\omega z}{c_1} \right) \right]^{1/2}$$

$$\times \cos\left\{ \omega t + \tan^{-1}\left[\left(\frac{1 - R}{1 + R} \right) \tan\left(\frac{\omega z}{c_1} \right) \right] \right\} \qquad (4\text{-}9)$$

Now let $R = +1$, corresponding to a very stiff interface ($Z_2 \gg Z_1$). Equation (4-9) simplifies to

$$\mathrm{Re}[p(t,z)] = A_1 \left[2 \cos\left(\frac{\omega z}{c_1} \right) \right] \cos \omega t$$

$$= A_1 \left[2 \cos\left(\frac{2\pi z}{\lambda} \right) \right] \cos \omega t \qquad (4\text{-}10)$$

where $\lambda = c_1/f$.

Equation (4-10) describes a standing wave of pressure above the interface similar to the standing wave of voltage that exists on an open-circuited transmission line (see Section 2.1). The pressure magnitude has a maximum value of $|2A_1|$ at the interface and at other locations where z is an integral multiple of $\lambda/2$. That is,

$$|p(t,z)| = |2A_1| \qquad \text{for } z = \frac{n\lambda}{2}, \quad n = 0, 1, 2, \ldots \qquad (4\text{-}11)$$

The pressure is zero for z an odd multiple of $\lambda/4$.

$$p(t,z) = 0 \qquad \text{for } z = \frac{m\lambda}{4}, \quad m = 1, 3, 5, \ldots \qquad (4\text{-}12)$$

For $R = -1$, corresponding to a soft, or pressure release, surface ($Z_2 \ll Z_1$) an expression similar to (4-10) is obtained except that the pressure is

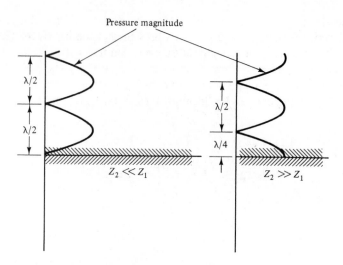

Figure 4-2   Pressure standing waves above a pressure release surface ($Z_2 \ll Z_1$) and a stiff surface ($Z_2 \gg Z_1$).

zero at the interface. Thus for $R = -1$

$$\text{Re}[p(t, z)] = A_1 \left[2 \sin\left(\frac{2\pi z}{\lambda}\right) \right] \sin \omega t$$

$$p(t, z) = 0 \qquad \text{for } z = \frac{n\lambda}{2}, \quad n = 0, 1, 2, \ldots \qquad (4\text{-}13)$$

$$|p(t, z)| = |2A_1| \qquad \text{for } z = \frac{m\lambda}{4}, \quad m = 1, 3, 5, \ldots$$

The magnitude of the pressure standing wave is sketched in Figure 4-2 for both a pressure release surface and a stiff surface.

4.1.1 Normal Modes of Vibration

Assume now a stiff interface with a standing wave of pressure as shown on the right side of Figure 4-2. Consider the consequence of placing a zero impedance pressure release surface one-quarter wavelength above the stiff boundary. The standing wave pressure amplitude is zero at the location of the pressure release surface, thereby satisfying the *required* boundary condition for this type of surface. This same result would hold for the pressure release surface located at distances $3\lambda/4, 5\lambda/4, \ldots$ above the lower boundary. On the other hand, if the pressure release surface is placed at any other location it is impossible simultaneously to satisfy both the upper and lower boundary conditions with a standing (or stationary) pressure amplitude distribution. The two-boundary situation is shown in Figure 4-3.

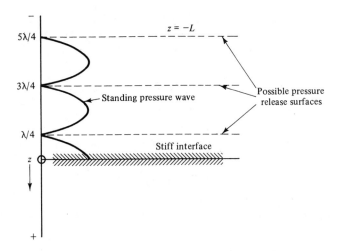

Figure 4-3   Pressure distribution between pressure release surface and stiff surface.

This two-boundary model is similar to the situation in the ocean, with the upper surface representing the air–water interface and the lower surface the ocean bottom. Representing the ocean bottom as a stiff interface is an oversimplification that nevertheless provides useful insights for our present purpose.

Let the depth be fixed at L and define the *allowed* values for the wavelengths that are consistent with a stationary (standing wave) pressure amplitude distribution. These are given by

$$\lambda_j = \frac{4L}{j}, \qquad j = 1, 3, 5, \ldots \tag{4-14}$$

The allowed stationary states are called the *normal modes of vibration* for the channel of depth L and the indicated boundary conditions. For the jth mode, with the pressure given by

$$p_j(t, z) = A \left[2 \cos \left(\frac{2\pi z}{\lambda_j} \right) \right] \sin \omega t$$

the z-dependent portion of the expression is called the *characteristic function*. For the assumed boundary conditions, the characteristic functions are defined as

$$\Psi_j(z) = 2 \cos \left(\frac{2\pi z}{\lambda_j} \right), \qquad j = 1, 3, 5, \ldots \tag{4-15}$$

We shall now show that the characteristic functions are obtained as the stationary solutions to the plane wave equation, assuming a sinusoidal plane pressure wave normal to the boundaries. Assume a pressure distribution given by

$$p(t, z) = \Psi(z) \cos (\omega t - \theta) \tag{4-16}$$

where Ψ is an arbitrary function of z and θ is an arbitrary phase. The assumed boundary conditions require that the pressure amplitude be a maximum at the bottom and zero at the surface. Letting $z = 0$ at the bottom and $-L$ at the surface, this means that

$$\Psi(0) = \Psi_{max}$$
$$\Psi(-L) = 0 \tag{4-17}$$

The plane wave equation for motion in the z-direction is

$$\frac{\partial^2 p}{\partial t^2} = c^2 \frac{\partial^2 p}{\partial z^2} \tag{4-18}$$

Partial differentiation of (4-16) twice with respect to t and twice with respect to z gives

$$\frac{\partial^2 p}{\partial t^2} = -\omega^2 \Psi(z) \cos(\omega t - \theta)$$

$$\frac{\partial^2 p}{\partial z^2} = \frac{\partial^2 \Psi}{\partial z^2} \cos(\omega t - \theta) \tag{4-19}$$

Substitute (4-19) into (4-18) to obtain

$$\frac{\partial^2 \Psi}{\partial z^2} + \frac{\omega^2}{c^2} \Psi = 0$$

or, letting $\omega/c = 2\pi/\lambda$,

$$\frac{\partial^2 \Psi}{\partial z^2} + \frac{4\pi^2}{\lambda^2} \Psi = 0 \tag{4-20}$$

Equation (4-20) is the well-known differential equation for simple harmonic motion with the general solution

$$\Psi(z) = a \cos\left(\frac{2\pi z}{\lambda}\right) + b \sin\left(\frac{2\pi z}{\lambda}\right) \tag{4-21}$$

setting $\Psi(0) = \Psi_{max}$, we obtain

$$\Psi_{max} = a \tag{4-22}$$

Because the maximum value of Ψ is also $(a^2 + b^2)^{1/2}$, this implies that $b = 0$. Hence

$$\Psi(z) = \Psi_{max} \cos\left(\frac{2\pi z}{\lambda}\right) \tag{4-23}$$

For this particular problem, Ψ_{max} is twice the magnitude of either the incident or reflected waves.

The requirement that $\Psi(-L) = 0$ means that (4-23) can only be a proper solution if

$$\frac{2\pi L}{\lambda} = j\frac{\pi}{2}, \qquad j = 1, 3, 5, \ldots$$

or
(4-24)

$$\lambda_j = \frac{4L}{j}$$

which is identical to the result in (4-14).

Because each normal mode satisfies the boundary conditions, the sum of two or more normal modes also represents a stationary solution. Thus, an arbitrary pressure distribution may be expressed as an infinite sum of normal modes. For the boundary conditions of Figure 4-3, the general solution is

$$p(t,z) = \sum_j a_j \cos\left(\frac{2\pi z}{\lambda_j}\right) \cos\left(2\pi f_j t - \theta_j\right)$$
(4-25)

For other boundary conditions, the characteristic functions may include both the sine and cosine components of (4-21). The arbitrary constants associated with the normal mode expansion are evaluated by methods related to the Fourier expansion of functions discussed in Chapter 6.

The expansion of the pressure field as an infinite series of normal mode terms is a very useful analytic tool, especially when the wavelength is not insignificant relative to the water depth. The discussion of normal modes in this section considers only a very simple and restrictive case (that is, a plane wave normal to the surface and bottom with very simple boundary conditions). The generalized normal mode expansion starts with the general wave equation and provides a three-dimensional expansion of the pressure field with arbitrary boundary conditions. Since this added complexity is beyond our present scope, it is hoped that the simplified development presented provides some understanding of the normal mode concept. For a comprehensive treatment of normal mode theory, see for instance, the excellent discussion in [2].

4.2 OBLIQUE INCIDENCE

Consider a plane acoustic wave incident at a fluid boundary at an arbitrary grazing angle θ_i relative to the plane boundary, as shown in Figure 4-4. In optics, the angles are generally measured relative to the normal to the boundary. In ocean acoustics, the important boundaries are horizontal and the acoustic propagation is often near-horizontal. It is therefore more natural to measure direction of propagation relative to the horizontal. By convention, positive angles are associated with downward propagation, while upward propagation is assigned negative angles. Similarily, the positive z-axis is directed downward.

With the plane wave traveling in the direction θ_i relative to the x-axis, a distance s in the direction of propagation may be related to its x and z components by

$$s = x \cos \theta_i + z \sin \theta_i$$

Therefore, the incident plane wave in complex notation becomes

$$p_i(t, x, z) = A_1 \exp \left\{ j\omega \left[t - \frac{(x \cos \theta_i + z \sin \theta_i)}{c_1} \right] \right\} \quad (4\text{-}26)$$

Similarly, the reflected and transmitted waves are

$$p_r(t, x, z) = B_1 \exp \left\{ j\omega \left[t - \frac{(x \cos \theta_r + z \sin \theta_r)}{c_1} \right] \right\}$$

$$p_t(t, x, z) = A_2 \exp \left\{ j\omega \left[t - \frac{(x \cos \theta_t + z \sin \theta_t)}{c_2} \right] \right\} \quad (4\text{-}27)$$

At the fluid boundary, the pressure must be continuous. Thus

$$p_i + p_r = p_t \quad (\text{at } z = 0)$$

or

$$A_1 \exp \left(\frac{-j\omega x \cos \theta_i}{c_1} \right) + B_1 \exp \left(\frac{-j\omega x \cos \theta_r}{c_1} \right)$$

$$= A_2 \exp \left(\frac{-j\omega x \cos \theta_t}{c_2} \right) \quad (4\text{-}28)$$

As an additional boundary condition, we require that the particle velocity in the z-direction be continuous. This results in the relationship

$$U_i \sin \theta_i + U_r \sin \theta_r = U_t \sin \theta_t \quad (4\text{-}29)$$

where $\quad U_i = \dfrac{A_1}{Z_1}$

$$U_r = \frac{B_1}{Z_1}$$

$$U_t = \frac{A_2}{Z_2}$$

Two results from optics are helpful in defining the relationships among the incident, reflected, and transmitted (refracted) waves at the fluid boundary. First, the angle of reflection is the negative of the angle of incidence. That is

$$\theta_r = -\theta_i \quad (4\text{-}30)$$

Second, Snell's law relates the angles of incidence and transmission in terms of

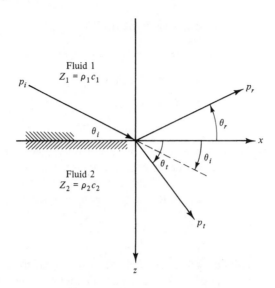

Figure 4-4   Transmission through a boundary at oblique incidence.

the sound speeds in the two media. For the angles as defined in Figure 4-4, Snell's law requires that

$$\frac{\cos\theta_i}{c_1} = \frac{\cos\theta_t}{c_2} \tag{4-31}$$

Applying these relationships to (4-28), we see that

$$\frac{\cos\theta_i}{c_1} = \frac{\cos\theta_r}{c_1} = \frac{\cos\theta_t}{c_2} \tag{4-32}$$

with the result that all exponents in (4-28) are equal. Therefore, we again obtain

$$A_1 + B_1 = A_2 \tag{4-33}$$

In a similar manner, using (4-30) and (4-33) in (4-29), we write

$$(U_i - U_r)\sin\theta_i = U_t\sin\theta_t$$

or

$$\left(\frac{A_1}{Z_1} - \frac{B_1}{Z_1}\right)\sin\theta_i = \frac{A_2}{Z_2}\sin\theta_t = \frac{A_1 + B_1}{Z_2}\sin\theta_t$$

Solving for the ratio of reflected to incident amplitude,

$$\frac{B_1}{A_1} = \frac{Z_2\sin\theta_i - Z_1\sin\theta_t}{Z_2\sin\theta_i + Z_1\sin\theta_t} \tag{4-34}$$

4.2.1 Conditions for Zero Reflection

The reflection coefficient defined by (4-34) is zero if, and only if

$$Z_2 \sin \theta_i = Z_1 \sin \theta_t$$

or

$$\rho_2 c_2 \sin \theta_i = \rho_1 c_1 \sin \theta_t \tag{4-35}$$

Notice that if $Z_1 = Z_2$, zero reflection is guaranteed only at normal incidence. If Z_1 does not equal Z_2, there *may* be an angle that results in zero reflection. Using Snell's law, (4-35) may be solved for the conditions necessary for zero reflection. Thus the incident angle for zero reflection is given by

$$\sin^2 \theta_i = \frac{(c_1/c_2)^2 - 1}{(\rho_2/\rho_1)^2 - 1} \tag{4-36}$$

The incident angle defined by (4-36), if it exists, is called the *angle of intromission*. Because the square of the sine must lie between zero and $+1$, this angle will exist only if the numerator and denominator in (4-36) have the same sign and if the denominator equals or exceeds the numerator in absolute value. These conditions require that

$$\frac{\rho_2}{\rho_1} \geq \frac{c_1}{c_2} \geq 1$$

or   (4-37)

$$\frac{\rho_2}{\rho_1} \leq \frac{c_1}{c_2} \leq 1$$

A trivial case occurs if $\rho_1 = \rho_2$ and $c_1 = c_2$, resulting in both conditions in (4-37) being met. In this case the two fluids are indistinguishable acoustically, so that in fact no boundary exists and there is no reflection under any condition.

4.2.2 Other Special Conditions

In Figure 4-4, the angle of transmission, θ_t is larger than θ_i, indicating a positive or downward refraction. From Snell's law we see that this condition results if $c_1 > c_2$ That is,

$$\cos \theta_i = \left(\frac{c_1}{c_2}\right) \cos \theta_t \tag{4-38}$$

Therefore,

$$\cos \theta_i > \cos \theta_t \qquad \text{for } \frac{c_1}{c_2} > 1$$

The ratio of sound speeds, c_1/c_2, is the *index of refraction* of the second fluid relative to the first, and in general may range in value from zero to infinity. In the limit as c_1/c_2 approaches infinity in (4-38), θ_t must approach 90°, or a direction normal to the boundary. As an example, assume that fluid 1 is water and fluid 2 is air. The index of refraction of air relative to water is approximately 5, with the result that as θ_i in water varies from zero to 90°, θ_t in air is restricted to the range 78 to 90°. To a person in a submerged submarine, a sound originating outside the pressure hull in the water seems always to originate in a direction normal to the hull, regardless of the actual bearing to the source of the sound. This is a direct result of the large index of refraction of the air inside the hull relative to the outside water. Because of the large impedance mismatch at the air–water boundary, the energy transmitted from the water into the air is a small portion of the total, and the reflection in the water is nearly complete at all incidence angles.

Now let $c_1 < c_2$. From (4-38) we see that θ_t must be less than θ_i, resulting in a negative refraction. Since θ_t cannot be less than zero, there is a critical value of θ_i below which the transmission into the second medium must be zero. The cosine of the critical incidence angle is equal to the index of refraction. That is,

$$\cos \theta_c = \frac{c_1}{c_2} \cos \theta_t = \frac{c_1}{c_2} \quad \text{for } \theta_t = 0 \qquad (4\text{-}39)$$

For θ_i less than θ_c, the reflection from the boundary is total.

4.2.3 Transmission Coefficient at Oblique Incidence

Equations (4-29), (4-30), and (4-33) may be used to obtain the ratio of the transmitted to incident pressure amplitudes, with the following result:

$$\frac{A_2}{A_1} = \frac{2Z_2 \sin \theta_i}{Z_2 \sin \theta_i + Z_1 \sin \theta_t} \qquad (4\text{-}40)$$

4.3 TWO-BOUNDARY PROBLEM

Transmission and reflection phenomena in multilayered media are considerably more complex than the single-boundary problem of previous sections. In this section we consider a two-boundary problem as shown in Figure 4-5. The two boundaries separate three fluids with characteristic impedances Z_1, Z_2, and Z_3. The middle fluid layer, with thickness h, separates the top and bottom fluids that are presumed to be infinitely thick.

At the boundary between fluid 1 and fluid 2, we define the reflection coefficient and transmission coefficient for an incident plane wave originating in

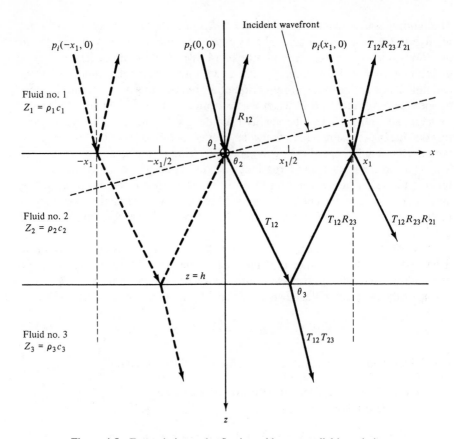

Figure 4-5 Transmission and reflection with two parallel boundaries.

fluid 1 as follows:

$$R_{12} = \frac{Z_2 \sin \theta_1 - Z_1 \sin \theta_2}{Z_2 \sin \theta_1 + Z_1 \sin \theta_2}$$

$$T_{12} = \frac{2Z_2 \sin \theta_1}{Z_2 \sin \theta_1 + Z_1 \sin \theta_2} \tag{4-41}$$

We shall also be interested in reflection and transmission for a plane wave originating in fluid 2 incident at the boundary between fluids 1 and 2. A little thought would show that

$$R_{21} = -R_{12}$$

$$T_{21} = \frac{2Z_1 \sin \theta_2}{Z_2 \sin \theta_1 + Z_1 \sin \theta_2} \tag{4-42}$$

The transmission and reflection coefficients are related by

$$T_{12}T_{21} = 1 - R_{12}^2 \tag{4-43}$$

Similar expressions could be developed at the boundary between fluids 2 and 3 for R_{23} and T_{23}.

We now define the incident, reflected, and transmitted components of pressure at the locations $(x = 0, z = 0)$, $(x = x_1/2, z = h)$, and $(x = x_1, z = 0)$ resulting from the incident pressure at the origin, p_i. In complex notation, let the incident pressure in fluid 1 be

$$p_i(t, x, z) = \exp\left[j\omega\left(t - \frac{x\cos\theta_1 + z\sin\theta_1}{c_1}\right)\right] \qquad (4\text{-}44)$$

The incident pressure at the origin $(x = 0, z = 0)$ results in reflected and transmitted components as follows:

$$p_{r_{12}} = R_{12}\exp\left[j\omega\left(t - \frac{x\cos\theta_1 - z\sin\theta_1}{c_1}\right)\right]$$
$$p_{t_{12}} = T_{12}\exp\left[j\omega\left(t - \frac{x\cos\theta_2 + z\sin\theta_2}{c_2}\right)\right] \qquad (4\text{-}45)$$

The transmitted component travels in fluid 2, intersecting the lower boundary at $x = x_1/2$, $z = h$. At this location the reflected and transmitted components become

$$p_{r_{23}} = T_{12}R_{23}\exp\left\{j\omega\left[t - \frac{x\cos\theta_2}{c_2} - \frac{(2h - z)\sin\theta_2}{c_2}\right]\right\}$$
$$p_{t_{23}} = T_{12}T_{23}\exp\left\{j\omega\left[t - \frac{x\cos\theta_3}{c_3} - \frac{z\sin\theta_3}{c_3} - h\left(\frac{\sin\theta_2}{c_2} - \frac{\sin\theta_3}{c_3}\right)\right]\right\} \qquad (4\text{-}46)$$

The reflected component from this location travels upward intersecting the upper boundary at $x = x_1$, $z = 0$. At that point we have

$$p_{r_{21}} = T_{12}R_{23}R_{21}\exp\left\{j\omega\left[t - \frac{x\cos\theta_2}{c_2} - \frac{(2h + z)\sin\theta_2}{c_2}\right]\right\}$$
$$p_{t_{21}} = T_{12}T_{21}R_{23}\exp\left\{j\omega\left[t - \frac{x\cos\theta_1}{c_1} + \frac{z\sin\theta_1}{c_1} - \frac{2h\sin\theta_2}{c_2}\right]\right\} \qquad (4\text{-}47)$$

We now form the ratio of the term $p_{t_{21}}$, traveling upward in fluid 1, to the pressure at the location $(x_1, 0)$ caused by the original downward plane wave in fluid 1. From (4-44), the incident pressure at $(x_1, 0)$ is

$$p_i(t, x_1, 0) = \exp\left[j\omega\left(t - \frac{x_1\cos\theta_1}{c_1}\right)\right] \qquad (4\text{-}48)$$

Letting $x = x_1$ and $z = 0$, $p_{t_{21}}$ becomes

$$P_{t_{21}}(t, x_1, 0) = T_{12}T_{21}R_{23}\exp\left[j\omega\left(t - \frac{x\cos\theta_1}{c_1} - \frac{2h\sin\theta_2}{c_2}\right)\right] \qquad (4\text{-}49)$$

from which

$$\frac{p_{t_{21}}(x_1, 0)}{p_i(x_1, 0)} = T_{12}T_{21}R_{23} \exp\left(-j\frac{\omega 2h \sin \theta_2}{c_2}\right) \qquad (4\text{-}50)$$

Equation (4-50) shows that the phase delay between the incident pressure at x_1 and the term arriving at x_1 via a single reflection in the middle layer is a function of the layer thickness h, frequency, and the angle θ_2.

We note that at the location $x = x_1$, $z = 0$, in addition to the upward term $p_{t_{21}}$, there is a direct reflection of $p_i(x_1, 0)$ with a reflection coefficient R_{12}. In fact, there will be an infinity of upward traveling terms at that location arriving as a result of multiple reflections in the middle layer from incident signals at $z = 0$, $x = x_1, 0, -x_1, -2x_1, \ldots$. The relative amplitude and phase of each of these terms are obtained by the same procedure used to arrive at (4-50). The ratio of the sum of these terms at any given location to the incident pressure at that location is defined as the *resultant reflection coefficient*, R_{13}, assuming a continuous sinusoidal incident wave under steady-state conditions. Thus

$$R_{13} = R_{12} + T_{12}T_{21}R_{23} \exp(-j2\phi_2) + T_{12}T_{21}R_{23}^2R_{21} \exp(-j4\phi_2) + \cdots \quad (4\text{-}51)$$

where $\phi_2 = (\omega h \sin \theta_2)/c_2$. With some rearranging, and taking advantage of the relationships in (4-42) and (4-43), (4-51) can be recognized as a geometric series with the closed-form solution

$$R_{13} = \frac{R_{12} + R_{23} \exp(-j2\phi_2)}{1 + R_{12}R_{23} \exp(-j2\phi_2)} \qquad (4\text{-}52)$$

By a similar procedure, the resultant transmission coefficient T_{13} from fluid 1 into fluid 3 can be shown to be

$$T_{13} = \frac{T_{12}T_{23} \exp(-j\phi_2)}{1 + R_{12}R_{23} \exp(-j2\phi_2)} \qquad (4\text{-}53)$$

In contrast with the reflection and transmission coefficients for a single boundary, the presence of multiple boundaries, or layers, results in complex coefficients. The magnitude and phase of both R_{13} and T_{13} are functions of frequency, the thickness of the middle layer, and the angle θ_2. As an example, let the upper and lower fluids in Figure 4-5 be water, and assume that fluid 2 has a characteristic impedance and sound velocity equal to two-thirds that of water. That is, let

$$Z_1 = Z_3 = 1.5Z_2$$

and   (4-54)

$$c_1 = c_3 = 1.5c_2$$

For simplicity we assume normal incidence ($\theta_1 = \theta_2 = 90°$), resulting in the single-boundary coefficients

$$R_{12} = \frac{Z_2 - Z_1}{Z_2 + Z_1} = -0.2$$

$$R_{23} = \frac{Z_3 - Z_2}{Z_3 + Z_2} = +0.2$$

$$T_{12} = \frac{2Z_2}{Z_2 + Z_1} = 0.8 \qquad (4\text{-}55)$$

$$T_{23} = \frac{2Z_3}{Z_3 + Z_2} = 1.2$$

Remember that these coefficients represent pressure ratios, so that it is quite permissible for the transmission coefficient to exceed unity without violating the principle of energy conservation.

Substitute the values from (4-55) in (4-52) and (4-53) to obtain

$$R_{13} = \frac{0.2[\exp(-j2\phi_2) - 1]}{1 - 0.04\,\exp(-j2\phi_2)}$$

$$T_{13} = \frac{0.96\,\exp(-j\phi_2)}{1 - 0.04\,\exp(-j2\phi_2)} \qquad (4\text{-}56)$$

where $\phi_2 = \omega h / c_2 = 2\pi h / \lambda_2$.

The transmission coefficient, T_{13}, is a maximum when the denominator is a minimum. This requires that

$$\exp(-j2\phi_2) = \exp\left(\frac{-j4\pi h}{\lambda_2}\right) = +1$$

$$\frac{4\pi h}{\lambda_2} = 2\pi n$$

$$h = \frac{n\lambda_2}{2} \qquad (4\text{-}57)$$

When the transmission is a maximum, the reflection coefficient R_{13} is a minimum, which in this example is zero. The magnitudes of the transmission and reflection coefficients are plotted in Figure 4-6 as a function of h/λ_2.

Repeated application of the procedures used to obtain T_{13} and R_{13} may be used to obtain the reflection and transmission through multilayered space. In general, this results in more complex dependence on frequency and the thickness of the various layers. One interesting result is obtained if a total reflection occurs at the bottom layer. Assuming that all layers are free of viscous losses, conservation of energy requires that the resultant reflection coefficient at the top layer for this case has unit amplitude with a phase shift varying with frequency and angle of incidence [5].

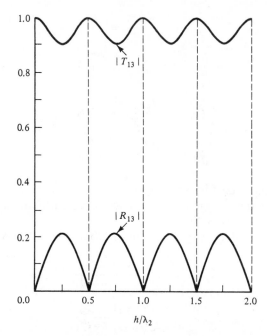

Figure 4-6 Reflection and transmission coefficients for two-boundary example $Z_1 = Z_3 = 1.5 Z_2$, $c_1 = c_3 = 1.5c_2$.

4.4 RAY ACOUSTICS: REFRACTION WITH A CONSTANT SOUND SPEED GRADIENT

In previous sections a directed arrow was used to indicate the direction of plane wave propagation. The arrow is normal to the wave front, where the wave front is defined as a line (or plane) connecting points of equal time phase in the traveling wave. In the plane wave case, a single directed arrow, or ray, is sufficient to define the direction of acoustic energy flow. Additional parallel rays may be used to indicate acoustic intensity. For plane waves, these rays are equally spaced with the separation inversely proportional to the intensity.

Consider a point source radiating acoustic energy equally in all directions in an infinite homogeneous medium, as shown in Figure 4-7. In the homogeneous medium, the energy propagates at the same rate in all directions, so that a wavefront at any point in time is a spherical surface. The rays are drawn perpendicular to the wave front, and therefore are straight lines with equal angular separation passing through the source location.

The number of rays originating at the source is proportional to the total acoustic power. At some reference distance (such as 1 m) from the source, the number of rays per unit area passing through the spherical surface is proportional to the acoustic intensity at that range. Because the spherical surface area increases in proportion to the square of range, the number of rays per unit area at

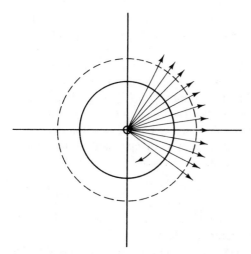

Figure 4-7 Point source radiating uniformly in a homogeneous medium.

range r meters is reduced by $1/r^2$, as compared with a reference range of 1 m. Thus, for the simple spherical acoustic wave, the *ray diagram* as shown in Figure 4-7 provides a useful model for depicting the local direction and intensity of the acoustic propagation.

In Figure 4-8 a point source is shown radiating uniformly in a nonhomogeneous medium. In the case shown, the acoustic sound speed is assumed to decrease with depth. At the source, the rays start out with equal angular separation. However, because the wave expands at different rates in different directions, the points of equal phase on the expanding wave lie on a nonspherical surface. The individual rays, always normal to the local wavefront, must follow curved paths with differing radii of curvature. This results in a density of rays that is not constant over the surface of an expanding wavefront. However, the ray density still provides a measure of local acoustic intensity. The job of ray

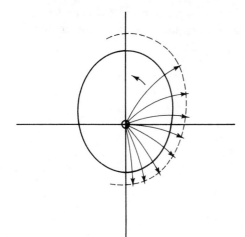

Figure 4-8 Point source radiating uniformly in a nonhomogeneous medium.

acoustics is to describe mathematically the paths taken by the acoustical rays and provide techniques for calculating the local acoustic intensity at any point in the medium.

In Section 4.2 we considered refraction at a plane boundary between two fluids with different sound speeds. The acoustic ray at oblique incidence experiences a discrete shift in direction at the fluid boundary, continuing in a straight line at the new angle in the second fluid. We now examine the refraction of acoustic rays in a medium in which the sound speed is a linear function of depth. We assume that at a fixed depth, the sound speed is independent of horizontal position. The sound speed may therefore be represented by

$$c(z) = c_0 + gz \qquad (4\text{-}58)$$

where c_0 is the speed at the reference depth $(z = 0)$, g is the sound speed gradient, and z is the depth below the reference depth.

With no loss in generality, we may choose the reference depth and c_0 to coincide with the point at which a sound ray is horizontal. Then by Snell's law, the angle associated with any other point along the ray is given by

$$\frac{\cos\theta}{c} = \frac{\cos\theta_0}{c_0} = \frac{1}{c_0} \qquad (4\text{-}59)$$

or

$$\cos\theta = \frac{c}{c_0}$$

Thus, at *any* point in the medium, the angle a ray makes with the horizontal is the arc cosine of the ratio of the local sound speed to the speed at the depth where the ray is horizontal. It follows that the local sound speed on the acoustic ray is always *less than* the value of c_0 because the acoustic ray must always bend toward the region of decreasing sound speed.

By differentiating (4-59), we obtain the relationship between an incremental change in sound speed and the resulting incremental change in direction:

$$\frac{dc}{c_0} = -\sin\theta \, d\theta \qquad (4\text{-}60)$$

Now differentiate (4-58) and obtain

$$dc = g \, dz \qquad (4\text{-}61)$$

Substitution of (4-61) in (4-60) gives

$$\frac{g \, dz}{c_0} = -\sin\theta \, d\theta$$

or

$$dz = -\frac{c_0}{g} \sin \theta \, d\theta \qquad (4\text{-}62)$$

Equation (4-62) defines the type of curve followed by an acoustic ray in a constant sound speed gradient medium. To recognize the type of curve, consider the vertical motion of a point moving on the arc of a circle of radius R as shown in Figure 4-9. From the figure, the relationship among the angle θ, incremental arc length ds, and incremental vertical distance dz is

$$dz = \sin \theta \, ds$$

But

$$ds = R \, d\theta$$

Therefore,

$$dz = R \sin \theta \, d\theta \qquad (4\text{-}63)$$

The form of (4-63) is identical to (4-62) if we let

$$R = -\frac{c_0}{g} = -\frac{c}{g \cos \theta} \qquad (4\text{-}64)$$

The ray path is therefore a circular arc with a radius of curvature defined by (4-64). The negative sign in (4-64) determines whether the ray path is curving downward or upward. If the gradient g is negative (sound speed decreasing with increasing depth), the radius of curvature is positive and the path curves downward. Conversely, with positive sound speed gradient, R is negative and the path curves upward.

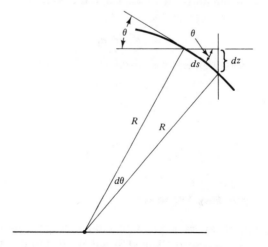

Figure 4-9   Vertical motion of a point moving on a circular arc.

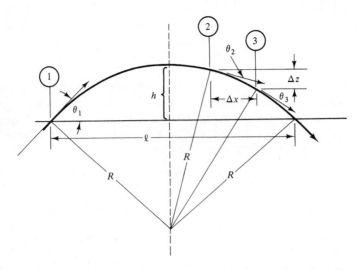

Figure 4-10 Circular ray path in medium with negative sound speed gradient.

From the geometry of a circle, a number of useful relationships are available to simplify the plotting of a ray path in a constant gradient region. Figure 4-10 shows a portion of a circular arc representing a ray path in a negative gradient region.

With a ray starting at point 1, directed upward at an angle θ_1, the maximum height reached by the ray above the starting point is

$$h = R(1 - \cos \theta_1) \tag{4-65}$$

where $R = -c_1/g \cos \theta_1$.

The cord length, ℓ, is the horizontal distance to the point where the ray returns to the starting depth and is given by

$$\ell = 2R \sin \theta_1 \tag{4-66}$$

For the arbitrary points 2 and 3, we have

$$\Delta z = R(\cos \theta_2 - \cos \theta_3)$$

$$= \frac{c_3 - c_2}{g} \tag{4-67}$$

$$\Delta x = R(\sin \theta_3 - \sin \theta_2) \tag{4-68}$$

4.4.1 Example: Ray Tracing

As an example of ray tracing in a medium with linear sound speed versus depth characteristics, we use the sound speed–depth profile shown in Figure 4-11. This profile results from a layer with a negative gradient of -0.05 from

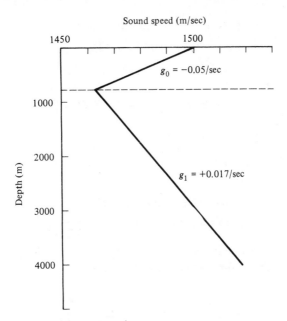

Figure 4-11   Sound speed–depth profile.

the surface to a depth of 750 m, above a region with a positive gradient of +0.017 from 750 m to the bottom.

Assume a source radiating omnidirectionally at a depth of 100 m. We now determine the information necessary to sketch the acoustic ray originating at the source with an initial angle $\theta_0 = 0$. Because of the negative gradient, the ray will curve downward with a radius of curvature

$$R_0 = -\frac{c_0}{g_0}$$

where c_0 is the sound speed at 100 m and g_0 is the gradient in the upper layer. From (4-58)

$$c_0 = 1500 - (0.05)(100) = 1495 \text{ m/sec}$$

and from (4-64)

$$R_0 = \frac{1495}{0.05} = 29{,}900 \text{ m}$$

The ray path in the top layer follows a circular path downward with radius R_0 to a depth of 750 m. The angle θ_1 that the ray makes with the boundary between the two layers can be determined by Snell's law. Thus

$$\cos \theta_1 = \frac{c_1}{c_0}$$

where c_1 is the sound speed at 750 m with the value

$$c_1 = 1500 - (0.05)(750) = 1462.5 \text{ m/sec}$$

Therefore,

$$\cos \theta_1 = \frac{1462.5}{1495} = 0.97826$$

$$\theta_1 = 11.97°$$

The horizontal distance traveled by the ray to reach the depth of 750 m may be calculated using (4-68):

$$x_1 = R_0(\sin \theta_1 - \sin \theta_0) = R_0 \sin \theta_1$$

$$= 29,900 \sin (11.97°) = 6200.57 \text{ m}$$

In the lower layer, the sound speed gradient is positive, resulting in a negative, or upward, curvature of the acoustic ray. The radius of curvature in the lower layer is given by

$$R_1 = -\frac{c_0}{g_1} \quad \text{or} \quad -\frac{c_1}{g_1 \cos \theta_1}$$

$$= -\frac{1495}{0.017} = -87,941.18 \text{ m}$$

The acoustic ray follows the circular arc with negative radius R_1 to a maximum depth determined by the point where the sound speed equals that at the source depth of 100 m. At this point, Snell's law requires that the angle, θ_2, of the ray relative to the horizontal equal that at the source depth. That is, θ_2 must equal zero because

$$\cos \theta_2 = \frac{c_2}{c_0} \cos \theta_0 = 1$$

This depth may be determined from

$$c_2 = c_0 = 1495 = c_1 + (0.017)(z_2 - 750)$$

$$= 1462.5 + (0.017)(z_2 - 750)$$

which gives

$$z_2 = 2661.76 \text{ m}$$

The horizontal distance from the source to the point of maximum depth is obtained with the help of (4-68).

$$x_2 - x_1 = R_1(\sin \theta_2 - \sin \theta_1) = -R_1 \sin \theta_1$$

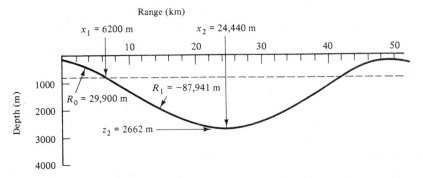

Figure 4-12   Acoustic ray path in a medium with a negative gradient layer above a positive gradient layer.

or

$$x_2 = x_1 - R_1 \sin \theta_1$$
$$= 6200.57 - (-87,941.18) \sin (11.97°)$$
$$= 24,439.57 \text{ m}$$

By symmetry, the acoustic ray will return to the source depth at a range $2x_2$ by a path identical to that traveled from the source to the point of maximum depth. One complete cycle of the ray path for this example is sketched in Figure 4-12. The depth and range in this figure are plotted on different scales to reveal more clearly the vertical excursions of the acoustic ray. Note that the actual total depth excursion of the ray is only about 5% of the range for one complete horizontal cycle. Assuming the characteristics of the medium are everywhere the same, the acoustic ray cycle shown in Figure 4-12 will repeat indefinitely in all horizontal directions away from the source.

4.4.2 Use of the Ray Pattern to Calculate Acoustic Intensity

In the preceding section the path of a single acoustic ray was determined. A principal purpose of constructing ray diagrams is to present a visual picture of the acoustic intensity at different points in the medium. This is accomplished by plotting many rays, separated at the source by small equal angular increments. The acoustic intensity at some point in the medium is then inferred by the separation of adjacent rays. For two rays with initial separation $\Delta\theta$ it can be shown [6] that the transmission loss at range r is given approximately by

$$\text{TL} = 10 \log \left(\frac{r \, \Delta h}{\Delta \theta} \right) \tag{4-69}$$

where Δh is the vertical separation of the rays at range r. In the absence of refraction (4-69) reduces to the spherical spreading law because for straight-line propagation $\Delta h \simeq r\Delta\theta$. In a layer with a constant gradient sound speed–depth characteristic, the loss will not depart significantly from the spherical spreading law if we can neglect boundary effects.

The presence of multiple layers, including both positive and negative sound speed gradients, can produce regions of ray convergence or divergence, resulting in a considerable departure from the spherical spreading law. For instance, in the example in Section 4.4.1 rays separated by a small angle at the source will intersect repeatedly, as shown in Figure 4-13(a). At the point of intersection, the intensity as predicted by simple ray theory is infinite, which is of course impossible. More rigorous solutions of the wave equations must be used to obtain accurate results in such situations. Although less than infinity, the intensity may significantly exceed that predicted for spherical spreading in regions containing intersecting rays.

In Figure 4-13(b) the top layer has a positive sound speed gradient followed by a negative gradient in the lower layer. This results in a region of diverging rays at the depth at which the gradient reverses polarity. Neglecting the effects of reflection from nearby boundaries, this will result in a lower intensity than for spherical spreading.

In addition to purely refractive effects, resulting from sound speed variation with depth, the acoustic intensity at any point in the medium may be dramatically affected by the surface and bottom boundaries. The total intensity at any point may include contributions from direct propagation by one or more

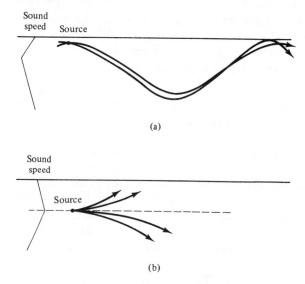

(a)

(b)

Figure 4-13   Sound speed profiles resulting in converging and diverging acoustic ray paths.

purely refractive paths and from paths involving one or more reflections from the surface and/or the bottom. In some cases the presence of the boundaries, together with the sound speed characteristics, precludes the possibility of direct propagation to certain regions of the medium. In these situations, reflected paths provide the only means of acoustic propagation and the intensity at any point is very dependent on the acoustic characteristics of the boundaries.

Mathematical methods are available for computing intensity based on acoustic ray density at any point in the medium given the sound speed profile and the boundary conditions. In most practical situations the computations are extensive and best suited to solution by computer. A detailed analytical discussion of all possibilities is beyond our scope. In the next chapter, a qualitative discussion is presented for the more commonly encountered acoustic situations in the ocean.

4.4.3 Limitations of Ray Acoustic Methods

Because the methods of ray acoustics do not represent exact solutions of the acoustic wave equations, a few words are in order concerning the range of validity of this approach. In optics, the ray approach is generally abandoned when features to be illuminated have dimensions of the same order of magnitude as the wavelength of light. The same is true for acoustics. Because the acoustic wavelength may vary from hundreds of feet to a small fraction of a foot, it is apparent that the ray acoustic method may involve more risk than the ray optic method for applications within the realm of our everyday experience. In ray acoustics there is cause for concern in the following situations:

1. The acoustic channel dimensions are not very large relative to the acoustic wavelength.
2. The speed of sound varies considerably over a distance of one wavelength or less.
3. The ray acoustic method predicts a large change in intensity in a distance small compared with a wavelength.

If accurate numerical results are required and any of the situations listed above exist, the more rigorous approaches using normal mode solutions or equivalent should be considered. Some computer programs for calculating acoustic intensity use a combination of ray acoustic and normal mode techniques to take advantage of the best features of each. Alternatively, some programs introduce approximate correction factors to reduce the errors inherent in ray acoustic methods in some particular situations.

In spite of its inherent limitations, the method of ray acoustics is very useful for the visualization of the acoustic situation in a given medium. Given the sound speed profile and the surface and bottom characteristics, the general behavior of acoustic intensity with range and depth can be rapidly estimated using relatively simple ray tracing techniques. Often the resulting accuracy is sufficient for preliminary design studies.

PROBLEMS

4.1. Assume plane wave sound in water normal to a water–steel plane interface. The density of the steel is 7700 kg/m³ and the sound speed is 6000 m/sec. Assuming that the steel thickness is effectively infinite, determine the reflection and transmission coefficients. Assume that the density and velocity for seawater are 1000 kg/m³ and 1500 m/sec, respectively.

4.2. A layer of water is on top of a fluid layer with a density of 1400 kg/m³ and a sound velocity of 1400 m/sec. Calculate and sketch the reflection coefficient for an acoustic plane wave in water as a function of the grazing angle, θ_i. Calculate specifically the angle of intromission resulting in zero reflection. Determine the transmission angle, θ_t, corresponding to the angle of intromission.

4.3. A layer of steel 0.01 m thick separates two layers of water. A plane acoustic wave at a frequency of 1000 Hz is normally incident at the water–steel boundary. Determine the transmission coefficient, T_{13}, for transmission of sound through the steel plate. Maximum transmission is obtained if the steel thickness is made one-half of the acoustic wavelength in steel. In this example, does the improvement obtained warrant the added thickness required to optimize the transmission?

4.4. A simple acoustic lens may be formed by filling a thin spherical shell with a liquid that has an index of refraction greater than that of the surrounding water. Acoustic rays in a plane wave incident at the surface will converge by refraction inside the spherical container. By properly choosing the index of refraction, the rays will converge in a small region on the lens surface or just inside the spherical surface. The acoustic intensity in this focal region exceeds the intensity in the incident plane wave. Referring to Figure P4-1, determine the required index of refraction to cause the refracted ray to pass through point O on the lens surface for ray a, with grazing incidence. Repeat for ray b with θ_1 equal to 30° and 60°. With a constant index of refraction equal to 1.8, sketch the ray paths for the rays with grazing incidence and for grazing angles of 30 and 60°. Assume that Snell's law can be applied at the spherical surface with angles measured relative to a tangent to the spherical surface at the point the ray intercepts the surface.

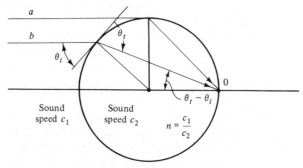

Figure P4-1 Simple liquid lens with constant index of refraction.

4.5. An acoustic source is located at a depth of 100 m in a medium with a sound speed of 1500 m/sec at the surface and a negative sound speed gradient of -0.2 m/sec/m. Determine the angle above the horizontal at the source for the ray that just grazes the surface. Determine the range at which the surface grazing ray returns to the depth of the source.

SUGGESTED READING

1. *Physics of Sound in the Sea,* Part I: *Transmission;* originally issued as Division 6, Vol. 8, NDRC Summary Technical Reports. Wakefield Printing Co., Wakefield, Mass.; reprinted in 1969 by the U.S. Government Printing Office, Washington, D.C.

2. Tolstoy, I., and C. S. Clay, *Ocean Acoustics.* New York; McGraw-Hill Book Company, 1966.

3. Officer, C. B., *Introduction to the Theory of Sound Transmission.* New York: McGraw-Hill Book Company, 1958.

4. Kinsler, L. E., and A. R. Frey, *Fundamentals of Acoustics,* 2nd ed. New York: John Wiley & Sons, Inc., 1962.

5. Clay, C. S. and H. Medwin, *Acoustical Oceanography.* New York: John Wiley & Sons, 1977.

6. Urick, R. J., *Principles of Underwater Sound for Engineers,* 2nd ed. New York: McGraw-Hill Book Company, Chap. 5.

5

Sound Transmission in the Ocean

~~~~~~

5.0 INTRODUCTION

The detailed characteristics of the ocean and its boundaries that affect sound transmission are very complex. Sound speed is a function of temperature, depth, and salinity. Temperature, in turn, is a function of depth, time, location and weather conditions. The ocean surface varies from a glossy smooth reflector to a very rough and turbulent surface that scatters sound in a random fashion. The ocean bottom has a wide variety of compositions, slopes, and roughness, all of which affect sound transmission. The effects of sound speed and the ocean surface and bottom boundaries interact to produce the final acoustic transmission characteristic.

In spite of this rich variety of detailed characteristics, it is possible to recognize predictable patterns related to environmental conditions and geographic locations. Thus typical sound speed profiles are often available for a given geographic location and season. Acoustic loss data for the boundaries, derived from a combination of experience and theoretical considerations, cover the expected ranges of wind speeds, grazing angles, bottom characteristics, and frequencies. Using this type of information, propagation prediction routines based on ray acoustics or more rigorous techniques are available to produce the average, or expected, transmission characteristics for a given situation. The results will agree in general character with actual on-site measurements. However, if accurate detailed results are required, it is necessary to measure the sound speed profile carefully and establish the surface and bottom conditions at the actual location and time of the acoustic test.

In this chapter we consider a number of commonly encountered acoustic situations. The salient features of the acoustic field in these cases are revealed by sketching the important ray paths and estimating the effect of losses at the surface and bottom boundaries

5.1 SOUND SPEED VARIATION IN THE OCEAN

In Chapter 2 we learned that the speed of sound in water is related to density and the bulk modulus of elasticity. In particular,

$$c = \sqrt{\frac{B}{\rho}}$$

In the ocean, neither the bulk modulus, B, nor the density, ρ, are constant. The density varies slightly with the chemical composition of the water, and the bulk modulus varies with both temperature and pressure. Of these effects, the variation in bulk modulus is by far the most important.

From experimental results and theoretical considerations, a number of equations have been proposed to represent sound speed as a function of temperature, salinity, and depth (or pressure) [1–3]. A typical example of these equations is as follows [4]:

$$c = 1449 + 4.6T - 0.055T^2 + 0.0003T^3$$
$$+ (1.39 - 0.012T)(S - 35) + 0.017z \qquad (5\text{-}1)$$

where c = sound speed, m/sec
$\quad\quad T$ = temperature, °C
$\quad\quad S$ = salinity, parts per thousand
$\quad\quad z$ = depth, m

In this equation, temperature is of course a function of depth, varying from something near the ambient air temperature at the surface to near freezing at depths of a few thousand feet.

At a point far removed from the mouth of a river or other source of fresh water, the salinity may vary by only a fraction of a part per thousand from the surface to the bottom. The average ocean salinity is often assumed to be 35 parts per thousand, in which case the effect on sound speed is slight. The average salinity in oceans and seas around the world may actually differ by several parts from the average value of 35, resulting primarily in a small bias effect on sound speed relative to the value obtained neglecting the salinity term. Near the mouths of rivers or in the vicinity of melting ice, the salinity is quite variable and may have a significant effect on sound speed. In the following discussions we assume the effect of salinity is negligible. This leaves the temperature and depth as the primary factors affecting sound speed.

The characteristics of sound speed variation with depth are best discussed by considering the ocean as divided into a series of horizontal layers as shown in Figure 5-1. The depths and thicknesses of the layers defined in the figure vary

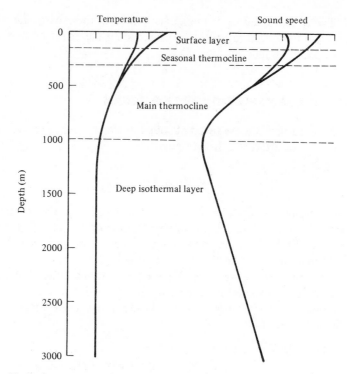

Figure 5-1 Typical sound speed versus depth profile at middle latitudes.

significantly with latitude and other factors [5]. The example shown is typical for middle latitudes in the North Atlantic Ocean.

The surface layer extends from the surface to perhaps 150 m and is the layer most affected by local weather conditions and even the time of day. In calm weather, the water temperature decreases rapidly with depth in the surface layer, resulting in a strong negative sound speed gradient. In stormy weather there is a strong mixing action in this layer that tends to reduce the temperature gradient to zero. The result is a positive sound speed gradient of +0.017 per second caused entirely by the last term in (5-1). Once thoroughly mixed, the surface layer may retain the isothermal condition for an appreciable time period following a storm.

In calm weather the top 10 m or so at the ocean surface may exhibit a changing sound speed characteristic during the course of a day. The surface acquires heat from the sun, resulting in negative temperature and sound speed gradients by late afternoon. During the night there is some mixing action caused by normal wave activity as well as heat lost by radiation from the surface. These effects cause the negative temperature gradient to weaken considerably or possibly disappear completely.

Below the surface layer the water temperature is affected less by transient effects such as storms or the day–night cycle, but there is a noticeable change with seasons. This layer, called the *seasonal thermocline,* extends down to perhaps 300 m and is characterized (in middle latitudes) by a negative temperature gradient.

The third layer has a stable temperature versus depth characteristic with a negative gradient. This layer is called the *main thermocline.* As the depth increases, the temperature gradient decreases until the temperature reaches its minimum value near freezing. The sound speed decreases in this layer to a minimum value determined by the point where the decrease in sound speed caused by decreasing temperature is balanced by the increase caused by increasing depth. This occurs typically at about 1000 m at middle latitudes and marks the top of the final layer.

The final layer is called the *deep isothermal layer* because of the nearly uniform temperature. The sound speed in the deep isothermal layer increases gradually with depth, with a positive sound speed gradient that asymptotically approaches +0.017 per second.

The depth associated with the minimum sound speed decreases as we proceed to more northerly latitudes. In the North Atlantic Ocean at latitudes above about 55°, and in areas not affected by the Gulf Stream, the minimum sound velocity often occurs at or near the surface. In this case, aside from some seasonal variation very near the surface, isothermal conditions prevail from the surface to the bottom. Figure 5-2 presents a series of sound speed profiles for the

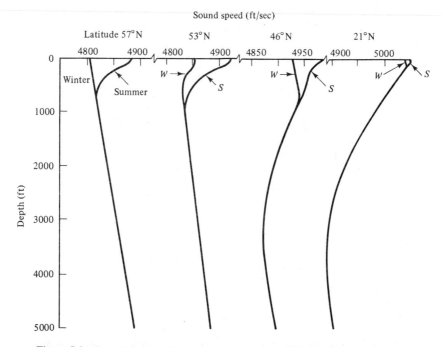

Figure 5-2   Sound speed profile variation with latitude in the North Atlantic Ocean. (From Podeszwa [6].)

North Atlantic, showing the nature of the variation that occurs with latitude. Because of the presence of ocean currents and other natural factors, there can be considerable variation from the "typical" characteristics shown in Figure 5-2.

5.2 ACOUSTIC LOSS AT THE OCEAN SURFACE

In Figure 5-3(a), reflection of acoustic rays from a smooth air–water interface is shown. Because of the large impedance mismatch at the boundary, the reflection coefficient is very nearly −1. The acoustic signal at location R, resulting from the reflected path, may be thought of as originating in a source at the image location, with equal amplitude but opposite phase compared with the actual source. Obviously, with a smooth surface there is no loss in acoustic image intensity caused by the surface reflection.

In Figure 5-3(b), the interface is not perfectly smooth. The acoustic rays are reflected in a random manner from the rough surface, with the result that the image of the source, as viewed from R, is smeared over an area such that the apparent image intensity is reduced compared with the result for a smooth surface. As the interface becomes very rough, diffuse reflection occurs, resulting in a large loss in intensity at R.

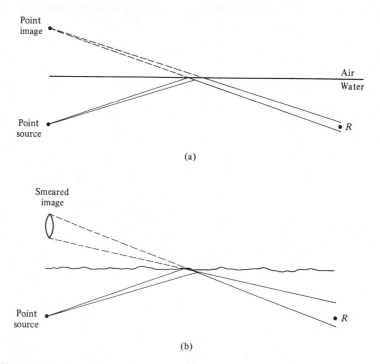

Figure 5-3 (a) Reflection from a smooth surface; (b) reflection from a rough surface.

The question arises as to what may be considered a "smooth" surface and what constitutes a "rough" surface. Intuitively, if the surface deviates from a plane by an amount that is very small compared with a wavelength, the surface may be considered smooth. Conversely, if the surface deviations are large compared with a wavelength, the surface may be considered rough. Specifically, according to the Rayleigh criterion, the surface is considered smooth if $h \sin \theta < \lambda/8$, where h is the height of the roughness feature, θ is the grazing angle, and λ is the acoustic wavelength. Between these extremes exists a variety of conditions, depending on wind speed, constancy and duration of wind conditions, ocean depth, and even the size of the ocean basin. With respect to acoustic transmission, the alignment of the ocean wave structure relative to the acoustic direction of interest is also of importance.

Rigorous mathematical modeling of all possible surface conditions with their effects on acoustic transmission is not practical. Lord Rayleigh considered the problem of the scattering of sound normally incident on a simple corrugated surface. Marsh [7] and Marsh et al. [8] extended this work to scattering from randomly irregular surfaces.

From oceanographic data, statistical properties of the sea surface, such as mean-squared wave height and average trough-to-crest wave height, have been determined and related to wind speed, assuming certain standard conditions. Marsh et al. [8] derived an expression for acoustic loss caused by reflection from the sea surface as a function of the product of wave height and acoustic frequency. Their expression is

$$\alpha_s = -10 \log[1 - 0.0234(fH)^{3/2}] \tag{5-2}$$

where α_s = surface reflection loss, dB
$\quad\quad f$ = acoustic frequency, kHz
$\quad\quad H$ = average trough-to-crest wave height, ft

Equation (5-2) was derived assuming a small grazing angle. In the form shown it should not be used for loss values exceeding 3 dB. Consideration of higher-order terms that were neglected in (5-2) leads to a graphical representation of acoustic loss per surface reflection, or "bounce," over a wider range as shown in Figure 5-4. As an example, assume a frequency of 5 kHz and a wave height of 3 ft. The expected surface loss per bounce is then approximately 5.5 dB according to Figure 5-4. Notice that at a frequency of 500 Hz, with the acoustic wavelength large compared to the wave height, the loss is essentially zero. From (5-2), the loss is less than 0.5 dB if the wave height is less than half the acoustic wavelength.

It is convenient to relate the condition of the sea surface to wind speed, assuming that the wind speed has been steady for a considerable time over many miles of open sea. Figure 5-5 presents the relationship of wind speed to wave height in graphical form. Also included on the figure is a division of the wind speed–wave height regime into numbered bands, in each of which the sea surface condition, or *sea state,* is characterized by a simple adjective. Thus for

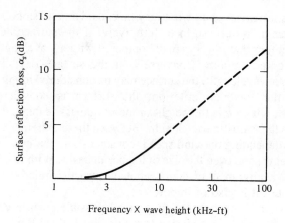

Figure 5-4 Sea surface reflection loss at low grazing angles. (From Marsh et al. [8].)

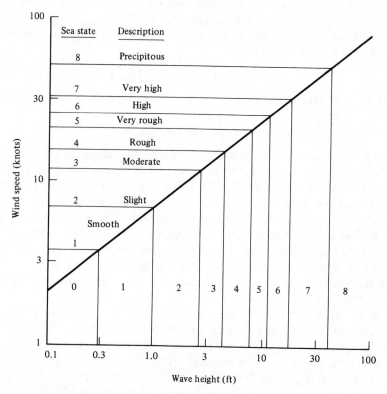

Figure 5-5 Sea state, wind speed, and wave height relationships.

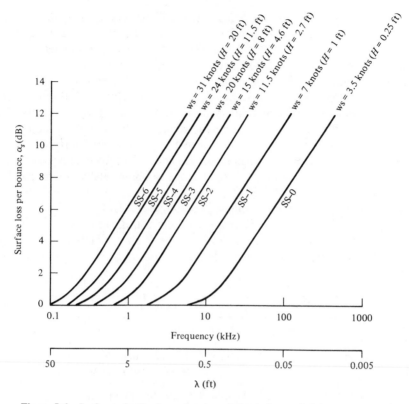

Figure 5-6 Surface reflection loss related to wind speed, wave height, sea state, and frequency.

wind speeds between 7 and 11 knots the sea surface is described as having slight waves with a maximum height less than 3 ft. The surface condition for this wind speed range is designated *sea state 2*.

Combining the information contained in Figures 5-4 and 5-5, we obtain Figure 5-6, relating surface loss to wind speed, wave height, sea state, and frequency. Because the sea surface is in constant motion, measured acoustic intensity from a surface reflection is subject to wide and rapid variation. This time variation, as well as the basic complexity of the process, makes experimental verification of loss predictions, such as presented in Figure 5-6, difficult.

5.3 REFLECTION LOSS AT THE OCEAN BOTTOM

The complexity of the surface reflection loss process results primarily from the nonplanar shape of the surface and the fact that this shape varies with time. The nature of the acoustic interaction at the air–water interface is actually quite simple because of the severe impedance mismatch. If the surface is calm, a nearly perfect reflection occurs.

In contrast, the reflection process at the ocean bottom is quite complex even if the bottom is perfectly flat. The impedance mismatch between water and bottom sediment materials is much less severe than that between water and air at the sea surface. Thus a portion of the incident acoustic energy at the bottom may be transmitted into the bottom materials and a portion may be reflected. The energy transmitted into the bottom may encounter layers of different solid materials, resulting in reflection and transmission at each layer boundary. Energy reflected within the bottom eventually returns to the water to combine with the acoustic wave reflected directly from the water–bottom interface. As discussed in Section 4.3, the resultant reflection coefficient from a layered bottom involves both a loss in amplitude and a change in phase relative to the incident wave. The layered nature of the bottom sediments causes the complex reflection coefficient to vary with both frequency and angle of incidence.

The type of bottom materials varies widely with geographic location. For our purposes, it is convenient to consider the highly simplified descriptive classification of bottom materials as defined by the U.S. Navy Hydrographics Office in the following table.

Bottom Type	Description
Mud	90% or more particles less than 0.062 mm in diameter
Mud and sand	50 to 90% $d < 0.062$ mm
Sand and mud	10 to 50% particles with $d < 0.062$ mm
Sand	Less than 10% with $d < 0.062$ mm; at least 90% with $d < 2.0$ mm
Stony	Mainly pebbles and cobbles
Rock	Mainly rock outcrop or boulders

A sandy bottom is typically smooth and often found in relatively shallow seas, such as the North Sea, and on continental shelves. Both the density and sound speed of a sandy bottom exceed those of water. For coarse sand the density is approximately twice that of water and the sound speed is about 1830 m/sec. With these parameters, the characteristic impedance of the sandy bottom is approximately 2.4 times that for seawater.

Assuming a nonlayered sandy bottom, the reflection coefficient, as given by (4-34), can be put in the form

$$R = \frac{(Z_2/Z_1) \sin \theta_i - \sin \theta_t}{(Z_2/Z_1) \sin \theta_i + \sin \theta_t}$$

where Z_2/Z_1 is the ratio of characteristic impedances of sand and water. With an impedance ratio of 2.4, for normal incidence this gives

$$R = \frac{2.4 - 1}{2.4 + 1} = 0.4118$$

The reflection loss is defined as

$$\alpha_b = -20 \log_{10} R = 7.7 \text{ dB} \qquad (5\text{-}3)$$

Because the sound speed in water is less than that in the sandy bottom, there will be a critical grazing angle below which the reflection should be total and the reflection loss, therefore, should theoretically be zero. In accordance with (4-39), the critical angle is

$$\theta_c = \text{arc cos} \left(\frac{c_1}{c_2} \right)$$

where c_1/c_2 is the ratio of sound speeds in water and the sandy bottom. For the parameters given for coarse sand, this gives a critical angle of about 35°. The theoretical reflection loss for the smooth sandy bottom is shown in Figure 5-7.

In practice, the theoretical characteristic of Figure 5-7 is modified by acoustic losses in the bottom material and possible roughness of the bottom relative to the acoustic wavelength. These effects increase with increasing frequency and tend to increase the loss relative to that shown by the solid curve in the figure. Thus even at low grazing angles the reflection loss from a sandy bottom is not zero. The range of expected reflection loss for the smooth sandy bottom for frequencies from below 500 Hz to approximately 3.5 kHz is shown by the dashed curves in Figure 5-7. Notice that the effect of nonzero viscous

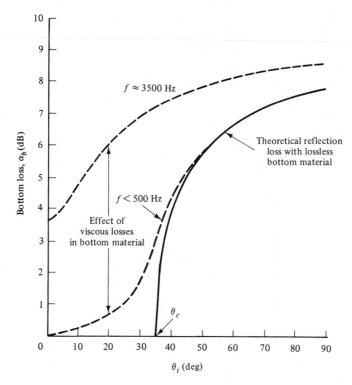

Figure 5-7 Reflection loss for a smooth, coarse sandy bottom.

losses in the bottom material is relatively more pronounced at low grazing angles than near normal incidence.

Mud bottoms have a density slightly greater than water and a sound speed that ranges from slightly above to slightly below the speed of sound in water. As an example, assume a bottom density of 1400 kg/m^3 and a sound speed of 1535 m/sec. With a sound speed in water of 1500 m/sec, this results in a characteristic impedance ratio of 1.433. Proceeding as before, the reflection coefficient and reflection loss at normal incidence, and the critical angle are

$$R = \frac{1.433 - 1}{1.433 + 1} = 0.178$$

$$\alpha_b = -20 \log (0.178) \simeq 15 \text{ dB}$$

$$\theta_c = \text{arc cos} \left(\frac{1500}{1535} \right) = 12°$$

The reflection loss as a function of grazing angle for a mud bottom is shown in Figure 5-8. The reflection loss for the mud bottom is considerably larger than for the sandy bottom at all angles of incidence.

Some soft mud bottoms have a sound speed that is slightly less than the

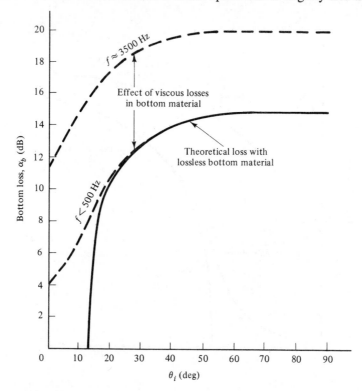

Figure 5-8   Reflection loss for a mud bottom.

speed in water, resulting in the possibility of the existence of an angle of intromission. Let the bottom material density and sound speed be 1420 kg/m³ and 1490 m/sec. We then obtain at normal incidence

$$R = \frac{1.41 - 1}{1.41 + 1} = 0.17$$

$$\alpha_b = -20 \log (0.17) = 15.4 \text{ dB}$$

and the angle of intromission is

$$\theta_i = \arc\sin \left[\frac{(c_1/c_2)^2 - 1}{(\rho_2/\rho_1)^2 - 1} \right]^{1/2} = 6.6°$$

At the angle of intromission, the reflection coefficient is zero and therefore the reflection loss is infinite. The complete loss versus grazing angle is plotted in Figure 5-9.

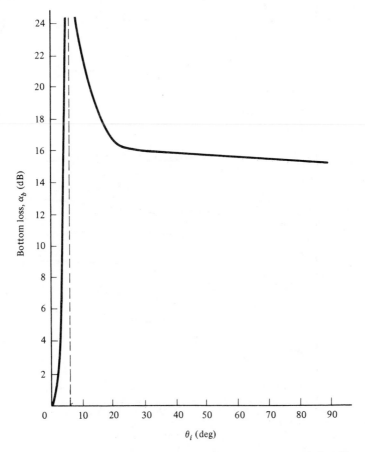

Figure 5-9   Reflection loss for a bottom material that supports an angle of intromission.

As an aid in acoustic analysis, the bottom types may be arranged and numbered in order of increasing reflection loss. One common classification system, developed by the U.S. Naval Fleet Numerical Weather Center (FNWC), identifies nine bottom classes starting with a low loss sandy bottom as number one. The bottom loss versus grazing angle for these nine bottom types is shown in Figure 5-10 at a frequency of 1 kHz. Oceanographic surveys provide descriptions of bottom sediments that are used to assign the appropriate acoustic bottom type number. It must be remembered that the compression of the great variety of bottom characteristics into such a limited number of bottom classes results in a loss in precision with respect to the prediction of acoustic propagation that

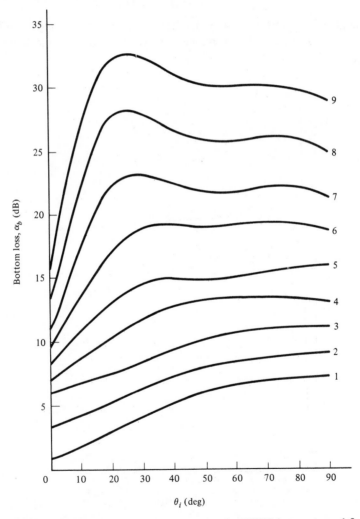

Figure 5-10 Bottom reflection loss per bounce for FNWC bottom types 1-9 at 1 kHz. (From Weinberg [9].)

involves interaction with the bottom. It is often justified on the basis that a more rigorous characterization is not practical, and that the precision obtained is adequate. This is generally true if the bottom characteristics are not the dominant factor in acoustic propagation.

5.4 ABSORPTION OF SOUND IN THE OCEAN

Up to this point, the water medium has been assumed lossless. That is, the integral of acoustic intensity over any closed surface including the source has been assumed constant regardless of distance from the source. Actually, as each volume element of the medium is subjected to the compression and rarefaction caused by an acoustic wave, some of the acoustic energy is lost to the volume element in the form of heat.

If the energy lost per unit volume in the medium is a constant fraction of the incident energy, it is easy to show that the intensity loss caused by the effect is an exponential function of range. Assuming a lossy homogenous medium, the intensity of a spherical wave expanding about a point source has the form

$$I(r) = I_0 r^{-2} \exp(-br) \qquad (5\text{-}4)$$

where I_0 is the intensity at unit range for a lossless medium and b is the exponential loss factor. The transmission loss, TL, is the ratio, $I_0/I(r)$, converted to decibels. Thus

$$\text{TL} = 10 \log\left[\frac{I_0}{I(r)}\right] = 20 \log r + 10 \log[\exp(br)]$$

This is the usual spherical spreading loss, $20 \log r$, plus the exponential *absorption loss* term. The absorption loss is usually written in the following form:

$$10 \log[\exp(br)] = \alpha_a r \, 10^{-3} \qquad (5\text{-}5)$$

where α_a = absorption loss in dB/km = $(b \, 10^{-3}) 10 \log_{10}(\epsilon)$
  r = range in meters
  ϵ = base of natural logarithms

Some of the causes of absorption loss are well known. In fresh water, the measured losses are adequately explained by consideration of viscous effects. In seawater, the measured loss at frequencies below 100 kHz is considerably in excess of that anticipated from viscous effects alone. An ionic relaxation process involving magnesium sulfate has been identified as a significant contributor to absorption loss below 100 kHz. This involves the disassociation–reassociation of the magnesium sulfate ion under the influence of acoustic pressure.

A similar effect occurs at frequencies below 5 kHz with the boric acid ion. At frequencies below 100 Hz, the absorption loss exceeds that predicted from viscosity and known relaxation effects. The reasons for this discrepancy are not well understood. Figure 5-11 gives the absorption loss as a function of frequency

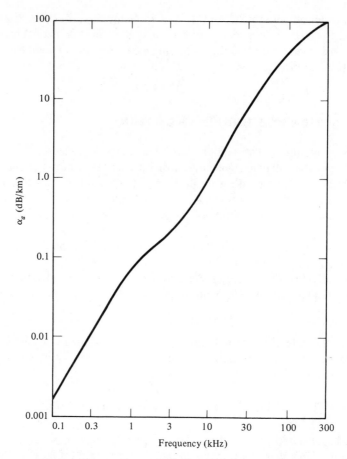

Figure 5-11   Absorption loss in the ocean. (After Thorp [10].)

as derived by Thorp [10], using data measured by a number of different investigators.

  The nature of absorption loss is such that once it becomes appreciable it
soon dominates the transmission loss equation. For instance, assume that at
range r the absorption loss is 6 dB. At range $10r$ the absorption loss would be
60 dB. In this same range interval, the loss caused by spherical spreading
increases by only 20 dB. Figure 5-12 gives the 6-dB absorption loss range
contour as a function of frequency. This figure provides some guidance in the
selection of operating frequency for various applications. Although the boundary
indicated in Figure 5-12 should not be considered as a hard limit, it is evident
that applications requiring very long range (hundreds of miles) must use very
low frequencies. Conversely, the frequency range above 30 kHz should not be
considered unless the required range is on the order of 1 km or less.

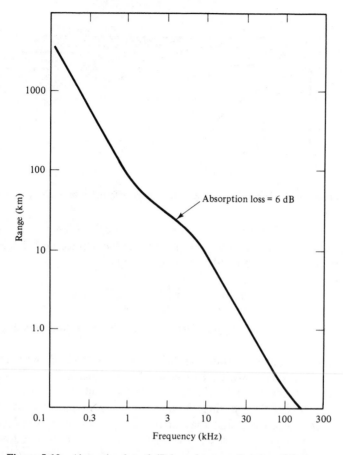

Figure 5-12   Absorption loss 6-dB boundary as a function of frequency.

5.5 ACOUSTIC CHARACTERISTICS OF THE SURFACE LAYER

The surface layer of the ocean is at once the region of greatest sound speed variability and the region often containing both the source and receiving sensor of an acoustic system. At the extremes, the surface layer may have either a positive or negative sound speed gradient. In shallow water, the surface layer characteristics may extend to the ocean bottom.

5.5.1 Negative Gradient

Consider acoustic transmission with a simple negative sound speed gradient as shown in Figure 5-13. Because of the negative gradient, all rays are

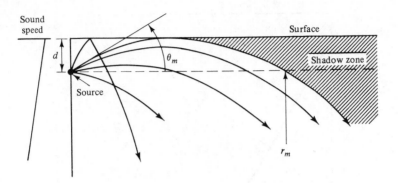

Figure 5-13   Transmission in a surface layer with negative gradient.

refracted downward. This means that for a source depth d there is an angle θ_m above the horizontal for which the ray just grazes the surface before refracting toward deeper water. Using (4-65), θ_m is given by

$$\theta_m = \text{arc cos}\left(1 + \frac{dg}{c_0}\right) \qquad (5\text{-}6)$$

where c_0 is the sound speed at the surface and g is the sound speed gradient (negative in this case).

Any rays that leave the source at an angle higher than θ_m will strike the surface and be reflected toward deep water. Notice that no rays leaving the source can reach directly into the shaded area to the right of the surface grazing ray. According to ray theory, this *shadow zone* is a region of zero acoustic intensity, except for energy that may arrive by way of bottom reflection or scattering from a rough sea surface. Actually, a more rigorous solution of the wave equation shows that the intensity cannot suddenly drop to zero at the shadow zone boundary. Nevertheless, in deep water the acoustic intensity may drop by as much as 10 to 60 dB in the shadow zone.

From (4-66), the horizontal range from the source to the shadow zone boundary is

$$r_m = -\left(\frac{2c_0}{g}\right)\sin\theta_m \qquad (5\text{-}7)$$

In deep water, it may be difficult to detect an acoustic signal if the source and receiver are at the same depth, and if the range exceeds r_m. As a numerical example, let

$$c_0 = 5000 \text{ ft/sec}$$

$$g = -0.2/\text{sec}$$

$$d = 400 \text{ ft}$$

Then

$$\theta_m = \text{arc cos}\left[1 - (400)\left(\frac{0.2}{5000}\right)\right] = 10.26°$$

$$r_m = (2)\left(\frac{5000}{0.2}\right)\sin 10.26° = 8905 \text{ ft} = 2969 \text{ yd}$$

Thus, for this example, the co-depth performance may be severely limited if source and receiver are separated by more than 3000 yd. At ranges less than r_m, the transmission loss will very nearly follow the spherical spreading law.

Assume now a water depth of 5000 yd and consider the effect of bottom reflection on the co-depth acoustic intensity in the region immediately to the right of the shadow zone boundary in Figure 5-14. Let the acoustic frequency be 3 kHz and assume a medium-loss bottom. At the steep grazing angles resulting from the geometry of this example, the bottom-reflection loss may be on the order of 14 dB. In addition, an absorption loss of approximately 0.2 dB/kyd must be included.

In Figure 5-15, the transmission loss from the source out to the shadow zone boundary is shown as a spherical spreading loss. At the shadow zone boundary the direct path intensity drops to zero in accordance with ray theory.

Now consider the bottom-reflected signal. At zero horizontal range, the bottom-reflected signal travels a round-trip distance of approximately 9740 yd. Assuming spherical spreading for this path, this results in a 79.8-dB loss. To this must be added the bottom-reflection loss of 14 dB and an absorption loss of 2 dB, for a total of 95.8 dB. Compared with the direct path signal, the bottom-reflected signal is obviously negligible for ranges shorter than r_m.

For simplicity, straight-line propagation is assumed for the bottom-reflected path. Thus the path length for the reflected ray that reaches the source depth at the horizontal range r_m is 10,192 yd. The spherical spreading loss is

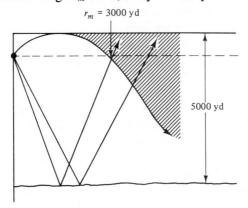

Figure 5-14 Bottom-reflected paths in deep water.

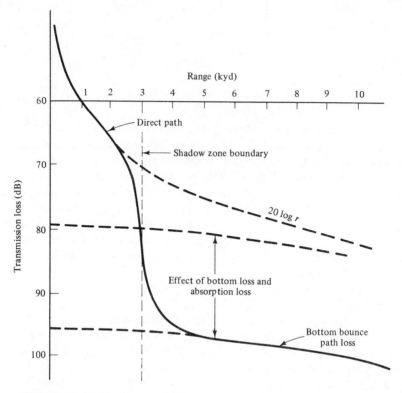

Figure 5-15 Estimated transmission loss in shadow zone region resulting from bottom reflection in deep water. Source and receiver depth = 400 ft, water depth = 5000 yd.

therefore 80.16 dB, giving a total loss of 96.16 dB, including the bottom loss and absorption loss. Notice that at horizontal ranges less than twice the water depth, the transmission loss for the bottom-reflected path changes very slowly. At a horizontal range of 10,000 yd, the bottom-reflected path length is 13,960 yd giving a spreading loss of 83 dB, an absorption loss of 2.8 dB and a bottom loss of 14 dB, for a total of 99.8 dB.

The resultant total transmission loss is shown in Figure 5-15 as the combination of the direct path loss out to 3 Kyds and the bottom path loss beyond 3 Kyds. The sharp transition at the shadow zone boundary predicted by ray theory is modified in the figure to give a more realistic picture.

In shallow water the negative sound speed gradient may extend to the ocean bottom. In this case the bottom loss becomes a dominant factor in acoustic transmission. In Figure 5-16, an omnidirectional source is shown at a depth d in a channel of total depth h. Rays that leave the source at angles of $\pm \theta_m$ will just graze the surface and be reflected at the bottom at intervals r_s. Assuming that the water depth is small compared with the radius of curvature for these grazing

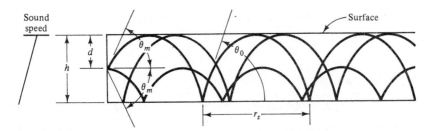

Figure 5-16 Acoustic propagation in a shallow channel with a negative temperature gradient.

rays, both θ_m and θ_0 (the maximum angle of the grazing ray) are small. Using small-angle approximations, the equations presented in Section 4.4 simplify to give the following relationships [5]:

$$\theta_m = \left(\frac{2d}{R}\right)^{1/2} \quad \text{rad}$$

$$\theta_0 = \left(\frac{2h}{R}\right)^{1/2} \quad \text{rad} \tag{5-8}$$

$$r_s = (8hR)^{1/2}$$

$$R = -\frac{c_0}{g}$$

All rays at angles less than $\pm\theta_m$ do not reach the surface, but suffer bottom reflections at intervals less than r_s. Rays that leave the source at angles exceeding θ_m reflect off both the surface and bottom at intervals that decrease rapidly as the angle increases.

Recognizing that a loss is suffered at each bottom and surface reflection, rays that leave the source at angles outside the range $\pm\theta_m$ attenuate rapidly because of these reflections. The energy that propagates to long ranges in the channel is primarily that which leaves the source within the angular range $\pm\theta_m$. The intensity at unit range from the source is inversely proportional to the area on a unit sphere intercepted by the angle $2\theta_m$. This area is easily shown to be $4\pi \sin\theta_m$. The energy in this area expands radially from the source and is confined by the surface and bottom boundaries to a cylindrical section of height h. The area of the cylindrical surface at some long-range r is therefore $2\pi rh$. Neglecting for a moment the bottom-reflection losses and absorption loss, the reduction in intensity from unit range to range r is proportional to the ratio of areas. Thus

$$TL = 10 \log\left(\frac{2\pi rh}{4\pi \sin\theta_m}\right)$$

$$= 10 \log\left(\frac{rh}{2 \sin\theta_m}\right) \tag{5-9}$$

Now define a transition range, r_0, as

$$r_0 = \frac{h}{2 \sin \theta_m} \qquad (5\text{-}10)$$

and (5-9) becomes

$$TL = 10 \log(rr_0) = 10 \log r_0 + 10 \log r \qquad (5\text{-}11)$$

This expression indicates that at long range, that is, $r > r_0$, the intensity decreases inversely with r rather than with r^2. This transmission characteristic is referred to as *cylindrical spreading*. At $r = r_0$ the loss is inversely proportional to r_0^2. Thus spherical spreading loss prevails from the source to the transition range, followed by cylindrical spreading at longer ranges.

The bottom-reflection losses and absorption loss must now be added to the loss given by (5-11). For the surface grazing rays, the number of bottom bounces at some long range r is approximately

$$n = \frac{r}{r_s} = \frac{r}{(8hR)^{1/2}} \qquad (5\text{-}12)$$

Assume a nominal sound speed of 5000 ft/sec and (5-12) can be put in the form

$$n = 1.5r(10^{-2})\left(\frac{-g}{h}\right)^{1/2} \qquad (5\text{-}13)$$

with r in yards and h in feet. Rays leaving the source at angles less than θ_m will suffer more bottom bounces than indicated by (5-13). Using (5-13) as the minimum number, the *minimum* expected transmission loss for the shallow channel at ranges beyond r_0 becomes

$$TL = 10 \log(r_0) + 10 \log(r) + 10^{-3}\left(\alpha_a + 15\alpha_b \sqrt{\frac{-g}{h}}\right)r \qquad (5\text{-}14)$$

where r_0 = transition range, yd
  r = range, yd
  α_a = absorption loss, dB/kyd
  α_b = bottom-reflection loss per bounce, dB
  g = sound speed gradient
  h = water depth, ft

As an example, assume a channel depth of 400 ft, source depth of 200 ft, a gradient of -0.2 per second, and a surface sound speed of 5000 ft/sec. Using the relationships in (5-8), we obtain

$$\theta_m = 0.1265 \text{ rad} = 7.25°$$

and from (5-10)

$$r_0 = 1585 \text{ ft} = 528 \text{ yd}$$

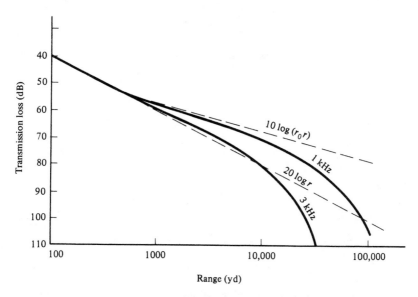

Figure 5-17 Transmission loss in a shallow channel with a negative sound speed gradient.

Equation (5-14) now becomes

$$TL = 10 \log(528) + 10 \log(r) + 10^{-3}(\alpha_a + 0.335\,\alpha_b)r$$

The transmission loss at 1 kHz and 3 kHz is plotted in Figure 5-17, assuming the following values for α_a and α_b:

$$\alpha_a = 0.07 \text{ dB/kyd at 1 kHz}$$

$$= 0.2 \text{ db/kyd at 3 kHz}$$

$$\alpha_b = 0.5 \text{ dB/bounce at 1 kHz}$$

$$= 3 \text{ dB/bounce at 3 kHz}$$

Notice the dramatic frequency dependence caused primarily by the bottom loss term. At 1 kHz the transmission loss exceeds that for spherical spreading at a range of 100 kyd. At 3 kHz the loss is equal to spherical spreading at 10 kyd. At frequencies below 500 Hz the bottom-reflection loss becomes negligible, resulting in a cylindrical spreading characteristic. However, at very low frequencies the methods of ray acoustics are not adequate for predicting the acoustic intensity in a shallow channel. Normal-mode solutions at low frequencies may give results departing significantly from the simple cylindrical spreading law shown on Figure 5-17.

The frequency response of the shallow channel in this example is plotted in Figure 5-18 at several ranges. The response at each range is shown relative

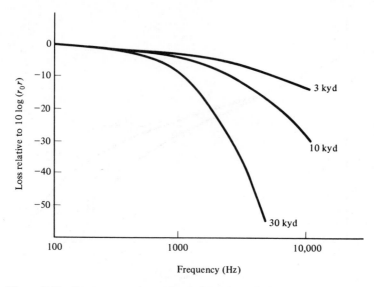

Figure 5-18   Frequency response of a shallow channel with a negative gradient.

to the cylindrical spreading loss curve in Figure 5-17. The shallow channel behaves as a low-pass filter with a corner frequency that decreases as the range from the source increases. If the acoustic source is a broadband source, it is evident that the shallow channel will modify the shape of the source spectral density in a manner that varies with range.

The frequency response of the shallow channel is reduced as the channel depth decreases, or as the negative gradient increases, or as the bottom loss per bounce increases. Because of the wide range of bottom types, the proper choice of bottom loss data is important if accurate results are required. If possible, on-site acoustic measurements should be used to confirm the loss properties of the bottom.

5.5.2 Positive Surface Gradient

In many areas of the world an isothermal surface layer over water with a negative temperature gradient is a common occurrence. In the North Atlantic such conditions exist perhaps 50% of the time in the winter, with layer depths exceeding 200 ft [5]. Such a situation is shown in Figure 5-19 together with a typical ray diagram for a source located in the layer. With a constant temperature in the layer, the sound speed gradient is +0.017 per second.

From the figure it is evident that rays that leave the source at angles exceeding θ_m either reflect off the surface and refract to deep water or refract directly to deep water. Note the shadow zone created below the layer, beyond the range where the first grazing ray reaches the bottom of the layer.

If the surface layer perfectly contains the energy radiated by the source in the angle $\pm\theta_m$, the transmission loss in the layer will tend to follow a cylindrical

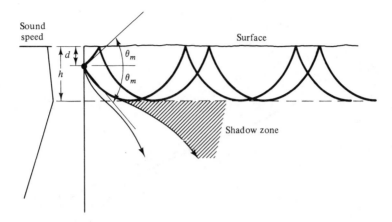

Figure 5-19 Acoustic transmission with an isothermal surface layer over a negative gradient region.

spreading law as described by (5-11). As with the negative gradient channel, several effects act to modify the cylindrical spreading characteristic. These include reflection losses at the sea surface, absorption loss, and leakage of sound out of the layer at the lower boundary as a result of irregularities at the interface between the surface layer and the water below. At low frequencies the ray acoustic technique may be inadequate to describe the acoustic properties of the layer if the wavelength is not small compared with the layer depth.

The surface reflection loss results in a frequency-dependent transmission characteristic similar to that observed with the shallow channel negative gradient condition. However, because of the very large radius of curvature in the isothermal layer, the number of bounces per kiloyard is less than normally encountered in a strong negative gradient.

With an isothermal surface layer depth of 400 ft and a source depth of 200 ft, we obtain from (5-8) and (5-10)

$$R = \frac{5000}{0.017} = 294,117 \text{ ft}$$

$$\theta_m = \left[\frac{2(h - d)}{R} \right]^{1/2} = 0.037 \text{ rad or } 2.1°$$

$$r_s \simeq 10 \text{ kyd}$$

$$r_0 = 1800 \text{ yd}$$

Assuming the number of surface reflections is inversely proportional to r_s, the transmission loss for the surface layer will be at least

$$TL = 10 \log(1800) + 10 \log(r) + (10^{-4})r\alpha_s + (10^{-3})r\alpha_a$$

where α_s is the surface-reflection loss per bounce. At 1 kHz and 3 kHz assume the surface-reflection loss to be 0.5 dB and 3 dB, respectively. From Figure 5-6,

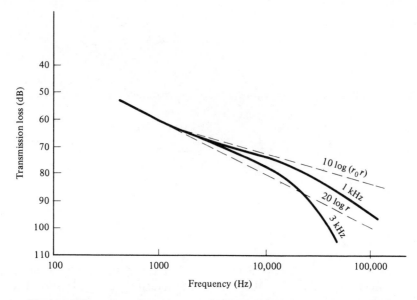

Figure 5-20 Acoustic transmission loss in isothermal surface layer, $h = 400$ ft, $d = 200$ ft, SS–3.

this is seen to correspond approximately to sea state 3. The resulting transmission loss is plotted in Figure 5-20. At 1 kHz the transmission loss is considerably better than spherical spreading even at ranges exceeding 100 kyd. However, at 3 kHz the loss increases very rapidly at ranges exceeding 30 kyd. At higher sea states the performance of the surface layer as an acoustic channel deteriorates rapidly. For very calm surface conditions, the transmission loss more closely approaches the cylindrical spreading law.

5.6 THE DEEP SOUND CHANNEL

The transition region between the main thermocline and the deep isothermal layer gives rise to an acoustic lens effect in the ocean. At middle latitudes this region is located at depths between 3000 and 4000 ft and contains the point of minimum sound speed.

With an acoustic source located exactly at the point of minimum sound speed, rays originating at the source will oscillate back and forth across the source depth as shown in Figure 5-21. The minimum sound speed depth is called the *axis* of the *deep sound channel*. As shown in the figure, sound originating on this axis tends to be focused along the axis so that the acoustic intensity at this depth diminishes with a cylindrical, rather than spherical, spreading loss. Of course, the cylindrical characteristic is modified by absorption loss as a function of range and frequency.

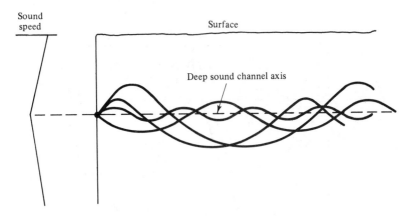

Figure 5-21   Acoustic ray paths for the deep sound channel.

The characteristics of the deep sound channel have been used for rescue operations of downed pilots at sea. A small depth charge, set to explode at the sound channel depth, is dropped by the downed pilot. By measuring the relative time of arrival of the received signal at several receiving stations, a fix can be obtained on the pilot's location. Signals from such charges have been detected at ranges of 2000 miles or more. The deep sound channel is sometimes called the SOFAR duct, for SOund Fixing And Ranging.

5.7 CONVERGENCE ZONE TRANSMISSION

With a sound speed profile of the type shown in Figure 5-21, a sound source located above or below the axis of the deep sound channel also generates rays that oscillate about the channel axis. In deep water, a case of practical interest involves a source and receiver relatively close to the surface. Consider the ray diagram for a source near the surface, as shown in Figure 5-22. This example is similar to that considered in Section 4.4.1. If all rays are plotted that do not intersect either the surface or bottom, we find that they periodically return to the near-surface region in narrow-range bands called *convergence zones*. At middle latitudes the distance between convergence zones is typically 30 to 35 miles, and the width of each convergence zone at the surface is 5 to 10% of range. Notice that for convergence zones to exist, the water depth must be sufficient for the sound speed at the bottom to equal or exceed that at the surface.

Now consider the transmission loss versus range for a receiver at the same depth as the source. At close range the loss obeys the spherical spreading law out to the boundary of the shadow zone. In the shadow zone, the acoustic intensity drops considerably to a level determined by the bottom-reflection characteristics. This low intensity persists out to the convergence zone. In the convergence zone the intensity increases abruptly to a value that typically ex-

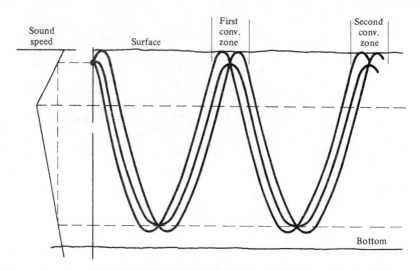

Figure 5-22 Convergence zone transmission.

ceeds that predicted by spherical spreading. Following the first convergence zone, the intensity again drops to a level determined by bottom characteristics. It is easy to show that two bottom bounces are required to reach the receiver from the region between the first and second convergence zones. A typical propagation loss characteristic for conditions supporting convergence zone transmission is shown in Figure 5-23.

5.8 RELIABLE ACOUSTIC PATH

The existence of convergence zones suggests another mode of operation using sonobuoys in the deep ocean. Suppose that a sensor is located at great depth at the point where the sound speed equals the maximum speed near the surface. A ray that is horizontal at the sensor is again horizontal at the surface. All rays above the horizontal at the sensor intersect the surface, as will all rays below the horizontal down to the ray that just grazes the bottom.

The ray paths will appear as in Figure 5-24, showing that solid coverage is obtained out to a range in excess of one-half the convergence zone range. The maximum range in this case is determined by the depth of water below the sensor. The transmission loss at the surface will approach that for spherical spreading (with range measured from the deep sensor to the surface).

Actually, the water does not have to be deep enough to support convergence zone paths to take advantage of this mode. If the sensor is mounted near the bottom in deep water, solid coverage will be obtained at least out to the range determined by the bottom grazing ray. This propagation path is called the *reliable acoustic path,* or RAP mode, because of its relatively stable characteristics.

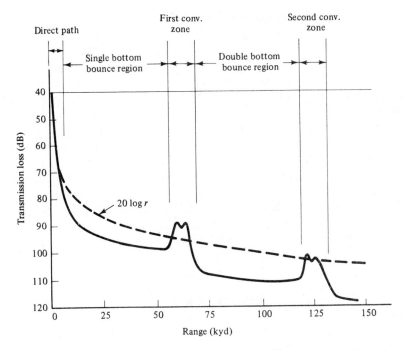

Figure 5-23 Transmission loss in deep water with convergence zones.

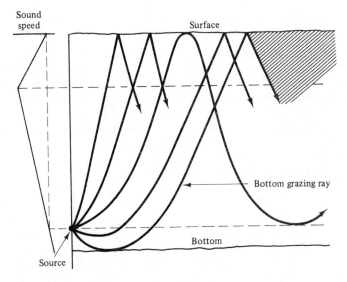

Figure 5-24 Reliable acoustic path (RAP mode) transmission.

5.9 FLUCTUATION OF SOUND IN THE SEA

The acoustic intensity measured at some distance from a steady source is not constant, but fluctuates with time. The obvious activity of the sea surface accounts for part of this fluctuation. A thermal microstructure in the top layers of the ocean is also responsible for intensity fluctuations. Sensitive temperature measurements at a fixed point over a period of time, or measurements at a fixed depth as a function of horizontal distance, reveal temperature variations of a fraction of a degree. Although small, the temperature variations result in sound speed variations that, in turn, introduce fluctuations into the refractive process. The cumulative effect over the acoustic transmission paths results in appreciable intensity variations.

Transmission loss predictions based on linear or smoothly varying sound speed profiles give, at best, the mean value of the measured results. Intensity fluctuations about the mean of as much as 50% may be expected as a result of the thermal microstructure [14].

PROBLEMS

5.1. Assuming a salinity of 35 parts per thousand, use (5-1) to calculate the sound speed–temperature gradient at $T = 0°C$ and $+10°C$.

5.2. The axis of the deep sound channel is at the depth of minimum sound speed. Assume a negative temperature gradient and a temperature near 0°C at this depth with a salinity of 35 parts per thousand. Starting with (5-1), calculate the temperature–depth gradient at the depth of minimum sound speed.

5.3. Using the Rayleigh criterion, with a grazing angle of 10°, calculate the maximum wave heights found in a "smooth" surface at acoustic frequencies of 1, 10, and 30 kHz. Using these wave heights, evaluate the resulting surface reflection loss according to (5-2).

5.4. Assume an isothermal surface layer with a depth of 100 m and an acoustic source at a depth of 30 m. Calculate θ_m, r_0, and r_s. Determine the number of surface bounces per kilometer for the grazing ray leaving the source at an angle of θ_m. Compare this with the number of surface bounces per kilometer for a ray that leaves the source at an angle of zero. Assume a nominal sound speed of 1500 m/sec.

5.5. For the isothermal surface layer and source described in Problem 5.4, estimate the transmission loss at 30 km for a receiver located in the layer. Assume a frequency of 3 kHz with a surface loss of 3 dB per bounce and an absorption loss of 0.2 dB/km. Assume first that the number of surface bounces is equal to the number for rays leaving the source at θ_m. Compare the loss obtained with that using the number of bounces associated with rays leaving the source at an initial angle of zero.

5.6. Assume a sound speed profile as given in Figure 4-11. With a bottom depth of 4000 m and a RAP mode receiver at a depth of 3000 m, calculate the horizontal range to a source at the surface for which the acoustic ray path just grazes the bottom.

SUGGESTED READING

1. Kuwahara, S., "Velocity of Sound in Sea Water and Calculation of the Velocity for Use in Sonic Sounding," *Hydrogr. Rev.*, Vol. 16, p. 123 (1939).

2. DelGrosso, V. A., "Velocity of Sound in Sea Water at Zero Depth," U.S. Naval Research Laboratory Report 4002, 1952.

3. Wilson, W. D., "Speed of Sound in Sea Water as a Function of Temperature, Pressure and Salinity," *J. Acoust. Soc. Am.* Vol. 32, p. 641 (1960).

4. Kinsler, L. E., and A. R. Frey, *Fundamentals of Acoustics*, 2nd ed. New York: John Wiley & Sons, Inc., 1962.

5. Urick, R. J., *Principles of Underwater Sound for Engineers*, 2nd ed. New York: McGraw-Hill Book Company, 1975, Chaps. 5 and 6.

6. Podeszwa, E. M., "Sound Speed Profiles for the North Atlantic Ocean," Naval Underwater Systems Center, Technical Document 5447, 1976.

7. Marsh, H. W., "Exact Solution of Wave Scattering by Irregular Surfaces," *J. Acoust. Soc. Am.*, Vol. 33, p. 330 (1961).

8. Marsh, H. W., M. Schulkin, and S. G. Kneale, "Scattering of Underwater Sound by the Sea Surface," *J. Acoust. Soc. Am.*, Vol. 33, p. 334 (1961).

9. Weinberg, H., "Generic Sonar Model," Naval Underwater Systems Center, Technical Document 5971-A, Feb. 27, 1980.

10. Thorp, W. H., "Deep Ocean Sound Attenuation in the Sub- and Low-Kilocycle-per-Second Region," *J. Acoust. Soc. Am.*, Vol. 38, p. 648 (1965).

11. Brekhovskikh, L. M., *Waves in Layered Media*, New York: Academic Press, Inc., 1960.

12. Mackenzie, K. V., "Long-Range Shallow-Water Transmission," *J. Acoust. Soc. Am.*, Vol. 33, p. 1505 (1961).

13. Weston, D. E., "Propagation of Sound in Shallow Water," *J. Br. IRE*, Vol. 26, p. 329 (1963).

14. Whitmarsh, D. C., E. Skudryzk, and R. J. Urick, "Forward Scattering of Sound in the Sea and Its Correlation with Temperature Microstructure," *J. Acoust. Soc. Am.*, Vol. 29, p. 1124 (1957).

6

Fourier Methods

6.0 INTRODUCTION

The analysis of underwater acoustic systems involves the consideration of signals, noise, and their modification by various system operations, such as filtering and detection. The transformation of complicated time-domain functions into the frequency domain by the methods of Fourier analysis proves to be a very useful technique in this type of analysis. The response of linear systems to complicated input signals may be deduced directly from their response to sinusoidal signals of different frequencies. The calculation of the output signal from such a system is often simplified if the input signal is expressed in these same terms.

In this chapter we cover the expansion of periodic functions as Fourier series, leading to the more general Fourier transform representation of functions. Basic operations required in system analysis, such as multiplication, convolution, differentiation, and integration, are discussed in both the time and frequency domains. Using the concepts and notation developed in this chapter, the sampling theorem is derived. Finally, the analytic signal is discussed in relation to the complex representation of real waveforms and the concept of the envelope function.

6.1 EXPANSION OF FUNCTIONS

Many complicated functions can be represented over some finite interval, $[x_1, x_2]$, as

$$f(x) = \sum_n a_n \phi_n(x) \qquad (6\text{-}1)$$

where the functions $\phi_n(x)$ are called the *basis functions* of the expansion, and the constants, a_n, must be determined. In general, an infinite number of terms are required for the mean-squared difference between the original function and the series expansion to approach zero.

A series expansion familiar to most engineers is the *Taylor series* expansion of a function, using powers of the independent variable as the basis functions. A continuous function $f(x)$ having finite derivatives at a point x_0 may be represented about the point x_0 by the Taylor series

$$f(x) = \sum_{n=0}^{\infty} a_n(x - x_0)^n \tag{6-2}$$

where

$$a_n = \left.\frac{d^n f(x)}{n!\, dx^n}\right|_{x=x_0} = \frac{f^{(n)}(x_0)}{n!}$$

Although the Taylor series is very useful in many applications, it is not well suited to most signal and system analysis problems.

A very useful class of basis functions is one in which the member functions of a given set are mutually orthogonal. A group of functions, $\phi_n(x)$, is orthogonal over a region $[x_1, x_2]$. with respect to a nonnegative weighting function $r(x)$, if

$$\int_{x_1}^{x_2} r(x)\phi_m(x)\phi_n^*(x)\, dx = \begin{cases} A_n & \text{if } m = n \\ 0 & \text{if } m \neq n \end{cases} \tag{6-3}$$

where A_n is a constant and $\phi_n^*(x)$ is the complex conjugate of $\phi_n(x)$. A relationship such as (6-3) is sometimes written in more compact form using a function called the *Kronecker delta function,* defined as

$$\delta_{mn} = \begin{cases} 1 & \text{if } m = n \\ 0 & \text{if } m \neq n \end{cases} \tag{6-4}$$

Using (6-4), (6-3) becomes

$$\int_{x_1}^{x_2} r(x)\phi_m(x)\phi_n^*(x)\, dx = \delta_{mn} A_n \tag{6-5}$$

The value of orthogonal functions as the basis for expansion lies in the ease with which the coefficients a_n can be evaluated. For instance, assume that the basis functions in (6-1) are orthogonal and multiply both sides of the expression by $r(x)\phi_n^*(x)$, where $\phi_n(x)$ is one particular member of the set of basis functions. Now integrate over the interval of orthogonality and obtain

$$\int_{x_1}^{x_2} r(x)f(x)\phi_n^*(x)\, dx = \int_{x_1}^{x_2} r(x)[a_0\phi_0(x)\phi_n^*(x) + \cdots$$

$$+ a_m\phi_m(x)\phi_n^*(x) + a_n\,|\phi_n(x)|^2 + \cdots]\, dx \tag{6-6}$$

Because of the orthogonality relationship of (6-3) or (6-5), all terms on the right in (6-6) are zero except the one involving $|\phi_n(x)|^2$. Hence

$$\int_{x_1}^{x_2} r(x)f(x)\phi_n^*(x)\,dx = a_n \int_{x_1}^{x_2} r(x)|\phi_n(x)|^2\,dx = a_n A_n \tag{6-7}$$

from which

$$a_n = \frac{1}{A_n} \int_{x_1}^{x_2} r(x)f(x)\phi_n^*(x)\,dx \tag{6-8}$$

6.2 PERIODIC FUNCTIONS: COMPLEX FOURIER SERIES

A function is defined as periodic if for any integer n

$$f(x) = f(nT + x)$$

where T is called the period. It follows that if the function is known in any interval $[x_1, x_1 + T]$, it is known over the entire range of x.

The expansion of periodic functions is usually accomplished using basis functions that are themselves periodic, with periods related to the period of the function to be expanded. For instance, the Fourier series expansion uses the trigonometric sine and cosine functions as the basis functions. It is convenient to combine the sine and cosine terms, using the Euler identity, to form a complex basis function. Thus

$$\text{complex basis function} = \exp(j\omega_n x) = \cos \omega_n x + j \sin \omega_n x \tag{6-9}$$

where $\omega_n = n\omega_1 = 2\pi n f_1$. The complex exponential basis function results in a complex form of the Fourier series coefficients. The resulting expansion of a periodic function is identical to that obtained using the real trigonometric basis functions.

The basis functions defined by (6-9) are orthogonal over the fundamental period, $T = 1/f_1$, with respect to the weighting function $r(x) = 1.0$. It is easy to verify that

$$\int_{x_1}^{x_1+T} \exp(jm\omega_1 x)\exp(-jn\omega_1 x)\,dx = \delta_{mn}T \tag{6-10}$$

The complex Fourier series expansion for an arbitrary function $f(x)$ may be written as

$$f(x) = \sum_{n=-\infty}^{+\infty} \alpha_n \exp(jn\omega_1 x) \tag{6-11}$$

where the constants α_n are obtained using (6-8) and (6-10). Thus

$$\alpha_n = \frac{1}{T} \int_T f(x)\exp(-jn\omega_1 x)\,dx \tag{6-12}$$

Notice that in (6-11) the summation index, n, is allowed to range from minus infinity to plus infinity. Also, notice that the constant α_n may in general be complex, even though $f(x)$ is real.

It is instructive to separate $f(x)$ into its even and odd components in order to investigate the properties of the complex Fourier coefficients. An even function is defined as one for which $f(x) = f(-x)$. The cosine function is an even function of its argument about the origin. For an odd function, $f(x) = -f(-x)$. The sine function is a typical example of an odd function. A real function $f(x)$ may be separated into its even and odd components as follows:

$$f_e(x) = \frac{f(x) + f(-x)}{2}$$

$$f_0(x) = \frac{f(x) - f(-x)}{2} \tag{6-13}$$

The product of the even and odd components of $f(x)$ gives

$$f_e(x)f_o(x) = \frac{f^2(x) - f^2(-x)}{4} \tag{6-14}$$

The integral of (6-14) over all x is easily shown to be zero, leading to the conclusion that the even and odd components of a function are orthogonal. This can be generalized to show that all even functions are orthogonal to odd functions in an interval symmetric about the origin.

The Fourier coefficients for $f(x)$ may now be expressed as

$$\alpha_n = \frac{1}{T} \int_T [f_e(x) + f_o(x)] \exp(-jn\omega_1 x)\, dx \tag{6-15}$$

Recognizing that $\exp(-jn\omega_1 x)$ may itself be separated into even and odd components by the Euler identity, and that the products of the form $f_e(x) \sin n\omega_1 x$ and $f_o(x) \cos n\omega_1 x$ will integrate to zero, (6-15) finally becomes

$$\alpha_n = \frac{1}{T} \int_T f_e(x) \cos n\omega_1 x\, dx - \frac{j}{T} \int_T f_o(x) \sin n\omega_1 x\, dx \tag{6-16}$$

This shows that the real part of α_n results from the even part of $f(x)$. Similarly, the odd part of $f(x)$ gives rise to the imaginary part of α_n.

Because of the even and odd symmetry of cosine and sine functions, the real part of α_n is unchanged by a change in sign of n and the imaginary part changes sign along with a sign change in n. That is, for $f(x)$ real

$$\text{Re}[\alpha_n] = \text{Re}[\alpha_{-n}]$$

$$\text{Im}[\alpha_n] = -\text{Im}[\alpha_{-n}]$$

from which

$$\alpha_n = \alpha_{-n}^* \tag{6-17}$$

Equation (6-17) indicates that the complex Fourier coefficients for a real function exhibit *conjugate symmetry*.

6.2.1 Example: Periodic Pulse Train

Consider a periodic train of rectangular pulses as shown in Figure 6-1. This could be, for instance, a voltage or pressure waveform as a function of time. The width of each pulse is t_p, the period is T, and the pulse amplitude is A. Because of the periodic nature of the waveform, the function is completely defined by specifying its value in the interval $[-T/2, +T/2]$. Thus

$$f(t) = \begin{cases} A & \text{for } \dfrac{-t_p}{2} \le t \le \dfrac{t_p}{2} \\ 0 & \text{elsewhere in the period} \end{cases}$$

The complex Fourier coefficients are defined as

$$\alpha_n = \frac{1}{T} \int_{-T/2}^{+T/2} f(t) \exp(-jn\omega_1 t)\, dt = \frac{A}{T} \int_{-t_p/2}^{+t_p/2} \exp(-jn\omega_1 t)\, dt$$

From elementary calculus we obtain

$$\alpha_n = -\frac{A}{jn\omega_1 T} \exp(-jn\omega_1 t)\Big|_{-t_p/2}^{+t_p/2}$$

$$= \frac{A}{jn\omega_1 T}\left[\exp\left(\frac{jn\omega_1 t_p}{2}\right) - \exp\left(\frac{-jn\omega_1 t_p}{2}\right) \right]$$

By using the Euler identity, this can be put in the form

$$\alpha_n = \frac{At_p}{T}\frac{\sin(n\omega_1 t_p/2)}{n\omega_1 t_p/2} = \frac{At_p}{T}\frac{\sin(\pi n f_1 t_p)}{\pi n f_1 t_p} \tag{6-18}$$

Notice that for this example the imaginary part of α_n is zero. This is because the function $f(t)$ is an even function. The Fourier expansion of $f(t)$ may now be written as

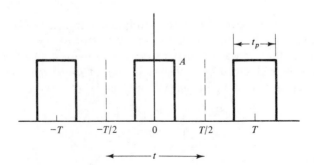

Figure 6-1 Rectangular pulse train.

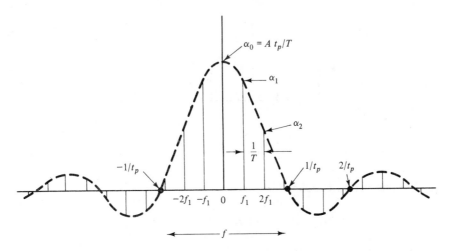

Figure 6-2   Amplitude spectrum for a rectangular pulse train centered at the origin.

$$f(t) = \frac{At_p}{T} \sum_{n=-\infty}^{+\infty} \frac{\sin(\pi n f_1 t_p)}{\pi n f_1 t_p} \exp(j2\pi n f_1 t) \qquad (6\text{-}19)$$

Because $f(t)$ is real in this example, the summation of the imaginary terms in (6-19) is zero.

The amplitude of α_n is plotted in Figure 6-2 as a function of frequency. This plot is called the amplitude spectrum of $f(t)$ and reveals the frequency content of the function in graphical form. The term at the origin, α_0, is the dc component of the signal, At_p/T. Notice that for $n = 0$, α_n is actually indeterminant and must be evaluated in the limit as the argument of α approaches zero. The sum of the components at $n = +1$ and -1 gives the amplitude of the component at the fundamental frequency, f_1, contained in the signal. Similarly, the content at each of the harmonics of the fundamental is given by the sum of the components at $n = \pm2, \pm3, \cdots$. The following key features may be noted from Figure 6-2:

1. The lines of the spectrum of a periodic signal are separated by $f_1 = 1/T$.
2. The spectrum of this particular signal is symmetrical about $f = 0$.
3. The spectrum for this signal is concentrated mainly in the region between $f = -1/t_p$ and $f = +1/t_p$.

If the period of the signal is increased, the separation between lines in the spectrum is decreased. However, the *envelope* of the spectrum remains unchanged provided the shape of the signal within one period remains the same.

As a variation on this example, assume that the periodic pulse train in Figure 6-1 is translated to the right by an amount t_0, as shown in Figure 6-3. The

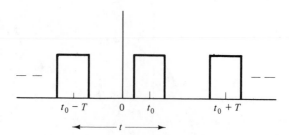

Figure 6-3 Periodic pulse train centered at $t = t_0$.

signal in one period is defined as

$$f(t) = A \qquad \text{for } t_0 - \frac{t_p}{2} \le t \le t_0 + \frac{t_p}{2}$$

and the Fourier coefficients are

$$
\begin{aligned}
\alpha_n &= \frac{A}{T} \int_{t_0 - t_p/2}^{t_0 + t_p/2} \exp(-jn\omega_1 t)\, dt \\
&= -\frac{A}{jn\omega_1 T} \exp(-jn\omega_1 t) \Big|_{t_0 - t_p/2}^{t_0 + t_p/2} \\
&= \frac{A}{jn\omega_1 T} \exp(-jn\omega_1 t_0) \left[\exp\left(\frac{+jn\omega_1 t_p}{2}\right) - \exp\left(\frac{-jn\omega_1 t_p}{2}\right) \right] \\
&= \left[\frac{A t_p}{T} \frac{\sin(n\omega_1 t_p/2)}{(n\omega_1 t_p/2)} \right] \exp(-jn\omega_1 t_0)
\end{aligned}
\tag{6-20}
$$

The translation to the location t_0 results in a spectrum that contains an imaginary term because the periodic waveform now contains both even and odd components. Notice, however, that the spectrum *amplitude* is identical to that in (6-18). The information relating to the time shift is contained entirely in the complex exponential term, $\exp(-jn\omega_1 t_0)$. The translation results in a phase shift, $n\omega_1 t_0$, applied to each line of the original spectrum. The phase is a linear function of frequency, and directly proportional to t_0.

6.2.2 Power Spectral Density: Parseval's Theorem

Let $s(t)$ be a time-domain periodic voltage waveform with a Fourier expansion as follows:

$$s(t) = \sum_n \alpha_n \exp(jn\omega_1 t) \tag{6-21}$$

The average power delivered to a load by this waveform is proportional to the average value of the square of the waveform, and inversely proportional to the effective parallel load resistance. It is convenient to assume a 1-Ω load resistance and equate power directly with the mean-squared voltage waveform. This convention is used frequently in this and subsequent chapters, where a 1-Ω load is tacitly assumed unless otherwise indicated. The average power in the waveform in (6-21) is therefore given by

$$\frac{1}{T} \int_T |s(t)|^2 \, dt = \frac{1}{T} \int_T \left| \sum_n \alpha_n \exp(jn\omega_1 t) \right|^2 dt$$

The squared summation may be written as a double summation to obtain

$$\frac{1}{T} \int_T |s(t)|^2 \, dt = \frac{1}{T} \int_T \left\{ \sum_m \sum_n \alpha_m \alpha_n^* \exp[j(m-n)\omega_1 t] \right\} dt$$

$$= \sum_m \sum_n \alpha_m \alpha_n^* \left\{ \frac{1}{T} \int_T \exp[j(m-n)\omega_1 t] \, dt \right\} \qquad (6\text{-}22)$$

Because of the orthogonality of the complex exponential functions, the integral on the right side of (6-22) is the Kronecker delta function as defined in (6-4). Hence

$$\frac{1}{T} \int_T |s(t)|^2 \, dt = \sum_m \sum_n \alpha_m \alpha_n^* \delta_{mn} = \sum_n |\alpha_n|^2 \qquad (6\text{-}23)$$

The average power in the waveform is thus given either by the time average of the squared magnitude of the waveform, or by the sum of the squared magnitudes of the Fourier coefficients. This relationship is Parseval's theorem. The magnitude-squared Fourier coefficients, considered as a function of frequency, give the distribution of waveform power in the frequency domain. A frequency plot of these coefficients is called the *power spectral density* of the waveform. The power contained in any limited frequency region is obtained by summation of the power spectral density over that region.

6.3 TIME-LIMITED WAVEFORMS: THE FOURIER INTEGRAL TRANSFORMATION

In contrast to continuous periodic waveforms, many functions are of limited extent. In the case of time-domain functions, such time-limited waveforms are often called transient signals.

Assume a time-limited waveform as shown in Figure 6-4. This function has the indicated shape in the interval $[t_1, t_2]$ and is zero elsewhere. If this waveform is a voltage or pressure waveform, it may be thought of as having finite total energy, but zero average power when averaged over all time. A continuous, or periodic waveform, on the other hand, has finite average power but infinite total energy. Hence a transient signal is sometimes referred to as an *energy signal*, while a periodic or continuous function is called a *power signal*.

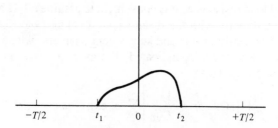

Figure 6-4 Time-limited waveform.

The function in Figure 6-4 may be expanded as a Fourier series over the interval $[t_1, t_2]$ by assuming that the function repeats with a period equal to or greater than the interval $[t_1, t_2]$. In Figure 6-5(a), the discrete amplitude spectrum is shown assuming that the period, T, is exactly $t_2 - t_1$. The separation between Fourier components in this case is $1/(t_2 - t_1)$. If the period is increased, as shown in Figure 6-5(b), the envelope of the spectrum remains the same (except

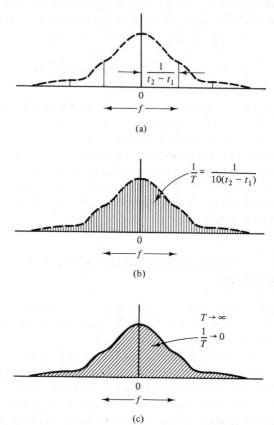

Figure 6-5 Effect on spectrum of increasing the interval of expansion for a time-limited waveform.

for scale factor), but the spacing between lines decreases. The actual time-limited waveform is approached as the period approaches infinity. As shown in Figure 6-5, this causes the separation between spectral lines to approach zero, resulting in a continuous spectrum.

As the period increases, the fundamental frequency, f_1, decreases. In the Fourier expansion of the waveform, we may change the summation index from n to nf_1. Thus

$$s(t) = \sum_{nf_1=-\infty}^{+\infty} \alpha_{(nf_1)} \exp(j2\pi nf_1 t)$$

where

$$\alpha_{(nf_1)} = f_1 \int_{-\frac{1}{2f_1}}^{+\frac{1}{2f_1}} s(t) \exp(-j2\pi nf_1 t) \, dt$$

Combining these expressions yields

$$s(t) = \sum_{nf_1=-\infty}^{+\infty} \left[f_1 \int_{-\frac{1}{2f_1}}^{+\frac{1}{2f_1}} s(t') \exp(-j2\pi nf_1 t') \, dt' \right] \exp(j2\pi nf_1 t) \qquad (6\text{-}24)$$

As the period approaches infinity, the discrete variable, nf_1, approaches a continuous variable, f. Also, we let

$$f_1 \rightarrow df$$

and

$$\frac{1}{2f_1} \rightarrow \infty$$

In the limit, the summation in (6-24) becomes an integral, and we have

$$s(t) = \int_{-\infty}^{+\infty} \left[\int_{-\infty}^{+\infty} s(t') \exp(-j2\pi ft') \, dt' \right] \exp(+j2\pi ft) \, df \qquad (6\text{-}25)$$

Equation (6-25) is the *Fourier integral* representation of the time-limited function, $s(t)$. This equation can be put in more compact form if we define

$$S(f) = \int_{-\infty}^{+\infty} s(t) \exp(-j2\pi ft) \, dt \qquad (6\text{-}26)$$

and write for (6-25)

$$s(t) = \int_{-\infty}^{+\infty} S(f) \exp(j2\pi ft) \, df \qquad (6\text{-}27)$$

Equation (6-26) is the *Fourier transform* of the time-domain function, $s(t)$, to a frequency-domain function $S(f)$. Conversely, (6-27) is the *inverse Fourier transform* of $S(f)$ back to the time domain. The functions $s(t)$ and $S(f)$ are called *Fourier transform pairs* and will be identified by the following symbolic representation:

$$s(t) \longleftrightarrow S(f)$$

The frequency-domain function, $S(f)$, is the continuous version of the Fourier spectrum of the function $s(t)$. As with the discrete function, α_n, if the time-domain function is real, the spectrum exhibits conjugate symmetry. That is,

$$S(f) = S^*(-f) \qquad \text{for } s(t) \text{ real} \qquad (6\text{-}28)$$

The real part of $S(f)$ is the transform of the even part of $s(t)$, and the imaginary part is the transform of the odd part of $s(t)$. It follows that if $s(t)$ is even, its spectrum is purely real.

As an example of the Fourier integral transform, consider the triangular signal shown in Figure 6-6. This signal is defined by

$$s(t) = \begin{cases} A\left(1 + \dfrac{t}{t_p}\right) & \text{for } -t_p \le t \le 0 \\[2ex] A\left(1 - \dfrac{t}{t_p}\right) & \text{for } 0 \le t \le t_p \\[2ex] 0 & \text{elsewhere} \end{cases}$$

The Fourier transform of $s(t)$ is

$$S(f) = \int_{-t_p}^{+t_p} s(t) \exp(-j2\pi ft)\, dt$$

Because $s(t)$ is an even function of t, the imaginary part of $S(f)$ is zero. Therefore,

$$S(f) = \int_{-t_p}^{+t_p} s(t)\cos 2\pi ft\, dt = 2 \int_0^{t_p} s(t)\cos 2\pi ft\, dt$$

$$= 2A \int_0^{t_p} \left(1 - \frac{t}{t_p}\right)\cos 2\pi ft\, dt$$

Using a table of elementary integrals, we obtain

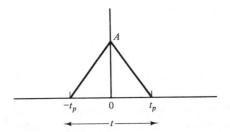

Figure 6-6   Triangular waveform.

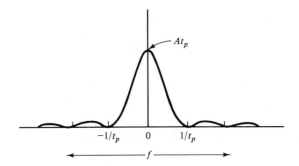

Figure 6-7   Triangular pulse spectrum.

$$S(f) = At_p \frac{\sin^2(\pi f t_p)}{(\pi f t_p)^2} \tag{6-29}$$

A plot of the spectrum of the triangular pulse is shown in Figure 6-7. Notice that the shape of this spectrum is the square of the shape of the spectrum of a rectangular pulse of width t_p. The significance of this will be made clear later in this chapter.

6.4 BASIC BUILDING BLOCK FUNCTIONS

Certain functions occur so frequently in signal and system analysis that they deserve special consideration and special notation. The rectangular pulse, considered in earlier examples, the unit impulse function, and the unit step function are in this category.

6.4.1 Rectangular Function and Its Transform

In previous examples, the rectangular pulse was defined by writing

$$s(t) = \begin{cases} A & \text{for } \frac{-t_p}{2} \le t \le \frac{+t_p}{2} \\ 0 & \text{elsewhere} \end{cases}$$

A standard rectangular pulse is one with unit amplitude and unit width. Thus

$$\text{standard rectangular pulse} = \begin{cases} 1.0 & \text{for } -\frac{1}{2} \le t \le +\frac{1}{2} \\ 0 & \text{elsewhere} \end{cases}$$

This standard pulse may be represented more compactly by the following notation [1, 2]:

$$\text{standard rectangular pulse} = \text{rect}(t) \tag{6-30}$$

A pulse of amplitude A and width t_p then becomes

$$A \operatorname{rect}\left(\frac{t}{t_p}\right) = \begin{cases} A & \text{for } -\tfrac{1}{2} \le \dfrac{t}{t_p} \le +\tfrac{1}{2} \\ & \text{or } \dfrac{-t_p}{2} \le t \le \dfrac{t_p}{2} \\ 0 & \text{elsewhere} \end{cases} \quad (6\text{-}31)$$

The transform of the rectangular function is easily obtained as

$$A \operatorname{rect}\left(\frac{t}{t_p}\right) \longleftrightarrow At_p \frac{\sin(\pi f t_p)}{\pi f t_p} \quad (6\text{-}32)$$

from which the standard pulse and its transform are

$$\operatorname{rect}(t) \longleftrightarrow \frac{\sin(\pi f)}{\pi f} \quad (6\text{-}33)$$

The transform of the rectangular pulse is called a *sinc function*, defined as

$$\operatorname{sinc}(f) = \frac{\sin(\pi f)}{\pi f}$$

or

$$At_p \operatorname{sinc}(ft_p) = At_p \frac{\sin(\pi f t_p)}{\pi f t_p} \quad (6\text{-}34)$$

The use of the rectangular function is not restricted to the time domain, nor is the sinc function restricted to the frequency domain. For instance, a time-domain sinc function with amplitude A and width to the first zero, t_p, together with its transform may be expressed as

$$A \operatorname{sinc}\left(\frac{t}{t_p}\right) \longleftrightarrow At_p \operatorname{rect}(ft_p) \quad (6\text{-}35)$$

The transform-pair relationships of (6-32) and (6-35) are sketched in Figure 6-8.

6.4.2 Impulse Function

The function in Figure 6-9(a) has an amplitude of zero or A, except in the region about the origin where it changes linearly from zero to A in the distance a. Figure 6-9(b) is the derivative of this function, which is a rectangular pulse of amplitude A/a and width a.

Notice that as a becomes very small, the function in Figure 6-9(a) becomes almost discontinuous. At the same time, the derivative becomes a vanishingly narrow pulse with an amplitude inversely proportional to the width. The area of the rectangular pulse remains constant, with a value equal to the magnitude of the amplitude change of the original function at the near-discontinuity.

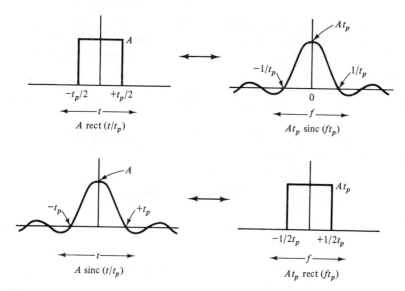

$A \, \text{rect} \, (t/t_p)$ $At_p \, \text{sinc} \, (ft_p)$

$A \, \text{sinc} \, (t/t_p)$ $At_p \, \text{rect} \, (ft_p)$

Figure 6-8   Rectangular functions and sinc functions in the time and frequency domains.

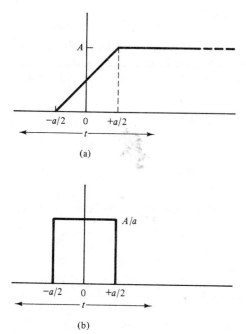

(a)

(b)

Figure 6-9   An almost-discontinuous function and its derivative.

The impulse function is defined using a function like the one shown in Figure 6-9(b), evaluated in the limit as a approaches zero. The *strength* of the impulse is equal to the area of the function. Thus an impulse of strength A is defined as

$$A\delta(t) = \lim_{a \to 0} \left[\frac{A}{a} \operatorname{rect}\left(\frac{t}{a}\right) \right] \tag{6-36}$$

The *unit impulse* is obtained if A is unity. The defining equations for the unit impulse can also be expressed as

$$\delta(t) = \begin{cases} \infty & \text{for } t = 0 \\ 0 & \text{elsewhere} \end{cases}$$

$$\int_{-\infty}^{+\infty} \delta(t)\, dt = 1.0 \tag{6-37}$$

$$\int_{-\infty}^{0} \delta(t)\, dt = \int_{0}^{+\infty} \delta(t)\, dt = \tfrac{1}{2}$$

The unit impulse may be located at some point other than the origin, such as t_0. We then write

$$\delta(t - t_0) = \begin{cases} \infty & \text{for } t = t_0 \\ 0 & \text{elsewhere} \end{cases} \tag{6-38}$$

$$\int_{-\infty}^{+\infty} \delta(t - t_0)\, dt = 1.0$$

Consider the result of multiplying an arbitrary signal $s(t)$ by a unit impulse located at t_0. The product will be zero at every point in time except at the location of the impulse. The result is an impulse with a strength equal to the amplitude of $s(t)$ at t_0. Thus

$$s(t)\delta(t - t_0) = s(t_0)\delta(t - t_0)$$

The integral of this product is

$$\int_{-\infty}^{+\infty} s(t)\delta(t - t_0)\, dt = s(t_0) \int_{-\infty}^{+\infty} \delta(t - t_0)\, dt = s(t_0) \tag{6-39}$$

Using the relationship in (6-39) we may determine the Fourier transform of the unit impulse. Let

$$s(t) = \delta(t)$$

Then

$$S(f) = \int_{-\infty}^{+\infty} s(t)\exp(-j2\pi ft)\, dt$$

$$= \int_{-\infty}^{+\infty} \delta(t)\exp(-j2\pi ft)\, dt = \exp(-j0) = 1.0$$

Hence

$$\delta(t) \longleftrightarrow 1.0 \tag{6-40}$$

The spectrum of the unit impulse located at the origin is equal to unity at all frequencies from minus infinity to plus infinity. For an impulse located at some other point, the transform is

$$S(f) = \int_{-\infty}^{+\infty} \delta(t - t_0) \exp(-j2\pi ft) \, dt = \exp(-j2\pi ft_0)$$

or

$$\delta(t - t_0) \longleftrightarrow \exp(-j2\pi ft_0) \tag{6-41}$$

The unit impulse may also be defined in the frequency domain as

$$\delta(f - f_0) = \begin{cases} \infty & \text{for } f = f_0 \\ 0 & \text{elsewhere} \end{cases} \tag{6-42}$$

$$\int_{-\infty}^{+\infty} \delta(f - f_0) \, df = 1.0$$

The frequency-domain impulse represents a single spectral line at $f = f_0$. The inverse transform of the frequency-domain impulse is easily obtained using the same approach as with the time-domain impulse. Thus

$$1.0 \longleftrightarrow \delta(f)$$
$$\exp(j2\pi f_0 t) \longleftrightarrow \delta(f - f_0) \tag{6-43}$$

The frequency-domain impulse at the origin is the Fourier transform of a constant, or dc, time-domain function.

From the transform pair in (6-43), the Fourier transforms for cosine and sine functions may be obtained. That is, because

$$\cos 2\pi f_0 t = \frac{1}{2}[\exp(j2\pi f_0 t) + \exp(-j2\pi f_0 t)]$$

$$\sin 2\pi f_0 t = \frac{1}{2j}[\exp(j2\pi f_0 t) - \exp(-j2\pi f_0 t)]$$

we may write

$$\cos 2\pi f_0 t \longleftrightarrow \frac{1}{2}[\delta(f - f_0) + \delta(f + f_0)]$$
$$\tag{6-44}$$
$$\sin 2\pi f_0 t \longleftrightarrow \frac{1}{2j}[\delta(f - f_0) - \delta(f + f_0)]$$

Some of the transform-pair relationships for unit impulses in the time and frequency domains are shown in Figure 6-10.

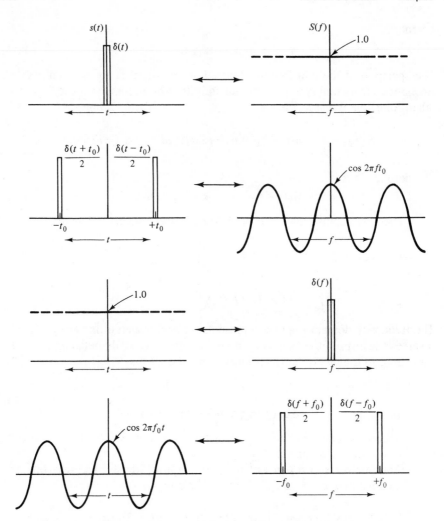

Figure 6-10 Time- and frequency-domain representations of impulse functions.

The unit impulse is helpful in deriving Parseval's theorem for energy-type signals. Let $s(t)$ be a finite-energy signal with a Fourier transform $S(f)$ such that

$$s(t) = \int_{-\infty}^{+\infty} S(f) \exp(j2\pi ft) \, df$$

The squared magnitude of $s(t)$ can be expressed as

$$|s(t)|^2 = \left| \int_{-\infty}^{+\infty} S(f) \exp(j2\pi ft) \, df \right|^2$$

$$= \int_{-\infty}^{+\infty} \int_{-\infty}^{+\infty} S(f) S^*(f') \exp[j2\pi(f - f')t] \, df \, df'$$

The energy in the signal is proportional to the integral of $|s(t)|^2$. Thus

$$\int_{-\infty}^{+\infty} |s(t)|^2 \, dt = \int_{-\infty}^{+\infty} \int_{-\infty}^{+\infty} S(f)S*(f') \left[\int_{-\infty}^{+\infty} \exp[j2\pi(f-f')t] \, dt \right] df \, df'$$

But the integral with respect to time on the right side of this expression is by definition [see, for instance, (6-43)] the impulse function $\delta(f - f')$. Hence

$$\int_{-\infty}^{+\infty} |s(t)|^2 \, dt = \int_{-\infty}^{+\infty} \int_{-\infty}^{+\infty} S(f)S*(f')\delta(f - f') \, df' \, df$$

Consider the integral on the right with respect to f'. The integral is zero except for $f = f'$. Using the relationship in (6-39), we obtain

$$\int_{-\infty}^{+\infty} S(f)S*(f')\delta(f - f') \, df' = |S(f)|^2$$

and finally,

$$\int_{-\infty}^{+\infty} |s(t)|^2 \, dt = \int_{-\infty}^{+\infty} |S(f)|^2 \, df \tag{6-45}$$

Equation (6-45) indicates that the energy in the signal is the integral of the squared magnitude of either the time-domain or the frequency-domain version of the signal. The function $|S(f)|^2$ is called the *energy spectral density* of the signal.

6.4.3 Unit Step Function

A *unit step function* has a value of zero from minus infinity to some point, and a value of unity from that point to plus infinity. The time-domain step function is

$$u(t - t_0) = \begin{cases} 0 & \text{for } t < t_0 \\ 1.0 & \text{for } t > t_0 \end{cases} \tag{6-46}$$

A little thought would verify that the unit step function may also be defined as the integral of the unit impulse. Thus

$$u(t - t_0) = \int_{-\infty}^{t} \delta(\tau - t_0) \, d\tau \tag{6-47}$$

Some difficulty is encountered when attempting a formal Fourier transformation of the unit step function because the integral of the function over all time is infinite. This difficulty is overcome by considering the unit step function as the limiting form of function with a more well behaved integral. For instance, for a positive, let

$$u(t) = \begin{cases} \lim_{a \to 0} [\exp(-at)] & \text{for } t \geq 0 \\ 0 & \text{for } t < 0 \end{cases} \tag{6-48}$$

The Fourier transform of $u(t)$ now becomes

$$U(f) = \lim_{a \to 0} \left\{ \int_0^{+\infty} \exp[(-a + j2\pi f)t] \, dt \right\}$$

$$= \lim_{a \to 0} \left[\frac{1}{a + j2\pi f} \right] \qquad (6\text{-}49)$$

But

$$\frac{1}{a + j2\pi f} = \frac{a}{a^2 + (2\pi f)^2} + \frac{2\pi f}{j[a^2 + (2\pi f)^2]} \qquad (6\text{-}50)$$

Considering the first term on the right, note that

$$\lim_{a \to 0} \left[\frac{a}{a^2 + (2\pi f)^2} \right] = \begin{cases} \infty & \text{for } f = 0 \\ 0 & \text{for } f \neq 0 \end{cases}$$

and

$$\int_{-\infty}^{+\infty} \frac{a}{a^2 + (2\pi f)^2} \, df = \frac{1}{2}$$

This term behaves like an impulse, and we may write

$$\lim_{a \to 0} \left[\frac{a}{a^2 + (2\pi f)^2} \right] = \frac{1}{2} \delta(f) \qquad (6\text{-}51)$$

The second term on the right in (6-50) is easily evaluated for $a = 0$, with the result that the transform of the unit step function is

$$u(t) \longleftrightarrow \frac{1}{2} \left[\delta(f) + \frac{1}{j\pi f} \right] \qquad (6\text{-}52)$$

By a similar procedure, a frequency-domain unit step function and its transform may be defined. Let

$$u(f) = \begin{cases} 1.0 & \text{for } f > 0 \\ 0 & \text{for } f < 0 \end{cases}$$

Then

$$u(f) \longleftrightarrow \frac{1}{2} \left[\delta(t) - \frac{1}{j\pi t} \right] \qquad (6\text{-}53)$$

6.4.4 Signum Function

A useful function closely related to the unit step function is the *signum function*, defined as

$$\text{sgn}\,(t - t_0) = \begin{cases} +1.0 & \text{for } t > t_0 \\ -1.0 & \text{for } t < t_0 \end{cases}$$

$$= 2u(t - t_0) - 1.0 \qquad (6\text{-}54)$$

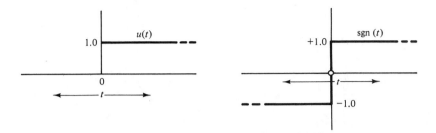

Figure 6-11   The unit step function and the signum function.

A frequency-domain signum function, sgn(f), may also be defined. With the help of (6-52) and (6-54), the transform-pair relationships for signum functions are

$$\text{sgn}(t) = 2u(t) - 1.0 \longleftrightarrow \frac{1}{j\pi f}$$

$$\frac{-1}{j\pi t} \longleftrightarrow 2u(f) - 1.0 = \text{sgn}(f)$$

(6-55)

The relationship between the signum function and the unit step function is demonstrated graphically in Figure 6-11.

6.5 TIME-DOMAIN RESPONSE OF LINEAR SYSTEMS

We now consider the response of a linear, time-invariant system to an arbitrary input signal. A linear system is one for which the output response to two or more input signals is equal to the sum of the responses to each of these signals acting alone. If the system is time invariant, the response to a given input signal is independent of the time of origin of the signal.

In Figure 6-12, a time-invariant linear system is shown, together with the output waveform, assuming that the input signal is a unit impulse. This figure also demonstrates that the output waveform shape is not affected by a change in the time of application of the input impulse.

If a series of impulses of different strengths and different time locations is applied at the input, the output signal is the sum of the responses to the individual impulses. Let the input signal be

$$s_1(t) = \sum_{n=1}^{N} a_n \delta(t - t_n)$$

(6-56)

The response of the system to a unit impulse at the origin is called the *impulse response, $h(t)$*. The output signal is, therefore,

$$s_2(t) = \sum_{n=1}^{N} a_n h(t - t_n)$$

(6-57)

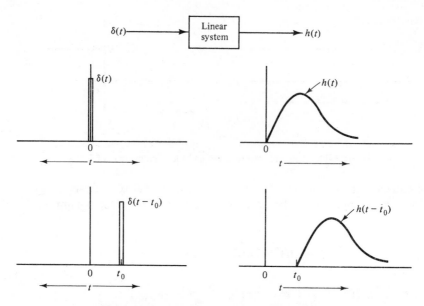

Figure 6-12   Impulse response of a linear system.

Now let the input signal pass to a continuous function, which can, nevertheless, be approximated by a contiguous series of narrow rectangular pulses. As the width of the rectangular pulses approaches zero, they become impulse-like, as shown in Figure 6-13. The continuous signal, $s_1(t)$, contains an impulsive element at $t = \tau$ such that

$$\text{impulse at } \tau = s_1(\tau)\delta(t - \tau)\, d\tau$$

The strength of this impulse is $s_1(\tau)\, d\tau$. The response of the system to this impulse is

$$\text{output} = s_1(\tau)h(t - \tau)\, d\tau \tag{6-58}$$

The total output resulting from all such impulsive elements in the input is

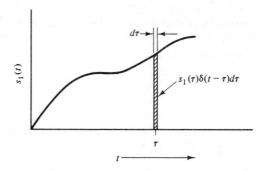

Figure 6-13   Impulsive element of arbitrary function $s_1(t)$.

obtained by summation of all components such as (6-58). Since τ is a continuous variable, this summation becomes an integral. Thus

$$s_2(t) = \int_{-\infty}^{+\infty} s_1(\tau)h(t - \tau)\, d\tau \qquad (6\text{-}59)$$

This integral is called the *convolution integral*. The output response of the linear system is obtained by the *convolution* of the system impulse response with the input function.

A little thought will show that (6-59) may also be written as

$$s_2(t) = \int_{-\infty}^{+\infty} h(\tau)s_1(t - \tau)\, d\tau \qquad (6\text{-}60)$$

The convolution integral is so frequently used that a shorthand notation is often used as follows:

$$s(t) \otimes h(t) = \int_{-\infty}^{+\infty} s(\tau)h(t - \tau)\, d\tau \qquad (6\text{-}61)$$

The convolution operation obeys many of the laws governing algebraic operations. Thus

$$s_1(t) \otimes [s_2(t) + s_3(t)] = s_1(t) \otimes s_2(t) + s_1(t) \otimes s_3(t) \qquad (6\text{-}62)$$

$$s_1(t) \otimes [s_1(t) \otimes s_3(t)] = [s_1(t) \otimes s_2(t)] \otimes s_3(t) \qquad (6\text{-}63)$$

and

$$s_1(t) \otimes s_2(t) = s_2(t) \otimes s_1(t) \qquad (6\text{-}64)$$

6.5.1 Additional Properties of Convolution

Consider the convolution of an arbitrary function, $s(t)$ with a unit impulse centered at the origin:

$$s(t) \otimes \delta(t) = \int_{-\infty}^{+\infty} s(\tau)\delta(t - \tau)\, d\tau \qquad (6\text{-}65)$$

Recalling the result given in (6-39), the integral in (6-65) is equal to the value of the function at $t = \tau$. Thus

$$s(t) \otimes \delta(t) = s(t) \qquad (6\text{-}66)$$

and we see that convolution with an impulse at the origin leaves a function unchanged.

In a similar fashion,

$$s(t) \otimes \delta(t - t_0) = s(t - t_0) \qquad (6\text{-}67)$$

If $s(t)$ is an impulse located at $t = t_1$, this becomes

$$\delta(t - t_1) \otimes \delta(t - t_0) = \delta[t - (t_1 + t_0)] \qquad (6\text{-}68)$$

Now assume three functions related by

$$s_3(t) = s_1(t) \otimes s_2(t)$$

Then

$$\begin{aligned}
s_1(t - t_1) \otimes s_2(t - t_2) &= s_1(t) \otimes \delta(t - t_1) \otimes s_2(t) \otimes \delta(t - t_2) \\
&= s_1(t) \otimes s_2(t) \otimes \delta[t - (t_1 + t_2)] \\
&= s_3(t) \otimes \delta[t - (t_1 + t_2)] \\
&= s_3[t - (t_1 + t_2)] \qquad\qquad\qquad\qquad \text{(6-69)}
\end{aligned}$$

Thus time delays are additive under the operation of convolution.

6.5.2 Convolution of Two Rect Functions

An operation frequently encountered is the convolution of a rect function with itself. Let

$$s(t) = \text{rect}\left(\frac{t}{t_p}\right)$$

Then

$$s(t) \otimes s(t) = \int_{-\infty}^{+\infty} \text{rect}\left(\frac{\tau}{t_p}\right) \text{rect}\left(\frac{t - \tau}{t_p}\right) d\tau$$

The product of the two rect functions inside the integral is zero except in the region of overlap. This is shown graphically in Figure 6-14.

For t negative, the limits of integration are from $-t_p/2$ to $t + t_p/2$. Therefore

$$s(t) \otimes s(t) = \int_{-t_p/2}^{t+t_p/2} d\tau = t + t_p \qquad \text{for } \frac{-t_p}{2} \le t \le 0$$

Similarly, for t positive

$$s(t) \otimes s(t) = \int_{t-t_p/2}^{t_p/2} d\tau = t_p - t \qquad \text{for } 0 \le t \le \frac{t_p}{2}$$

Combining these results yields

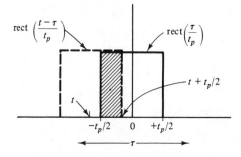

Figure 6-14 Intersection of two rect functions in convolution operation.

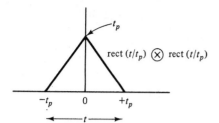

rect (t/t_p) $\otimes$ rect (t/t_p)

Figure 6-15 Result of self-convolution of a rect function.

$$s(t) \otimes s(t) = t_p\left(1 - \frac{|t|}{t_p}\right) \text{rect}\left(\frac{t}{2t_p}\right) \qquad (6\text{-}70)$$

Thus the self-convolution of a rectangular pulse results in a triangular pulse, as shown in Figure 6-15.

6.6 BASIC OPERATIONS IN FOURIER ANALYSIS

System analysis requires many basic operations on functions, such as addition, multiplication, integration, differentiation, and so on. We shall now consider the effect of these basic operations on the Fourier transform of the function.

6.6.1 Addition

The Fourier transform of a sum of two functions is the sum of the Fourier transforms. Thus

$$s_3(t) = s_1(t) + s_2(t) \longleftrightarrow S_1(f) + S_2(f) = S_3(f) \qquad (6\text{-}71)$$

6.6.2 Folding

The folded version of a function is obtained by replacing t with $-t$ in $s(t)$:

$$\text{folded } s(t) = s(-t)$$

To determine the effect of the folding operation on the Fourier transform, let

$$S_1(f) = \int_{-\infty}^{+\infty} s(t) \exp(-j2\pi ft)\, dt$$

$$S_2(f) = \int_{-\infty}^{+\infty} s(-t) \exp(-j2\pi ft)\, dt$$

Now let $t' = -t$ and substitute in the expression for $S_2(f)$.

$$S_2(f) = \int_{-\infty}^{+\infty} s(t') \exp[-j2\pi f(-t')]\, dt'$$

$$= \int_{-\infty}^{+\infty} s(t') \exp[-j2\pi(-f)t']\, dt'$$

The expression for $S_2(f)$ is now identical to that for $S_1(f)$ except for a change in variable from f to $-f$. Therefore,

$$S_2(f) = S_1(-f)$$

or, in general, if

$$s(t) \longleftrightarrow S(f)$$

then

$$s(-t) \longleftrightarrow S(-f)$$

6.6.3 Complex Conjugation

Let $s(t)$ be a complex function with the transform

$$S_1(f) = \int_{-\infty}^{+\infty} s(t) \exp(-j2\pi ft)\, dt$$

The complex conjugate of $s(t)$ has the transform

$$S_2(f) = \int_{-\infty}^{+\infty} s^*(t) \exp(-j2\pi ft)\, dt$$

$$= \left\{ \int_{-\infty}^{+\infty} s(t) \exp(+j2\pi ft)\, dt \right\}^*$$

$$= \left\{ \int_{-\infty}^{+\infty} s(t) \exp[-j2\pi(-f)t]\, dt \right\}^*$$

The integral inside the braces is the transform of $s(t)$ with respect to the variable $-f$. Therefore,

$$S_2(f) = S_1^*(-f)$$

or

$$s(t) \longleftrightarrow S(f)$$

$$s^*(t) \longleftrightarrow S^*(-f)$$

(6-73)

6.6.4 Scaling

Assume a transform-pair relationship

$$s(t) \longleftrightarrow S(f)$$

We now wish to change the scale of the variable t and determine the effect on the transform. Define a scaled variable t' such that for T a positive real constant, $t' = t/T$. The transform of the function with the new variable is

$$s_1(f) = \int_{-\infty}^{+\infty} s(t') \exp(-j2\pi ft)\, dt$$

but

$$t = t'T$$
$$dt = T\,dt'$$

so that

$$S_1(f) = T \int_{-\infty}^{+\infty} s(t') \exp[-j2\pi(fT)t']\,dt'$$

The integral is now identical to the transform of $s(t)$ with respect to the scaled frequency variable, fT. Thus

$$S_1(f) = TS(fT)$$

and

$$s(t/T) \longleftrightarrow TS(fT) \qquad (6\text{-}74)$$

Recall that a scaling operation was involved in Section 6.4.1 in defining the rect function. A standard rect function of unit width was changed to a pulse of width t_p by scaling the time variable, t. The transform-pair relationships were defined as

$$\text{rect}(t) \longleftrightarrow \text{sinc}(f)$$
$$\text{rect}\left(\frac{t}{t_p}\right) \longleftrightarrow t_p \,\text{sinc}(ft_p) \qquad (6\text{-}75)$$

which agrees with (6-74).

6.6.5 Translation

In the example in Section 6.2.1 it was demonstrated that a translation of the waveform in the time domain resulted in a linear phase versus frequency term in the spectrum. This will now be developed as a general relationship. Assume the transform-pair relationship

$$s(t) \longleftrightarrow S(f)$$

If $s(t)$ is translated in time by t_0, we may write

$$S_1(f) = \int_{-\infty}^{+\infty} s(t - t_0) \exp(-j2\pi ft)\,dt$$

Now let $t' = t - t_0$, so that

$$dt' = dt$$
$$t = t' + t_0$$

and substitute in the integral:

$$S_1(f) = \int_{-\infty}^{+\infty} s(t') \exp[-j2\pi f(t' + t_0)]\,dt$$

$$= \exp(-j2\pi ft_0) \int_{-\infty}^{+\infty} s(t') \exp(-j2\pi ft') \, dt'$$

$$= S(f) \exp(-j2\pi ft_0)$$

This gives the general transform-pair relationship

$$s(t - t_0) \longleftrightarrow S(f) \exp(-j2\pi ft_0) \tag{6-76}$$

Translation in the frequency domain is an operation commonly performed in systems to shift a signal to a more convenient frequency region. Because of the symmetrical nature of the Fourier transform, we may write immediately:

$$s(t) \longleftrightarrow S(f)$$

$$s(t) \exp(j2\pi f_0 t) \longleftrightarrow S(f - f_0) \tag{6-77}$$

From (6-77), the result of multiplication of the time-domain function by a real sinusoidal function is easily obtained. For instance,

$$\cos 2\pi f_0 t = \tfrac{1}{2}[\exp(j2\pi f_0 t) + \exp(-j2\pi f_0 t)]$$

Therefore,

$$s(t) \cos 2\pi f_0 t \longleftrightarrow \tfrac{1}{2}[S(f - f_0) + S(f + f_0)] \tag{6-78}$$

6.6.6 Convolution

The response of a linear system has been shown to be the result of the time-domain convolution of the input signal and the system impulse response. The Fourier transform of the result of a convolution operation in the time domain results in a particularly simple relationship in the frequency domain.

Let

$$s_1(t) \longleftrightarrow S_1(f)$$

$$s_2(t) \longleftrightarrow S_2(f)$$

$$s_3(t) = s_1(t) \otimes s_2(t) \longleftrightarrow S_3(f)$$

Then

$$S_3(f) = \int_{-\infty}^{+\infty} [s_1(t) \otimes s_2(t)] \exp(-j2\pi ft) \, dt$$

$$= \int_{-\infty}^{+\infty} \left[\int_{-\infty}^{+\infty} s_1(\tau)s_2(t - \tau) \, d\tau\right] \exp(-j2\pi ft) \, dt$$

Rearranging terms, we have

$$S_3(f) = \int_{-\infty}^{+\infty} \left[\int_{-\infty}^{+\infty} s_2(t - \tau) \exp(-j2\pi ft) \, dt\right] s_1(\tau) \, d\tau$$

The inner integral is the Fourier transform of $s_2(t - \tau)$. From the rule for translation

$$s_2(t - \tau) \longleftrightarrow S_2(f)\exp(-j2\pi f\tau)$$

Substitution in the expression for $S_3(f)$ gives

$$S_3(f) = S_2(f) \int_{-\infty}^{+\infty} s_1(\tau)\exp(-j2\pi f\tau)\,d\tau$$

The remaining integral is the Fourier transform of $s_1(t)$. The final result is

$$S_3(f) = S_1(f)S_2(f)$$

or

$$s_1(t) \otimes s_2(t) \longleftrightarrow S_1(f)S_2(f) \qquad (6\text{-}79)$$

Thus the relatively complicated operation of convolution in the time domain results in the simple multiplication operation in the frequency domain. As an example of the usefulness of this result, consider the time-domain convolution of two identical sinc functions.

$$s(t) = \text{sinc}\,(t) \otimes \text{sinc}\,(t) = \int_{-\infty}^{+\infty} \frac{\sin \pi\tau}{\pi\tau}\left[\frac{\sin \pi(t - \tau)}{\pi(t - \tau)}\right] d\tau$$

Although the integral in this case may be evaluated without great difficulty, considerable effort is saved by recognizing that

$$\text{sinc}\,(t) \longleftrightarrow \text{rect}\,(f)$$

Using (6-79), we obtain

$$\text{sinc}\,(t) \otimes \text{sinc}\,(t) \longleftrightarrow \text{rect}\,(f)\,\text{rect}\,(f) = \text{rect}\,(f)$$

Thus

$$s(t) = \text{sinc}\,(t) \otimes \text{sinc}\,(t) = \text{sinc}\,(t) \longleftrightarrow \text{rect}\,(f)$$

The self-convolution of a sinc function results in the same sinc function—a fact that is made immediately obvious using the transform-pair relationship in (6-79).

Convolution is an operation that is also often encountered in the frequency domain. By symmetry with the result in (6-79), frequency-domain convolution results in multiplication of the appropriate time-domain functions. Thus

$$s_1(t)s_2(t) \longleftrightarrow S_1(f) \otimes S_2(f) \qquad (6\text{-}80)$$

If $s_2(t)$ is the conjugate of $s_1(t)$, we obtain, with the help of (6-73),

$$|s_1(t)|^2 \longleftrightarrow S_1(f) \otimes S_1^*(-f) \qquad (6\text{-}81)$$

6.6.7 Differentiation

Consider the Fourier integral representation of $s(t)$:

$$s(t) = \int_{-\infty}^{+\infty} S(f)\exp(j2\pi ft)\,df$$

Now differentiate this expression with respect to time:

$$\frac{ds(t)}{dt} = s'(t) = \int_{-\infty}^{+\infty} [j2\pi f S(f)] \exp(j2\pi ft) \, df$$

The transform of the derivative of $s(t)$ is given by the bracketed term inside the integral. Thus, if

$$s(t) \longleftrightarrow S(f)$$

then

$$s'(t) \longleftrightarrow j2\pi f S(f) \tag{6-82}$$

In a similar manner, the transform of the derivative of the frequency-domain function can be shown to be

$$-j2\pi t s(t) \longleftrightarrow S'(f) \tag{6-83}$$

6.6.8 Integration

The effect of integration on transform relationships can be seen by considering the convolution of a function with a unit step function.

$$s(t) \otimes u(t) = \int_{-\infty}^{+\infty} s(\tau)u(t-\tau) \, d\tau$$

But

$$u(t-\tau) = \begin{cases} 1.0 & \text{for } -\infty \leq \tau \leq t \\ 0 & \text{for } \tau > t \end{cases}$$

Therefore,

$$s(t) \otimes u(t) = \int_{-\infty}^{t} s(\tau) \, d\tau$$

The transform of this integral is the product of the transforms of $s(t)$ and $u(t)$:

$$\int_{-\infty}^{t} s(\tau) \, d\tau \longleftrightarrow \frac{S(f)}{2} \left[\delta(f) + \frac{1}{j\pi f} \right] \tag{6-84}$$

In a similar manner, integration in the frequency domain gives

$$\frac{s(t)}{2} \left[\delta(t) - \frac{1}{j\pi t} \right] \longleftrightarrow \int_{-\infty}^{f} S(\nu) \, d\nu \tag{6-85}$$

6.7 REPETITION OPERATOR AND SAMPLING FUNCTION

In the first part of this chapter, the Fourier integral representation of limited-duration waveforms was developed by a limiting process on the Fourier series expansion of periodic waveforms. We shall now proceed in the reverse direction to arrive at the Fourier expansion of periodic waveforms by a limiting process on the expansion of limited-duration waveforms.

Consider the single rectangular pulse and its transform

$$s_1(t) = \text{rect}\left(\frac{t}{t_p}\right) \longleftrightarrow t_p \, \text{sinc}\,(ft_p)$$

A semiperiodic waveform can be created by adding a duplicate of $s_1(t)$ on either side of $s_1(t)$ at $-T$ and $+T$.

$$s(t) = s_1(t + T) + s_1(t) + s_1(t - T)$$

$$= \sum_{n=-1}^{+1} s_1(t - nT)$$

Using the properties of the unit impulse function, $s(t)$ can also be described as the convolution of $s_1(t)$ with a set of three impulses.

$$s(t) = s_1(t) \otimes \left[\sum_{n=-1}^{+1} \delta(t + nT) \right]$$

In either case, the transform of the three pulses is

$$s(t) \longleftrightarrow S_1(f)[\exp(+j2\pi fT) + 1 + \exp(-j2\pi fT)]$$

$$= t_p \, \text{sinc}\,(ft_p)[1 + 2 \cos 2\pi fT]$$

This spectrum is shown in Figure 6-16. Notice that the spectrum boundary has the form of $S_1(f)$, but with concentrated lobes spaced at intervals $1/T$. The

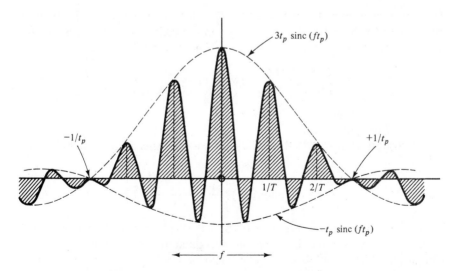

Figure 6-16   Spectrum of three equally spaced rectangular pulses.

characteristics typical of a periodic waveform are evident with the pulse repeated only three times.

Now add $N/2$ equally spaced pulses on each side for a total of $N + 1$ pulses. Then

$$s(t) = \sum_{n=-N/2}^{+N/2} s_1(t - nT) \tag{6-86}$$

and

$$S(f) = S_1(f) \sum_{n=-N/2}^{N/2} \exp(-j2\pi fnT) \tag{6-87}$$

The summation in (6-87) is a geometric series of the form

$$\sum_{n=-N/2}^{+N/2} (r)^n$$

with

$$r = \exp(-j2\pi fT)$$

The sum of a finite geometric series is readily available in closed form. With the help of trigonometric identities, the sum in (6-87) can be put in the following compact form:

$$\sum_{n=-N/2}^{+N/2} \exp(-j2\pi fnT) = \frac{\sin(N + 1)\pi fT}{\sin \pi fT} \tag{6-88}$$

This function is periodic with period $1/T$. One complete period is shown in Figure 6-17 for $N + 1 = 13$. Notice that in the vicinity of $f = 0$, the function has the general shape of a sinc function. That is, for $\pi fT \ll 1$,

$$\frac{\sin(N + 1)\pi fT}{\sin \pi fT} \simeq (N + 1) \frac{\sin(N + 1)\pi fT}{(N + 1)\pi fT} \tag{6-89}$$

The first zero occurs at $f = 1/(N + 1)T$, which is the inverse of the total duration of the pulse train.

Substitution of (6-88) into (6-87) gives for the spectrum of the finite pulse train

$$S(f) = t_p \operatorname{sinc}(ft_p) \left[\frac{\sin(N + 1)\pi fT}{\sin \pi fT} \right] \tag{6-90}$$

and is shown graphically in Figure 6-18. The spectrum is concentrated mostly in narrow spikes, separated by $1/T$, and weighted by the amplitude spectrum of a single pulse. The width of each spike is inversely proportional to the total width of the pulse train.

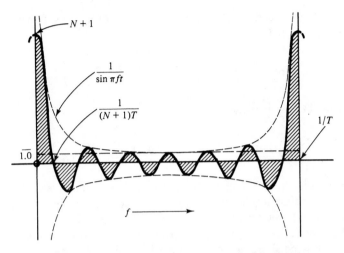

Figure 6-17   One period of $[\sin\,(N\,+\,1)\pi fT]/\sin\,\pi fT$.

From (6-86) and (6-87), a truly periodic waveform is obtained by letting N approach infinity.

$$s(t) = \sum_{n=-\infty}^{+\infty} s_1(t - nT) \longleftrightarrow S_1(f)\left\{ \lim_{N\to\infty} \left[\frac{\sin\,(N\,+\,1)\pi fT}{\sin\,\pi fT} \right] \right\} \quad (6\text{-}91)$$

The amplitude of the bracketed term in (6-91) approaches infinity at the locations $f = n/T$ as N approaches infinity. Furthermore,

$$\lim_{N\to\infty} \int_{-1/2T}^{+1/2T} \frac{\sin\,(N\,+\,1)\pi fT}{\sin\,\pi fT}\, df = \frac{1}{T} \quad (6\text{-}92)$$

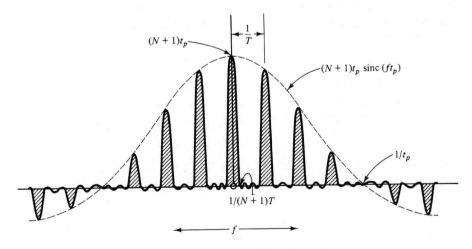

Figure 6-18   Spectrum of finite pulse train.

But these are the characteristics of a periodic train of impulses with strength $1/T$ and period $1/T$. Thus

$$\lim_{N \to \infty} \left[\frac{\sin (N + 1)\pi fT}{\sin \pi fT} \right] = \frac{1}{T} \sum_{n=-\infty}^{+\infty} \delta\left(f - \frac{n}{T}\right) \qquad (6\text{-}93)$$

Equation (6-91) now becomes

$$\sum_{n=-\infty}^{+\infty} s_1(t - nT) \longleftrightarrow \frac{1}{T} S_1(f) \sum_{n=-\infty}^{+\infty} \delta\left(f - \frac{n}{T}\right) \qquad (6\text{-}94)$$

The impulse train in (6-94) creates the line structure typical of a periodic waveform. The original spectrum, $S(f)$, is in effect sampled by the impulse train at intervals equal to the repetition frequency. By convention, we plot the *amplitude* of the spectral component as the *strength* of the corresponding impulse.

The following shorthand notation has been suggested by P.M. Woodward [1] for use with repeated waveforms. Define

$$\sum_{n=-\infty}^{+\infty} s(t - nT) = [\text{rep}_T s(t)] \qquad (6\text{-}95)$$

$$F \sum_{n=-\infty}^{+\infty} \delta(f - nF) = F \text{ comb}_F$$

$$F = \frac{1}{T} \qquad (6\text{-}96)$$

The repetition operation $[\text{rep} (\cdot)]$ indicates that the function within the brackets is to be repeated at the designated interval. Care must be exercised to set aside the rep operator and the function to be repeated with brackets. Suppose, for instance, that a periodic function is multiplied by a second function. This could be written as

$$[\text{rep}_T s_1(t)]s_2(t)$$

This indicates that the function $s_1(t)$ is repeated at intervals T, and the result multiplied by the function $s_2(t)$. This, in general, is not the same as multiplying $s_1(t)$ by $s_2(t)$, and then repeating the result at intervals, T. Thus, in general,

$$[\text{rep}_T s_1(t)]s_2(t) \neq [\text{rep}_T s_1(t)s_2(t)] \qquad (6\text{-}97)$$

The repetition operator, as defined by (6-95), may be recognized as a convolution operation, with one of the functions being an impulse train.

$$[\text{rep}_T s_1(t)] = s(t) \otimes \sum_{n=-\infty}^{+\infty} \delta(t - nT)$$

$$= s(t) \otimes \text{comb}_T \qquad (6\text{-}98)$$

The comb function, being a true function, has a Fourier transform as follows:

$$\text{comb}_T = \sum_{n=-\infty}^{+\infty} \delta(t - nT) \longleftrightarrow \sum_{n=-\infty}^{+\infty} \exp(-j2\pi nfT) = F \text{ comb}_F \qquad (6\text{-}99)$$

Thus a comb function in one domain transforms to a comb function in the opposite domain.

Using the relationships in (6-98) and (6-99), we develop the following transform pairs:

$$s(t) \longleftrightarrow S(f)$$

$$[\text{rep}_T \, s(t)] \longleftrightarrow FS(f)\,\text{comb}_F \qquad (6\text{-}100)$$

$$\text{comb}_T \, s(t) \longleftrightarrow F[\text{rep}_F \, S(f)] \qquad (6\text{-}101)$$

Equations (6-100) and (6-101) demonstrate that the sampling function (comb) and the repetition operator may be applied in either the time or the frequency domain.

6.8 TABULATION OF RULES AND PAIRS

For future reference, the rules governing useful operations in Fourier analysis, and some selected transform pairs, are tabulated in Table 6-1. Pair 6 in this table is the Gaussian waveform and its transform. Although this transform pair was not derived in this chapter, the Gaussian function is included because of its frequent occurrence in problems of practical interest. Notice that the transform of a Gaussian function is also Gaussian in shape.

The use of the rules and pairs in Table 6-1 will be demonstrated by means of several examples of practical interest.

6.8.1 High-Frequency Periodic Pulse Train

A single high-frequency rectangular pulse may be expressed as

$$s_1(t) = \text{rect}\left(\frac{t}{t_p}\right) \cos 2\pi f_0 t$$

where the carrier frequency, f_0, is assumed large compared with $1/t_p$. This signal may be converted to a periodic waveform, with period $T \gg t_p$, using the repetition operator.

$$s(t) = [\text{rep}_T s_1(t)] = \left[\text{rep}_T \, \text{rect}\left(\frac{t}{t_p}\right) \cos(2\pi f_0 t) \right]$$

The Fourier transform of the periodic waveform is now developed using the rules and pairs:

TABLE 6-1   Selected Rules and Pairs for Fourier Analysis

		Waveform	Transform	Notes
		$s(t)$	$S(f)$	Definition
Rule	1	$s_1(t) + s_2(t)$	$S_1(f) + S_2(f)$	Addition
	2	$s(-t)$	$S(-f)$	Folding
	3	$s*(t)$	$S*(-f)$	Complex conjugation
	4	$s(t/T)$	$TS(fT)$	Scaling of variable; T positive
	5	$s(t - t_0)$	$S(f)\exp(-j2\pi ft_0)$	Time translation
	6	$s(t)\exp(j2\pi f_0 t)$	$S(f - f_0)$	Frequency translation
	7	$s_1(t) \otimes s_2(t)$	$S_1(f)S_2(f)$	Time-domain convolution
	8	$s_1(t)s_2(t)$	$S_1(f) \otimes S_2(f)$	Time-domain multiplication
	9	$s'(t)$	$j2\pi fS(f)$	Differentiation
		$-j2\pi ts(t)$	$S'(f)$	
	10	$\int_{-\infty}^{t} s(\tau)\,d\tau$	$\dfrac{S(f)}{2}[\delta(f) + \dfrac{1}{j\pi f}]$	Integration
		$\dfrac{s(t)}{2}\left[\delta(t) - \dfrac{1}{j\pi t}\right]$	$\int_{-\infty}^{f} S(\nu)\,d\nu$	
	11	comb_T	$F\,\text{comb}_F$	Time-domain sample function $F = 1/T$
	12	$[\text{rep}_{TS}(t)]$	$F\,\text{comb}_F S(f)$	Time-domain repetition operator
	13	$\text{comb}_{TS}(t)$	$F[\text{rep}_F S(f)]$	Time-domain sampled waveform
Pair	1	$\delta(t)$	1	Unit impulse
	2	$u(t)$	$\dfrac{1}{2}\left[\delta(f) + \dfrac{1}{j\pi f}\right]$	Unit step functions
		$\dfrac{1}{2}\left[\delta(t) - \dfrac{1}{j\pi t}\right]$	$u(f)$	
	3	$\text{rect}\left(\dfrac{t}{t_p}\right)$	$t_p\,\text{sinc}(ft_p)$	Rectangular time-domain function
	4	$\text{sinc}\left(\dfrac{t}{t_p}\right)$	$t_p\,\text{rect}(ft_p)$	Time-domain sinc function
	5	$\cos 2\pi f_0 t$	$\dfrac{1}{2}[\delta(f - f_0) + \delta(f + f_0)]$	Sinusoidal functions
		$\sin 2\pi f_0 t$	$\dfrac{1}{2j}[\delta(f - f_0) - \delta(f + f_0)]$	
	6	$\exp\left(\dfrac{-t^2}{2\sigma^2}\right)$	$\sigma\sqrt{2\pi}\exp[-2(\pi f)^2\sigma^2]$	Gaussian waveform

Pair   3: $\text{rect}\left(\dfrac{t}{t_p}\right) \longleftrightarrow t_p\,\text{sinc}(ft_p)$

Pair   5: $\cos 2\pi f_0 t \longleftrightarrow \frac{1}{2}[\delta(f - f_0) + \delta(f + f_0)]$

Rule   8: $\text{rect}\left(\dfrac{t}{t_p}\right)\cos 2\pi f_0 t \longleftrightarrow \dfrac{t_p}{2}\left[\text{sinc}(f - f_0)t_p + \text{sinc}(f + f_0)t_p\right]$

Rule   12: $[\text{rep}_{TS_1}(t)] \longleftrightarrow \dfrac{t_p}{2}F\,\text{comb}_F[\text{sinc}(f - f_0)t_p + \text{sinc}(f + f_0)t_p]$

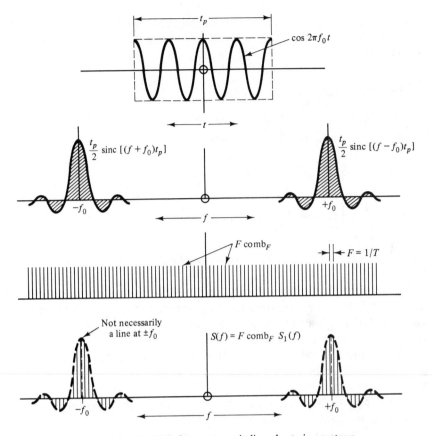

Figure 6-19   High-frequency periodic pulse train spectrum.

The time-domain waveform, $s_1(t)$, and the various steps in obtaining the final spectrum are shown in Figure 6-19. The spectrum of $s_1(t)$ is the sinc-function spectrum of the rectangular envelope, shifted to the locations $\pm f_0$. The final spectrum results from sampling the shifted sinc functions by the frequency-domain comb function. Notice that the final spectrum does not necessarily contain a line at $\pm f_0$ unless f_0 is related harmonically to $1/T$.

6.8.2 Phase-Coherent High-Frequency Pulse Train

Consider a signal formed as the product of a periodic rectangular pulse train and a continuous cosinusoidal carrier.

$$s(t) = \left[\text{rep}_T \, \text{rect}\left(\frac{t}{t_p}\right) \right] \cos 2\pi f_0 t$$

Notice that in this case the phase of the carrier in each pulse is determined by the phase of the continuous sinusoid. In the preceding example, the phase of the carrier was the same in each pulse, resulting in a truly periodic waveform.

Transforming term by term to the frequency domain, we obtain

Pair 3: $\quad \text{rect}\left(\dfrac{t}{t_p}\right) \longleftrightarrow t_p \text{sinc}\,(ft_p)$

Rule 12: $\quad \left[\text{rep}_T\ \text{rect}\left(\dfrac{t}{t_p}\right)\right] \longleftrightarrow t_p F \text{ comb}_F\ \text{sinc}\,(ft_p)$

Pair 5: $\quad \cos 2\pi f_0 t \longleftrightarrow \tfrac{1}{2}[\delta(f - f_0) + \delta(f + f_0)]$

Rule 8: $\quad \left[\text{rep}_T\ \text{rect}\left(\dfrac{t}{t_p}\right)\right]\cos 2\pi f_0 t \longleftrightarrow$

$$\frac{t_p}{2}\, F \text{ comb}_F \text{sinc}\,(ft_p) \otimes [\delta(f - f_0) + \delta(f + f_0)]$$

In this example, the sinc-function spectrum of the single rectangular pulse is first multiplied by the comb function to give the spectrum of a periodic pulse train. The resulting line spectrum is then translated to the locations $\pm f_0$. The time-domain signal and the development of the spectrum are shown in Figure 6-20. Notice that the spectrum of the pulse train envelope has a line at $f = 0$, representing the average value of the function. In the final spectrum, this line at the origin is translated to the locations $\pm f_0$. Thus, in contrast to the signal in the preceding example, the original carrier can be recovered by centering a narrow-band filter at the carrier frequency.

6.8.3 Triangular Pulse

In Section 6.5.2 the convolution of a rectangular pulse with itself was shown to result in a triangular pulse. In Section 6.3 the Fourier transform of a triangular pulse was shown to be the square of a sinc function. This result follows directly from rule 7 for convolution. That is,

Pair 3: $\quad \text{rect}\left(\dfrac{t}{t_p}\right) \longleftrightarrow t_p \text{sinc}\,(ft_p)$

Rule 7: $\quad \text{rect}\left(\dfrac{t}{t_p}\right) \otimes \text{rect}\left(\dfrac{t}{t_p}\right) = t_p\left(1 - \dfrac{|t|}{t_p}\right)\text{rect}\left(\dfrac{t}{2t_p}\right)$

$$t_p\left(1 - \frac{|t|}{t_p}\right)\text{rect}\left(\frac{t}{2t_p}\right) \longleftrightarrow t_p^2 \text{sinc}^2\,(ft_p)$$

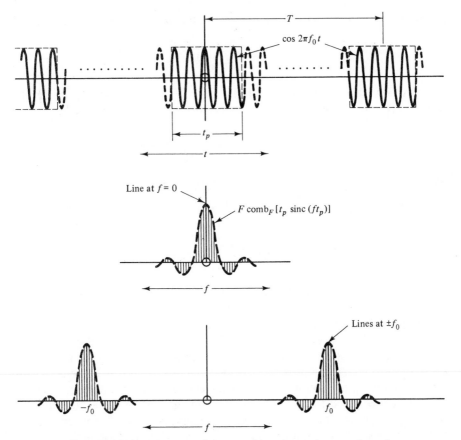

Figure 6-20   Development of spectrum for a phase-coherent pulse train.

6.8.4 Sampling Theorems

In practice, it is often convenient to deal with discrete periodic samples of waveforms rather than with the continuous functions. It is then necessary to determine how often the continuous function must be sampled to avoid the loss of information. It is also desirable to determine a method for reconstructing the original waveform from the sample values. The solution to these problems is the subject of the sampling theorem, which gives definitive answers provided that certain constraints (not actually realizable) are imposed on the signal. The use of the rules and pairs in Table 6-1 results in a particularly compact derivation of the sampling theorem [1, 5].

Assume a continuous waveform $s(t)$ with a spectrum limited to the frequency range $\pm F$. Let

$$s(t) \longleftrightarrow S(f)$$

Because $S(f)$ is zero outside the range $\pm F$, this may also be written

$$s(t) \longleftrightarrow S(f)\,\text{rect}\left(\frac{f}{2F}\right)$$

We may now cause $S(f)$ to repeat at intervals $2F$ without affecting the result, because the rect function will reject everything outside the range $\pm F$.

$$s(t) \longleftrightarrow [\text{rep}_{2F}S(f)]\,\text{rect}\left(\frac{f}{2F}\right) = S(f)$$

Now let $2F = 1/t_s$ and transform the spectrum to the time domain.

Rule 13: $t_s\,\text{comb}_{t_s}s(t) \longleftrightarrow \text{rep}_{2F}S(f)$

Pair   4: $\dfrac{1}{t_s}\,\text{sinc}\left(\dfrac{t}{t_s}\right) \longleftrightarrow \text{rect}\left(\dfrac{f}{2F}\right)$

Rule 7: $s(t) = \text{comb}_{t_s}s(t) \otimes \text{sinc}\left(\dfrac{t}{t_s}\right) \longleftrightarrow S(f)$ (6-102)

The product of comb_{t_s} and $s(t)$ is a sampled version of the continuous function with impulse samples at intervals $t_s = 1/2F$. Convolution with the sinc function is equivalent to passing the impulse samples through a rectangular filter, centered of $f = 0$ with a bandwidth $2F$. Notice that this result is in agreement with the Nyquist sampling requirement, which states that samples must be taken at a rate at least twice the highest frequency contained in the signal. In addition, (6-102) indicates that the original function is reproduced exactly by convolution of the samples with the appropriate sinc function. The time-domain sampling theorem may be stated as follows:

> A continuous function of time whose spectrum is limited to the band $\pm F$ is completely defined by time-domain samples taken at intervals $\frac{1}{2F}$. The original function can be exactly recovered provided that interpolation between samples is accomplished with the appropriate sinc function.

In Figure 6-21, a continuous function and its bandlimited spectrum are shown, together with the sampled function and its spectrum. The sample interval shown is slightly less than $1/2F$. This actually makes the job of recovery of the original signal easier because the separation is increased between the replicated versions at the signal spectrum. This relaxes the requirements on the filter used to reject all but the spectrum centered at the origin.

It is also possible to define a sampling theorem in the frequency domain. Let $s(t)$ be zero outside the interval $\pm T$. We shall now determine the required interval between samples in the frequency domain to define completely the spectrum of $s(t)$. Proceeding as before, let

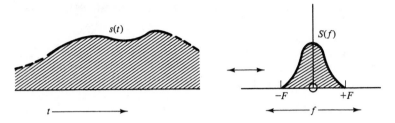

Original function and spectrum

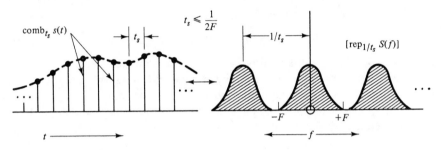

Sampled function and spectrum

Figure 6-21   Time-domain sampling of a bandlimited signal.

$$s(t) \longleftrightarrow S(f)$$

Because $s(t)$ is zero outside the interval $\pm T$, we may also write

$$s(t)\,\text{rect}\,(t/2T) \longleftrightarrow S(f) \otimes 2T\,\text{sinc}\,(2fT)$$

The function $s(t)$ may now be made repetitive at intervals $2T$. This gives

$$s(t) = [\text{rep}_{2T}\,s(t)]\,\text{rect}\left(\frac{t}{2T}\right) \longleftrightarrow f_s\,\text{comb}_{f_s}\,S(f) \otimes 2T\,\text{sinc}\,(2fT)$$

$$= \text{comb}_{f_s}\,S(f) \otimes \text{sinc}\,(2fT) = S(f)$$

where $f_s = 1/2T$.

The term $\text{comb}_{f_s}\,S(f)$ is a sampled version of $S(f)$ with samples taken at intervals $f_s = \frac{1}{2T}$. Convolution of the sampled spectrum with $\text{sinc}\,(2fT)$ permits recovery of the original continuous spectrum.

The frequency-domain sampling theorem may now be stated as follows:

The spectrum of a time function that exists only over a time interval $\pm T$ is completely defined by samples taken at intervals $f_s = \frac{1}{2T}$. The original spectrum can be exactly recovered provided interpolation between samples is accomplished with the appropriate sinc function.

6.9 COMPLEX REPRESENTATION OF SIGNALS: ENVELOPE FUNCTIONS

In previous chapters it has often been useful to replace real sinusoidal signals with the complex exponential function. The real signal was then identified as the real part of the complex signal. Thus

$$A \cos(\omega_0 t + \phi) = \text{Re}\{A \exp[j(\omega_0 t + \phi)]\} \qquad (6\text{-}103)$$

This procedure was justified on the basis that certain mathematical manipulations are simplified using the complex form. Notice, for instance, that a *complex envelope* function is easily separated from the carrier function when the signal is expressed in complex form. Thus

$$A \exp[j(\omega_0 t + \phi)] = [A \exp(j\phi)] \exp(j\omega_0 t) \qquad (6\text{-}104)$$

The envelope of a signal is intuitively considered to contain the slow amplitude and phase variations associated with the complex sinusoid, or carrier term. For the signal in (6-103), the amplitude and phase are constant.

The spectrum of the complex signal in (6-104) is confined entirely to the positive frequency region. Thus

$$A \exp(j\phi) \exp(j2\pi f_0 t) \longleftrightarrow A \exp(j\phi)\delta(f - f_0) \qquad (6\text{-}105)$$

We now let the amplitude and phase vary with time, and express the real signal as

$$x(t) = a(t) \cos[\omega_0 t + \phi(t)] \qquad (6\text{-}106)$$

Proceeding as before, a complex form of the signal is formed by writing

$$x(t) = \text{Re}\{a(t) \exp[j\phi(t)] \exp(j\omega_0 t)\} \qquad (6\text{-}107)$$

If the highest frequencies contained in the complex envelope, $a(t) \exp[j\phi(t)]$, are small compared with the carrier frequency, the spectrum of the total complex signal is again confined to the positive frequency region. The envelope function in this case behaves in an intuitively satisfying manner. If the envelope function has a broad frequency spectrum that includes the carrier frequency, the total complex signal spectrum will not be confined to the positive frequency region. Indeed, with a broad frequency spectrum, the concept of the envelope ceases to have meaning in the intuitive sense.

To develop a more general approach to deriving a complex signal from a real signal, we define the *analytic signal* [6]. The analytic signal is formed by suppressing the negative frequency region of the real signal spectrum, and doubling the positive frequency region. Thus, if $x(t)$ is a real signal, we write

$$x(t) \longleftrightarrow X(f)$$

and

$$\text{analytic signal} = x_a(t) \longleftrightarrow 2u(f)X(f) \qquad (6\text{-}108)$$

where $u(f)$ is the frequency-domain unit step function.

Notice that if $x(t)$ is a constant-amplitude sinusoid, the operation indicated in (6-108) consists simply of replacing the real sinusoid with a complex exponential. In the more general case, the real and imaginary parts of $x_a(t)$ may be identified by transforming the spectrum in (6-108) to the time domain.

$$\left[\delta(t) - \frac{1}{j\pi t}\right] \longleftrightarrow 2u(f)$$

$$x(t) \longleftrightarrow X(f)$$

$$x_a(t) = x(t) \otimes \left[\delta(t) - \frac{1}{j\pi t}\right] \tag{6-109}$$

The convolution of $x(t)$ with the impulse leaves $x(t)$ unchanged, so we may write

$$x_a(t) = x(t) + \frac{j}{\pi} \int_{-\infty}^{+\infty} \frac{x(\tau)}{t - \tau} d\tau \tag{6-110}$$

The real part of $x_a(t)$ is the original real function and the imaginary part is determined by the integral function involving $x(t)$ in (6-110). The integral in (6-110) is called the *Hilbert transform* of $x(t)$ and given the designation

$$\hat{x}(t) = \frac{1}{\pi} \int_{-\infty}^{+\infty} \frac{x(\tau)}{t - \tau} d\tau \tag{6-111}$$

A reciprocal relationship exists as follows:

$$x(t) = -\frac{1}{\pi} \int_{-\infty}^{+\infty} \frac{\hat{x}(\tau)}{t - \tau} d\tau \tag{6-112}$$

Thus $x(t)$ and $\hat{x}(t)$ constitute a *Hilbert-transform pair*.

The analytic signal may be written as

$$x_a(t) = x(t) + j\hat{x}(t) \tag{6-113}$$

It is easy to show that the function $\sin \omega_0 t$ is the Hilbert transform of $\cos \omega_0 t$. Therefore, the analytic signal corresponding to $\cos \omega_0 t$ is

$$x_a(t) = \cos \omega_0 t + j \sin \omega_0 t = \exp(j\omega_0 t)$$

It is convenient to express the general analytic signal in exponential form as

$$x_a(t) = |x_a(t)| \exp[j\Phi(t)] \tag{6-114}$$

where

$$|x_a(t)| = [x^2(t) + \hat{x}^2(t)]^{1/2}$$

$$\Phi(t) = \arctan\left[\frac{\hat{x}(t)}{x(t)}\right]$$

Now let $\Phi(t) = \omega_0 t + \phi(t)$ and write

$$x_a(t) = |x_a(t)| \exp[(j\phi(t)]\exp(j\omega_0 t)$$
$$= \mu(t)\exp(j\omega_0 t) \tag{6-115}$$

The complex envelope function, $\mu(t)$, is obtained by removing the complex carrier from the analytic signal.

$$\mu(t) = x_a(t)\exp(-j\omega_0 t) = |x_a(t)| \exp[j\phi(t)] \tag{6-116}$$

If $\mu(t)$ is a narrow-band function, relative to f_0, it will have properties we intuitively associate with an envelope. Otherwise, it is simply a convenient mathematical representation.

Recalling the discussion of the signum function in Section 6.4.4, the spectrum of $\hat{x}(t)$ is obtained as follows:

$$\frac{1}{\pi t} \longleftrightarrow -j\,\text{sgn}(f)$$

$$\hat{x}(t) = x(t) \otimes \frac{1}{\pi t} \longleftrightarrow -j\,\text{sgn}(f)X(f) \tag{6-117}$$

$$= -jX(f) \qquad \text{for } f > 0$$
$$= +jX(f) \qquad \text{for } f < 0$$

With the help of a generalization of Parseval's theorem, $x(t)$ and $\hat{x}(t)$ can be shown to be orthogonal. Given two functions $s_1(t)$ and $s_2(t)$, the generalized Parseval relationship states that

$$\int_{-\infty}^{+\infty} s_1(t)s_2^*(t)\,dt = \int_{-\infty}^{+\infty} S_1(f)S_2^*(f)\,df \tag{6-118}$$

from which

$$\int_{-\infty}^{+\infty} x(t)\hat{x}(t)\,dt = \int_{-\infty}^{+\infty} X(f)[-j\,\text{sgn}(f)X(f)]^*\,df \tag{6-119}$$

But

$$X(f)[-j\,\text{sgn}(f)X(f)]^* = +j|X(f)|^2\,\text{sgn}(f)$$

Because this is an odd function of frequency, the integral in (6-119) over the entire frequency range is zero. Therefore,

$$\int_{-\infty}^{+\infty} x(t)\hat{x}(t)\,dt = 0$$

The energy spectral density of $x(t)$ is identical to that for $\hat{x}(t)$. Therefore, the total energy in the analytic signal is twice that in the real signal.

PROBLEMS

6.1. With respect to the waveform in Figure P6-1:

    (a) Determine the Fourier transform by direct application of the Fourier integral.

    (b) Observe that the derivative of the waveform consists of two rectangular pulses symmetrically displaced with respect to the origin. Using pair 3 and rule 5 from Table 6-1, write the Fourier transform of the derivative of the waveform by inspection.

    (c) Use rule 10 of Table 6-1 to convert the result in part (b) to the transform of the original waveform.

    (d) The waveform in Figure P6-1 may be obtained by the convolution of two rectangular pulses. Define the amplitude and width of these pulses and use pair 3 and rule 7 to obtain the Fourier transform of the final waveform. The results obtained in parts (a), (c), and (d) should be identical.

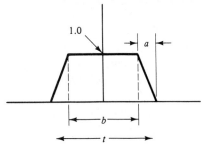

Figure P6-1

6.2. Pair 6 of Table 6-1 indicates that the transform of a Gaussian waveform also has a Gaussian shape. Derive this result by direct application of the Fourier integral.

6.3. Assume two Gaussian waveforms as follows:

$$s_1(t) = \frac{1}{\sigma_1\sqrt{2\pi}} \exp\left(\frac{-t^2}{2\sigma_1^2}\right)$$

$$s_2(t) = \frac{1}{\sigma_2\sqrt{2\pi}} \exp\left(\frac{-t^2}{2\sigma_2^2}\right)$$

Demonstrate by direct application of the convolution integral that the convolution of $s_1(t)$ with $s_2(t)$ results in a third Gaussian waveform. Demonstrate this result by using pair 6 and rule 7 from Table 6-1.

6.4. Given two time-domain functions $s_1(t)$ and $s_2(t)$, with Fourier transforms $S_1(f)$ and $S_2(f)$, prove that

$$\int_{-\infty}^{+\infty} s_1(t)s_2^*(t)\, dt = \int_{-\infty}^{+\infty} S_1(f)S_2^*(f)\, df$$

6.5. By use of (6-111) show that the Hilbert transform of $\cos \omega_0 t$ is $\sin \omega_0 t$. Demonstrate this fact using the Fourier transform relationship in (6-117).

6.6. A repetitive Gaussian pulse train may be expressed as

$$\left[\operatorname{rep}_T \exp\left(\frac{-t^2}{2\sigma_1^2}\right)\right], \qquad T \gg \sigma_1$$

Assume that this waveform is multiplied by a single Gaussian pulse as follows:

$$\exp\left(\frac{-t^2}{2\sigma_2^2}\right)\left[\operatorname{rep}_T \exp\left(\frac{-t^2}{2\sigma_1^2}\right)\right], \qquad \sigma_2 \gg T$$

Express the Fourier transform of this waveform in symbolic form. Sketch the resulting spectrum.

SUGGESTED READING

1. Woodward, P. M., *Probability and Information Theory, with Applications to Radar*. New York: McGraw-Hill Book Company/London: Pergamon Press, Inc., 1953, Chap. 2.

2. Burdic, W. S., *Radar Signal Analysis*. Englewood Cliffs, N. J.: Prentice-Hall, Inc., 1968, Chaps. 1, 2, 4.

3. Titchmarsh, E. C., *Introduction to the Theory of Fourier Integrals*. New York: Oxford University Press, Inc., 1948.

4. Miller, K. S., *Engineering Mathematics*, New York: Dover Publications, Inc., 1943.

5. Shannon, C. E., and N. Weaver, *The Mathematical Theory of Communication*. Urbana, Ill.: The University of Illinois Press, 1964, p. 86.

6. Dugundji, J., "Envelopes and Pre-envelopes of Real Waveforms," *IRE Trans. PGIT*, Vol. IT-4, No. 1, p. 53 (Mar. 1958).

7

Discrete Fourier Methods

7.0 INTRODUCTION

The Fourier series and integral transforms were developed in Chapter 6 assuming continuous functions, available over the entire range from plus to minus infinity, in both the time and frequency domains. When dealing with actual physical phenomena and practical systems, we are always constrained to consider only finite-length records of functions. For systems operating in real time, the record extends from some finite time past to the present. Similarly, practical spectral measurements of a function must be limited to a finite frequency range.

In addition to the finite-record constraint, systems using digital hardware require that the functions to be processed be converted to a discrete set of numbers, or sample values. The processing is then performed on these samples, rather than on the continuous functions. This leads directly to the development of a discrete form of the Fourier transform.

In this chapter, starting with continuous functions, the sampling theorems are used to describe sampled versions of both the time- and frequency-domain representations of functions. Rules are then developed for performing basic operations required in system analysis, and for relating the time-domain and frequency-domain sample values by means of the discrete Fourier transform (DFT). Finally, a particularly efficient method for computing the DFT, called the fast Fourier transform (FFT), is described in elementary form [1–3].

7.1 DISCRETE REPRESENTATION OF CONTINUOUS
FUNCTIONS

In Figure 7-1, a bandlimited continuous time function is shown, together with its sampled version in the time domain. As discussed in Chapter 6, if the time-domain sample interval, t_s, is the inverse of the total signal bandwidth, $2F$, the spectrum will repeat at intervals $2F$ with no overlap of the repeated spectra. In Figure 7-1(c), the sampled time-domain function is gated by the function rect (t/T). In the frequency domain, this requires convolution with a sinc function. Starting with the original continuous function, the operations depicted in Figure 7-1 may be written as:

Continuous function (bandlimited):

$$s(t) \longleftrightarrow S(f) \text{ rect}\left(\frac{f}{2F}\right)$$

Sampled function ($t_s = 1/2F$):

$$s(t) \text{ comb}_{t_s} \longleftrightarrow \left[2F \text{ rep}_{2F} S(f) \text{ rect}\left(\frac{f}{2F}\right)\right] \qquad (7\text{-}1)$$

Gated-sampled function:

$$[s(t) \text{ comb}_{t_s}] \text{ rect}\left(\frac{t}{T}\right) \longleftrightarrow \left[2F \text{ rep}_{2F} S(f) \text{ rect}\left(\frac{f}{2F}\right)\right] \otimes T \text{ sinc}(fT)$$

Notice that the repeated versions of the original spectrum are spread out by the convolution, resulting in overlap of the adjacent replicas. Indeed, because the sinc function is infinite in extent, the original bandlimited spectrum, $S(f)$, and each of its replicas are extended to an infinite frequency range. As a result, the original continuous function can no longer be recovered exactly by filtering the signal in Figure 7-1(c) with a rectangular filter of width $2F$ centered at the origin. The contamination of the original spectrum in the region $\pm F$ by overlap from adjacent regions results in an *aliasing error* when attempting to recover the original signal. This example serves to illustrate the fact that a function cannot be simultaneously limited in duration in both the time and frequency domains. However, if T is very large compared with $1/2F$, and if $S(f)$ decreases to a small value well before reaching the boundaries of $\pm F$, the aliasing error may be acceptably small.

Assume that the time gate in Figure 7-1(c) is centered at the origin, and that the total number of time samples in time T is

$$N + 1 = \frac{T}{t_s} = 2FT$$

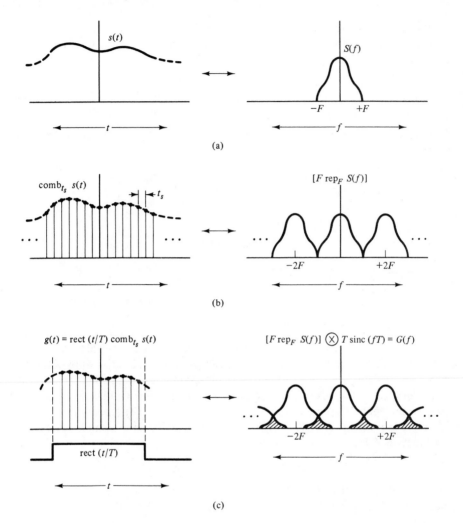

Figure 7-1 Effect of time-gating a sampled waveform.

The gate-sampled signal can be written as

$$g(t) = [s(t) \, \text{comb}_{t_s}] \, \text{rect}\left(\frac{t}{T}\right) = \sum_{n=-N/2}^{N/2} s(nt_s)\delta(t - nt_s) \qquad (7\text{-}2)$$

and the Fourier transform of $g(t)$ is

$$g(t) = \sum_{n=-N/2}^{+N/2} s(nt_s)\delta(t - nt_s) \longleftrightarrow \sum_{n=-N/2}^{+N/2} s(nt_s) \exp(-j2\pi fnt_s) = G(f)$$

$$(7\text{-}3)$$

Now create a sampled version of the spectrum with the help of the frequency-domain comb function. Because $g(t)$ is limited to the interval T, the frequency-domain sample interval should be $f_s = 1/T$. Therefore,

$$\text{comb}_{f_s}\, G(f) = \sum_{k=-\infty}^{+\infty} \left[\sum_{n=-N/2}^{+N/2} s(nt_s) \exp\left(-j2\pi fnt_s\right) \right] \delta(f - kf_s) \qquad (7\text{-}4)$$

The presence of the impulse function causes the entire expression to be zero except for $f = kf_s = k/T$. Hence

$$\text{comb}_{f_s}\, G(f) = \sum_{k=-\infty}^{+\infty} \left[\sum_{n=-N/2}^{+N/2} s(nt_s) \exp\left(\frac{-j2\pi knt_s}{T}\right) \right] \delta(f - kf_s)$$

$$= \sum_{k=-\infty}^{+\infty} G(kf_s)\delta(f - kf_s) \qquad (7\text{-}5)$$

where

$$G(kf_s) = \sum_{n=-N/2}^{+N/2} s(nt_s) \exp\left(\frac{-j2\pi knt_s}{T}\right) \qquad (7\text{-}6)$$

Sampling the spectrum, $G(f)$, causes $g(t)$ to repeat at intervals T as shown in Figure 7-2. Notice from (7-6) that the sample values, $G(kf_s)$, were obtained by a *discrete transformation* of the $N + 1$ time samples contained in the interval T. Because $G(f)$ extends over all frequencies, the frequency index, k, must also range from minus to plus infinity. However, because $G(f)$ repeats at intervals $2F$, there are only $2FT = 2F/f_s$ *distinct* samples required to completely define the function. Thus, for any integer r,

$$G(kf_s) = G[(k + 2FTr)f_s] \qquad (7\text{-}7)$$

Because the number of time samples, $N + 1$, is equal to $T/t_s = 2FT$, we see that the number of distinct frequency samples is identical to the number of time samples.

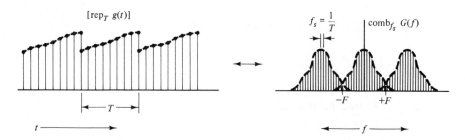

Figure 7-2 Effect of frequency-domain sampling on $g(t)$.

7.2 DISCRETE FOURIER TRANSFORM

The discrete sample values of the gated time-function and the corresponding discrete spectrum constitute a *discrete Fourier transform* (DFT) pair. Let $g(n)$ be the sample values, $s(nt_s)$, over the interval T, and $G(k)$ be the corresponding frequency-domain samples. With $T/t_s = N + 1$, (7-6) becomes

$$G(k) = \sum_{n=-N/2}^{+N/2} g(n) \exp\left(\frac{-j2\pi kn}{N + 1}\right) \tag{7-8}$$

For convenience in the following discussion, define

$$W = \exp\left(\frac{-j2\pi}{N + 1}\right)$$

Equation (7-8) may now be written more compactly as

$$G(k) = \sum_{n=-N/2}^{+N/2} g(n)W^{kn} \tag{7-9}$$

The discrete inverse transform of $G(k)$ over the interval $k = \pm N/2$ is

$$\sum_{k=-N/2}^{+N/2} G(k)W^{-kn} = \sum_{k=-N/2}^{+N/2} \left[\sum_{\ell=-N/2}^{+N/2} g(\ell)W^{k\ell}\right]W^{-kn}$$

$$= \sum_{k}\sum_{\ell} g(\ell)W^{(\ell-n)k}$$

$$= \sum_{\ell} g(\ell)\left[\sum_{k} W^{(\ell-n)k}\right] \tag{7-10}$$

Consider the nature of the bracketed term in (7-10). The exponential term is

$$W^{(\ell-n)k} = \exp\left[-j2\pi(\ell - n)\frac{k}{N + 1}\right]$$

$$= 1 \qquad \text{for } (n - \ell) = r(N + 1), \quad r = \text{any integer}$$

$$\text{or } n = \ell + r(N + 1)$$

A little thought will show that for any other value of n, summation of this term over k will give zero. Thus

$$\sum_{k=-N/2}^{+N/2} W^{(\ell-n)k} = N + 1 \qquad \text{for } n = \ell + r(N + 1) \tag{7-11}$$

$$= 0 \qquad \text{otherwise}$$

$$= (N + 1)\delta_{((\ell n))}$$

where $\delta_{((\ell n))}$ is the Kronecker delta function, with the double parenthesis used to indicate the repetitive nature of the required relationship between ℓ and n. Notice that n is *not* limited to the range $\pm N/2$. Equation (7-10) may now be written

$$\sum_{k=-N/2}^{+N/2} G(k)W^{-kn} = (N + 1) \sum_{\ell=-N/2}^{+N/2} g(\ell)\delta_{((\ell n))} \qquad (7\text{-}12)$$

The summation on the right side of (7-12) is the discrete equivalent of the integral of the product of a continuous function and an impulse function. The result is the value of $g(\ell)$ at $\ell = ((n))$. Thus

$$\sum_{k=-N/2}^{+N/2} G(k)W^{-kn} = (N + 1)g((n)) \qquad (7\text{-}13)$$

Equations (7-9) and (7-13) identify $g(n)$ and $G(k)$ as a discrete Fourier transform pair. An alternative representation of this transform pair involves the use of N points in the interval zero to $N - 1$.

$$G((k)) = \sum_{n=0}^{N-1} g(n)W^{kn}$$

$$g((n)) = \frac{1}{N} \sum_{k=0}^{N-1} G(k)W^{-kn} \qquad (7\text{-}14)$$

The N-point transform results in a discrete series in the opposite domain that repeats at interval N.

7.3 BASIC OPERATIONS ON DISCRETE FUNCTIONS

Basic operations required in analysis with discrete functions generally obey the same rules, in discrete form, as discussed in Chapter 6 for continuous functions. When more than one function is involved, care should be exercised to ensure that all functions have the same number of discrete elements, with the same sample interval. In working with discrete representations of continuous functions, we must remember that the discrete functions cannot perfectly represent the continuous functions in both the time and frequency domains. However, the transform relationships between the discrete functions are in themselves exact.

The rules governing operations with continuous functions and their transforms are tabulated in Table 6-1. We consider here only a few examples using discrete functions.

Let the discrete function, $g_3(n)$, be the sum of two other discrete functions. The DFT of $g_3(n)$ is then equal to the sum of the DFTs of the component functions.

$$g_3(n) = g_1(n) + g_2(n) \longleftrightarrow G_1(k) + G_2(k) = G_3(k) \qquad (7\text{-}15)$$

A discrete time function may be translated in time in discrete steps, resulting in the addition of a discrete linear phase term to the DFT. Let

$$g((n)) = \frac{1}{N} \sum_{k=0}^{N-1} G(k)W^{-nk}$$

$$g((n - \ell)) = \frac{1}{N} \sum_{k=0}^{N-1} G(k)W^{-(n-\ell)k}$$

$$= \frac{1}{N} \sum_{k=0}^{N-1} [G(k)W^{\ell k}]W^{-nk}$$

Therefore,

$$g((n - \ell)) \longleftrightarrow G(k)W^{\ell k} = G(k) \exp\left(\frac{-j2\pi\ell k}{N}\right) \qquad (7\text{-}16)$$

Translation in the frequency domain produces a similar effect on the time-domain function. If

$$G((k)) = \sum_{n=0}^{N-1} g(n)W^{nk}$$

then

$$g(n) \exp\left(\frac{+j2\pi qn}{N}\right) \longleftrightarrow G((k - q)) \qquad (7\text{-}17)$$

The product of discrete functions results in a type of discrete convolution of their DFTs in the opposite domain. Consider the product of the discrete functions $X(k)$ and $Y(k)$, which are the N-point DFTs of $x(n)$ and $y(n)$. We now determine the inverse DFT of the product. Define

$$v((n)) = \frac{1}{N} \sum_{k=0}^{N-1} X(k)Y(k)W^{-kn} \qquad (7\text{-}18)$$

Substitute for $X(k)$ and $Y(k)$ their equivalents in terms of the DFTs of x and y.

$$v((n)) = \frac{1}{N} \sum_{k=0}^{N-1} \left[\sum_{\ell=0}^{N-1} x(\ell)W^{\ell k} \right]\left[\sum_{m=0}^{N-1} y(m)W^{mk} \right]W^{-nk}$$

Rearranging, we obtain

$$v((n)) = \sum_{\ell=0}^{N-1}\sum_{m=0}^{N-1} x(\ell)y(m)\left[\frac{1}{N} \sum_{k=0}^{N-1} W^{-(n-\ell-m)k} \right] \qquad (7\text{-}19)$$

The bracketed term may be identified as the discrete impulse, or Kronecker delta function. That is,

$$\frac{1}{N} \sum_{k=0}^{N-1} W^{-(n-\ell-m)k} = \begin{cases} 1 & \text{for } m = n - \ell + rN \\ 0 & \text{otherwise} \end{cases} \tag{7-20}$$

$$= \delta_{((m(n-\ell)))}$$

Equation (7-19) is therefore zero unless $m = ((n - \ell))$. Hence

$$v((n)) = \sum_{\ell=0}^{N-1} x(\ell)y((n - \ell)) \tag{7-21}$$

Equation (7-21) is the discrete equivalent of the convolution integral for continuous functions. Notice that the resultant function, $v((n))$, repeats at intervals N. This is because of the repetitive nature of $y((n - \ell))$. To perform the convolution, the original function, $y(\ell)$, of length N, is first folded and caused

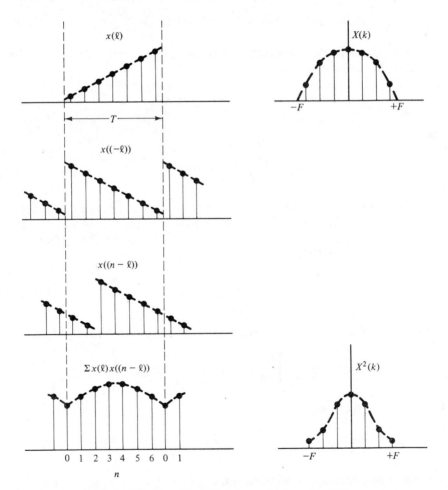

Figure 7-3 Example of discrete, or circular, convolution.

to repeat at intervals N, resulting in $y((-\ell))$. Consider the interval $0 \leq \ell \leq N - 1$, defined by the summation. As $y((-\ell))$ shifts to the right by n, sample values of y shift out of the summation interval at one end and reappear at the other end. Thus all N sample values of y are always present within the summation interval, circulating through the region with the index n. Because of this, the discrete convolution defined by (7-21) is often called a *circular convolution*, in contrast to the *linear convolution* normally associated with non-periodic functions.

As an example of discrete convolution, consider the convolution of a discrete function, $x(n)$, with itself. Let

$$v(n) = x(n) \otimes x((n)) \longleftrightarrow X^2(k)$$

In Figure 7-3 are shown $x(\ell)$ and its DFT, $X(k)$, together with $x((-\ell))$. The shifted function, $x((n - \ell))$, is multiplied by $x(\ell)$ and summed over the interval N to provide one point on the output function.

In some cases, it is desirable to avoid the overlap from adjacent intervals normally encountered with circular convolution. This can be accomplished by artificially extending the original time series with zeros. This procedure is demonstrated in Figure 7-4. In this example, the original series is doubled in

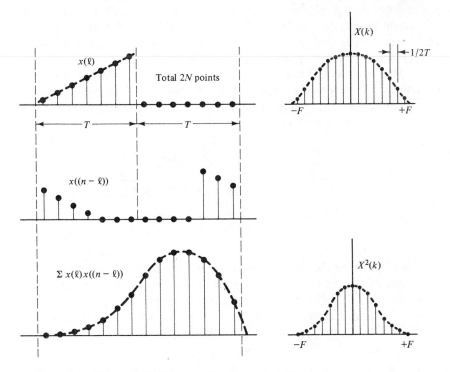

Figure 7-4   Effect of artificially extending time series with zeros to prevent overlap in convolution.

length by adding N zeros for a total time duration of $2T$. The additional zeros in the time series do not modify the shape or extent of the DFT, but the sample interval in the frequency domain is reduced to $1/2T$, resulting in $2N$ points in the interval $\pm F$. As shown in the figure, the presence of the additional points with zero amplitude prevents the overlap from adjacent repetitions of the original time series. The function resulting from the convolution repeats at intervals $2N$, rather than at intervals N as in the example in Figure 7-3.

7.4 THE FAST FOURIER TRANSFORM

Consider the DFT

$$G(k) = \sum_{n=0}^{N-1} g(n)W^{nk} \tag{7-22}$$

For each discrete frequency, k, the computation of $G(k)$ requires N complex multiplications and N complex additions. To obtain all N terms, therefore, requires N-squared complex multiplications and additions. For large values of N, this obviously becomes quite a computational load. Thus, for $N = 1000$, a million complex operations are required. The *fast Fourier transform* (FFT) algorithm has been developed to provide a more efficient method for computing the discrete Fourier transform [3]. Indeed, the FFT algorithm makes possible high-speed spectral analysis of waveforms, using modern digital techniques.

To demonstrate the principle of the FFT technique, we divide the summation defining $G(k)$ into two parts: one containing the odd terms in n, and one containing the even terms. Assume that N is divisible by 2, so that the resulting summations containing even and odd values of n contain equal numbers of components. Thus we may write

$$G(k) = \underbrace{\sum_{n=0}^{(N/2)-1} g(2n)W^{2kn}}_{(N/2) \ - \ \text{even terms}} + \underbrace{\sum_{n=0}^{(N/2)-1} g(2n+1)W^{k(2n+1)}}_{(N/2) \ - \ \text{odd terms}}$$

$$= \sum_{n=0}^{(N/2)-1} g(2n)W^{2kn} + W^k \sum_{n=0}^{(N/2)-1} g(2n+1)W^{2kn} \tag{7-23}$$

Each of the summations in (7-23) corresponds to a time function sampled at intervals $2t_s$, for a total of $N/2$ samples. Thus each summation can be used to generate only $N/2$ distinct values of $G(k)$, rather than N values. That is, for $k \geq N/2$, each summation will repeat a value already computed for a value $k \leq N/2 - 1$. For instance, for $k = 1$, we have

$$W^{2kn} = W^{2n}$$

For $k = (N/2) + 1$,

$$W^{2nk} = W^{nN}W^{2n}$$

but

$$W^{nN} = \exp\left(\frac{-j2\pi nN}{N}\right) = 1$$

In general,

$$W^{2nk} = W^{2n(k+N/2)} \tag{7-24}$$

Each summation in (7-23) requires $N/2$ complex multiplications for each value of k. In addition, one complex multiplication is required for the exponential term W^k outside the second summation. The total number of multiplications for *each* value of k is, therefore,

$$\text{multiplies for each } k = \left(\frac{N}{2}\right) + \left(\frac{N}{2}\right) + 1 = N + 1$$

For *all* values of k in the interval $[0, (N/2) - 1]$, the total number of complex multiplications is

$$\left(\frac{N}{2}\right)(N + 1) = \left(\frac{N^2}{2}\right) + \left(\frac{N}{2}\right)$$

Now consider the interval $(N/2) \le k \le (N - 1)$. Using (7-24), the values of $G(k)$ in this interval may be obtained using the values computed for k less than $N/2$. Let k_1 represent the values of k in the lower interval, and $k_2 = k_1 + N/2$. Then

$$G(k_2) = \sum_{n=0}^{(N/2)-1} g(2n)W^{2nk_1} + W^{(N/2)}W^{k_1}\sum_{n=0}^{(N/2)-1} g(2n + 1)W^{2nk_1} \tag{7-25}$$

but

$$W^{(N/2)} = \exp\left[\frac{-j2\pi(N/2)}{N}\right] = -1$$

Now let

$$X(k) = \sum_{n=0}^{(N/2)-1} g(2n)W^{2nk}$$

$$Y(k) = \sum_{n=0}^{(N/2)-1} g(2n + 1)W^{2nk}$$

and write

$$G(k_1) = X(k_1) + W^{k_1}Y(k_1)$$
$$G(k_2) = X(k_1) - W^{k_1}Y(k_1) \tag{7-26}$$

Hence, after performing the operations required to compute the DFT in the range $0 \le k \le (N/2) - 1$, the remainder of the range is obtained by reversing the sign of the $W^{k_1} Y(k_1)$ terms and performing $N/2$ complex additions. No additional complex multiplications are required. The total number of complex multiplications for the entire inverval $0 \le k \le N - 1$ is, therefore, equal to $N^2/2 + N/2$. For N large, this represents a reduction by almost a factor of 2, compared with a direct computation of the DFT.

Now assume that N is a power of 2. Each of the summations in (7-23) may be further subdivided into two groups with $N/4$ points each. This results in an additional reduction by almost a factor of 2 in the number of required complex multiplications. This procedure may be repeated until only two terms are left in each summation. The total number of complex multiplications then required is $(N/2) \log_2 N$. To demonstrate the magnitude of the computational savings this represents, let $N = 1024$. Using the FFT algorithm, the N-point DFT requires 5120 complex multiplications, whereas 1,048,576 are required by direct evaluation of (7-22).

As a simple example of the FFT technique, consider a discrete time series with $N = 8$. In accordance with (7-23), we divide the eight-point DFT into two four-point DFTs. Using (7-26), a functional flow diagram for computing the entire DFT is shown in Figure 7-5. The even-numbered terms in the time series provide the values X_0, X_1, X_2, and X_3, using a standard discrete transform. Similarly, the odd-numbered terms are used to generate Y_0, Y_1, Y_2, and Y_3. To obtain the final transform term, G_0, we add the product $W^0 Y_0$ to X_0. The output

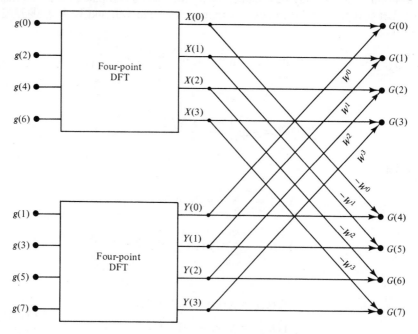

Figure 7-5   Reduction of eight-point DFT to a two four-point DFTs to illustrate FFT technique.

term, G_4, is obtained by subtracting the product $W^0 Y_0$ from X_0. Proceeding in this manner, all eight output terms are generated. The extension of this example to a group of four two-point DFTs is straightforward, although the functional flow diagram is more complex. The reader is referred to the Suggested Reading list for information on the organization of practical FFT devices.

PROBLEMS

7.1. A Kronecker delta function is defined by (7-11). For $N = 4$ and $(\ell - n) = 1, 2,$ and 4, demonstrate by geometric construction that the summation in (7-11) is zero. Demonstrate that the summation interval on k may be shifted in integral steps without affecting the result in (7-11). That is,

$$\sum_{k=m-N/2}^{m+N/2} W^{(\ell-n)k} = (N + 1)\delta_{((\ell n))} \qquad \text{for } m = \text{integer}$$

7.2. In Chapter 6 the effect of folding and complex conjugation on the Fourier transform relationship for continuous variables was derived. Using a similar approach for discrete variables, demonstrate that if

$$g((n)) \longleftrightarrow G(k)$$

then

$$g((-n)) \longleftrightarrow G(-k)$$
$$g^*((n)) \longleftrightarrow G^*(-k)$$

7.3. In Figure 7-5 the computational flow graph for an eight-point DFT computed by operating on the outputs of two four-point DFTs is shown. By a similar procedure construct the flow graph for an eight-point DFT using four two-point DFTs. As a suggestion, group the input samples in the order $g(0), g(4), g(2), g(6), g(1), g(5),$ $g(3), g(7)$.

SUGGESTED READING

1. Gold, B., and C. Rader, *Digital Processing of Signals*. New York: McGraw-Hill Book Company, 1969.
2. Oppenheim, A. V., and R. W. Schafer, *Digital Signal Processing*. Englewood Cliffs, N. J.: Prentice-Hall, Inc., 1975.
3. Cooley, J. W., and J. W. Tukey, "An Algorithm for the Machine Computation of Complex Fourier Series," *Math. Comp.*, Vol. 19, pp. 297–301. (Apr. 1965).

8 Correlation and Correlation Functions

8.0 INTRODUCTION

Correlation, in signal analysis, is the act or process of determining mutual relationships that exist among signals or functions. Signal characteristics that are often compared include shape, bandwidth, duration, and position, either in time or frequency.

In this chapter we develop the concept of correlation as the process of adjusting one function to minimize the mean-squared difference relative to another function. The amount of adjustment required is related to the degree of similarity between the two functions. The correlation function arises quite naturally in the analysis of systems, and provides a basis for measuring and optimizing system performance.

8.1 CORRELATION OF VECTORS

A geometric interpretation of correlation is obtained by considering the method by which we may compare two vectors. In Figure 8-1 are shown two vectors, V_1 and V_2, together with the resolution of each one into orthogonal components.[1]

We now wish to determine a quantitative measure of the degree to which V_1 is "like" V_2. One approach is to form the difference vector, V_d, as shown in

---

[1] The development of correlation coefficients presented here is patterned after Mason and Zimmerman, *Electronic Circuits, Signals and Systems,* Chapter 6, 1968; used with permission of MIT Press. (Originally published by John Wiley & Sons, Inc., 1960.)

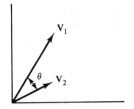

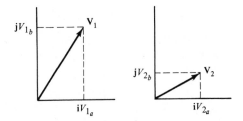

Figure 8-1 Resolution of two vectors into their orthogonal components.

Figure 8-2(a). If the difference vector is zero, $\mathbf{V}_1$ and $\mathbf{V}_2$ are identical. If $\mathbf{V}_d$ is not zero, suppose that we adjust the *magnitude* of $\mathbf{V}_2$ to minimize $\mathbf{V}_d$. Thus let

$$\mathbf{V}_d = \mathbf{V}_1 - c_{12}\mathbf{V}_2 \qquad (8\text{-}1)$$

where c_{12} is a real constant.

It is evident from Figure 8-2(b) that $\mathbf{V}_d$ is minimized when it is at right angles, or orthogonal, to $c_{12}\mathbf{V}_2$. The vector $\mathbf{V}_1$ is then composed of a component, $c_{12}\mathbf{V}_2$, that differs only in magnitude from $\mathbf{V}_2$, and a component, $\mathbf{V}_d$, that is orthogonal to $\mathbf{V}_2$.

By inspection of Figure 8-2(b), the constant c_{12} is identified as

$$c_{12} = \frac{|\mathbf{V}_1|}{|\mathbf{V}_2|} \cos \theta \qquad (8\text{-}2)$$

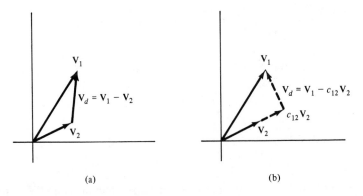

(a)   (b)

Figure 8-2 Adjustment of the magnitude of $\mathbf{V}_2$ to minimize the difference vector.

If the angle θ is 90° ($\mathbf{V}_1$ and $\mathbf{V}_2$ orthogonal), c_{12} is zero and $\mathbf{V}_1$ and $\mathbf{V}_2$ are as unalike as is possible. At the other extreme, if θ is zero, c_{12} is simply the ratio of the vector amplitudes, and would be unity if $|\mathbf{V}_1| = |\mathbf{V}_2|$.

An alternative form of (8-2) is obtained by recalling the definition of the *scalar* or *dot product* of two vectors. If θ is the angle between two vectors, then

$$\mathbf{V}_1 \cdot \mathbf{V}_2 = |\mathbf{V}_1||\mathbf{V}_2|\cos\theta$$

from which

$$c_{12} = \frac{\mathbf{V}_1 \cdot \mathbf{V}_2}{|\mathbf{V}_2|^2} \tag{8-3}$$

We shall now derive the value of c_{12} in a manner that will also be applicable to the comparison of functions other than simple two-dimensional vectors. Resolve $\mathbf{V}_1$ and $\mathbf{V}_2$ into rectangular components and write

$$\mathbf{V}_1 = \mathbf{i}V_{1a} + \mathbf{j}V_{1b}$$

$$\mathbf{V}_2 = \mathbf{i}V_{2a} + \mathbf{j}V_{2b}$$

$$\mathbf{V}_d = \mathbf{i}(V_{1a} - c_{12}V_{2a}) + \mathbf{j}(V_{1b} - c_{12}V_{2b}) \tag{8-4}$$

The squared magnitude of $\mathbf{V}_d$ is equal to the sum of the squares of its orthogonal components:

$$|\mathbf{V}_d|^2 = |\mathbf{V}_1 - c_{12}\mathbf{V}_2|^2 = (V_{1a} - c_{12}V_{2a})^2 + (V_{1b} - c_{12}V_{2b})^2 \tag{8-5}$$

In more general terms

$$|\mathbf{V}_d|^2 = \sum_{k=a}^{b} (V_{1k} - c_{12}V_{2k})^2 \tag{8-6}$$

The value of c_{12} that minimizes $|\mathbf{V}_d|^2$ also minimizes $|\mathbf{V}_d|$. Therefore, we may differentiate (8-6) with respect to c_{12}, equate to zero, and solve for c_{12}.

$$\frac{\partial |\mathbf{V}_d|^2}{\partial c_{12}} = \frac{\partial}{\partial c_{12}} \left[\sum_{k=a}^{b} (V_{1k}^2 - 2c_{12}V_{1k}V_{2k} + c_{12}^2 V_{2k}^2) \right] = 0$$

$$= 2c_{12} \sum_{k=a}^{b} V_{2k}^2 - 2 \sum_{k=a}^{b} V_{1k}V_{2k} = 0$$

or

$$c_{12} = \frac{\sum\limits_{k=a}^{b} V_{1k}V_{2k}}{\sum\limits_{k=a}^{b} V_{2k}^2} \tag{8-7}$$

The numerator in (8-7) is by definition the scalar product of $\mathbf{V}_1$ and $\mathbf{V}_2$, and the denominator is obviously the squared magnitude of $\mathbf{V}_2$. Thus (8-7) is identical to (8-3). The constant c_{12} is obtained by minimizing the *squared difference* between $\mathbf{V}_1$ and $\mathbf{V}_2$.

In this example, the magnitude of $\mathbf{V}_1$, rather than $\mathbf{V}_2$, could have been adjusted to minimize the difference vector. This would result in a constant, c_{21}, defined as

$$c_{21} = \frac{\sum\limits_{k=a}^{b} V_{1k} V_{2k}}{\sum\limits_{k=a}^{b} V_{1k}^2} \qquad (8\text{-}8)$$

If we write

$$\mathbf{V}_2 = c_{21} \mathbf{V}_1 + \mathbf{V}_d$$

we see that $\mathbf{V}_2$ can be resolved into a component that differs from $\mathbf{V}_1$ only in magnitude, and a component, $\mathbf{V}_d$, that is orthogonal to $\mathbf{V}_1$.

8.2 CORRELATION OF TIME WAVEFORMS

Consider the simple time functions shown in Figure 8-3. Each waveform is composed of two contiguous rectangular pulses of width Δt.

The total energy contained in $v_1(t)$ is

$$W_1 = \int_{-\infty}^{+\infty} |v_1(t)|^2 \, dt = (v_{1a}^2 + v_{1b}^2) \Delta t \qquad (8\text{-}9)$$

Because the rectangular components of $v_1(t)$ exist in different time intervals, they are by definition orthogonal. That is,

$$\int_{-\infty}^{+\infty} \left[v_{1a} \, \text{rect} \left(\frac{t - \Delta t/2}{\Delta t} \right) \right] \left[v_{1b} \, \text{rect} \left(\frac{t - 3\Delta t/2}{\Delta t} \right) \right] dt = 0 \qquad (8\text{-}10)$$

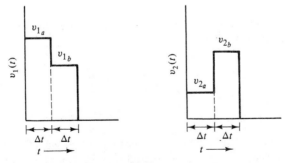

Figure 8-3 Two simple time functions.

The waveform energy is therefore proportional to the sum of the squares of its orthogonal components. This is analogous to identifying the squared magnitude of a vector with the sum of the squares of its orthogonal components.

A measure of similarity of two waveforms may be defined in a manner directly comparable to that used to compare vectors. For instance, multiply $v_2(t)$ by a constant c_{12} and form the difference waveform

$$v_d(t) = v_1(t) - c_{12}v_2(t) \tag{8-11}$$

The constant c_{12} is selected to minimize the *energy* in the difference waveform. Thus

$$\frac{\partial}{\partial c_{12}}\left\{\int_{-\infty}^{+\infty}[v_1^2(t) - 2c_{12}v_1(t)v_2(t) + c_{12}^2v_2^2(t)]\,dt\right\} = 0$$

or

$$c_{12} = \frac{\displaystyle\int_{-\infty}^{+\infty} v_1(t)v_2(t)\,dt}{\displaystyle\int_{-\infty}^{+\infty} v_2^2(t)\,dt} \tag{8-12}$$

For the simple rectangular waveforms in Figure 8-3, this can be put in the form

$$c_{12} = \frac{\displaystyle\sum_{k=a}^{b} v_{1k}v_{2k}\,\Delta t}{\displaystyle\sum_{k=a}^{b} v_{2k}^2\,\Delta t} \tag{8-13}$$

This is identical to the result obtained in (8-7) for the two vectors.

With c_{12} selected in accordance with (8-12), the waveform $v_1(t)$ can be separated into a component that differs only in magnitude from $v_2(t)$, and a component, $v_d(t)$, that is *orthogonal* to $v_2(t)$. The orthogonality at $v_d(t)$ and $v_2(t)$ is easily demonstrated using (8-11) and (8-12) to show that

$$\int_{-\infty}^{+\infty} v_2(t)v_d(t)\,dt = 0$$

The magnitude of $v_1(t)$ may be adjusted to minimize the energy in the difference waveform by defining a constant c_{21} as

$$c_{21} = \frac{\displaystyle\int_{-\infty}^{+\infty} v_1(t)v_2(t)\,dt}{\displaystyle\int_{-\infty}^{+\infty} v_1^2(t)\,dt} \tag{8-14}$$

Equations (8-12) and (8-14) are derived assuming real waveforms. In the more general case the functions being compared may be complex, in which case c_{12} and c_{21} may be complex and are defined as

$$c_{12} = \frac{\int_{-\infty}^{+\infty} v_1(t)v_2^*(t)\, dt}{\int_{-\infty}^{+\infty} |v_2(t)|^2\, dt}$$

$$c_{21} = \frac{\int_{-\infty}^{+\infty} v_1^*(t)v_2(t)\, dt}{\int_{-\infty}^{+\infty} |v_1(t)|^2\, dt}$$

(8-15)

8.3 NORMALIZED CORRELATION AND CORRELATION FUNCTIONS

The denominators in (8-15) relate to the energy in $v_1(t)$ and $v_2(t)$. The numerators contain the information concerning the similarity in *form* of the two functions. A useful modification of the expressions in (8-15) is obtained by making the denominator proportional to the product of the energies in the two functions. The result is the *normalized correlation coefficient* defined by

$$\rho_{12} = \frac{\int_{-\infty}^{+\infty} v_1(t)v_2^*(t)\, dt}{\left[\int_{-\infty}^{+\infty} |v_1(t)|^2\, dt \int_{-\infty}^{+\infty} |v_2(t)|^2\, dt\right]^{1/2}}$$

(8-16)

By normalizing with respect to the energy in both functions, ρ_{12} is constrained to fall within the range ± 1.0. That is, if $v_1(t) = kv_2(t)$, where k is any real constant, ρ_{12} is $+1$ or -1 depending on whether k is plus or minus.

Consider two signals, $s_1(t)$ and $s_2(t - \tau)$, where τ is an arbitrary time shift. The degree of correlation between the two signals is obviously a function of τ. Neglecting the normalization, the *cross-correlation functions* are

$$\mathscr{R}_{12}(\tau) = \int_{-\infty}^{+\infty} s_1(t)s_2^*(t - \tau)\, dt$$

(8-17)

$$\mathscr{R}_{21}(\tau) = \int_{-\infty}^{+\infty} s_1^*(t - \tau)s_2(t)\, dt$$

(8-18)

A little thought would show that we could also write

$$\mathscr{R}_{12}(\tau) = \int_{-\infty}^{+\infty} s_1(t + \tau)s_2^*(t)\, dt$$

(8-19)

$$\mathscr{R}_{21}(\tau) = \int_{-\infty}^{+\infty} s_1^*(t)s_2(t + \tau)\, dt$$

(8-20)

From (8-17) and (8-20) it can be shown that

$$\mathcal{R}_{12}(\tau) = \mathcal{R}_{21}^*(-\tau) \tag{8-21}$$

which for real signals becomes

$$\mathcal{R}_{12}(\tau) = \mathcal{R}_{21}(-\tau) \tag{8-22}$$

An upper bound on the value of the correlation function can be established with the help of the *Schwarz inequality,* which states that, in general,

$$\left| \int_{-\infty}^{+\infty} f_1(x)f_2(x)\, dx \right|^2 \le \int_{-\infty}^{+\infty} |f_1(x)|^2\, dx \int_{-\infty}^{+\infty} |f_2(x)|^2\, dx \tag{8-23}$$

Applying this relationship to (8-17) gives

$$|\mathcal{R}_{12}(\tau)| = \left| \int_{-\infty}^{+\infty} s_1(t)s_2^*(t-\tau)\, dt \right| \le \left[\int_{-\infty}^{+\infty} |s_1(t)|^2\, dt \int_{-\infty}^{+\infty} |s_2(t)|^2\, dt \right]^{1/2} \tag{8-24}$$

This shows that the correlation function is bounded by the square root of the product of the signal energies. Dividing through by the right-hand side of (8-24) verifies that the normalized correlation function cannot exceed unity.

If $s_1(t) = s_2(t)$, in (8-17) we obtain the *autocorrelation function:*

$$\mathcal{R}_{11}(\tau) = \int_{-\infty}^{+\infty} s_1(t)s_1^*(t-\tau)\, dt = \int_{-\infty}^{+\infty} s_1(t+\tau)s_1^*(t)\, dt \tag{8-25}$$

By the Schwarz inequality, the autocorrelation function is equal to or less than its value with $\tau = 0$.

$$\mathcal{R}_{11}(\tau) \le \int_{-\infty}^{+\infty} |s_1(t)|^2\, dt = \mathcal{R}_{11}(0) \tag{8-26}$$

The autocorrelation function exhibits conjugate symmetry.

$$\mathcal{R}_{11}(\tau) = \mathcal{R}_{11}^*(-\tau) \tag{8-27}$$

If $s_1(t)$ is real, the autocorrelation function is, therefore, an even function of τ.

8.3.1 Fourier Transform of Correlation Functions

Consider two signals, $s_1(t)$ and $s_2(t)$, with Fourier transforms $S_1(f)$ and $S_2(f)$. By definition the Fourier transform of the cross-correlation function is

$$S_3(f) = \int_{-\infty}^{+\infty} \mathcal{R}_{12}(\tau)\, exp\, (-j2\pi f\tau)\, d\tau$$

$$= \int_{-\infty}^{+\infty} \left[\int_{-\infty}^{+\infty} s_1(t)s_2^*(t-\tau)\, dt \right] exp\, (-j2\pi f\tau)\, d\tau$$

Rearranging terms and interchanging the order of integration, we obtain

$$S_3(f) = \int_{-\infty}^{+\infty} s_1(t) \left[\int_{-\infty}^{+\infty} s_2^*(t - \tau) \exp(-j2\pi f\tau)\, d\tau \right] dt \qquad (8\text{-}28)$$

The inner integral is the Fourier transform of $s_2^*(t - \tau)$ from the τ-domain to the f-domain. Using the rules from Table 6-1, we find that

$$s_2^*(t - \tau) \longleftrightarrow S_2^*(f) \exp(-j2\pi ft) \qquad (8\text{-}29)$$

Substitution of (8-29) in (8-28) gives

$$S_3(f) = \left[\int_{-\infty}^{+\infty} s_1(t) \exp(-j2\pi ft)\, dt \right] S_2^*(f)$$

The remaining integral is obviously the transform of $s_1(t)$. Therefore,

$$\mathcal{R}_{12}(\tau) \longleftrightarrow S_3(f) = S_1(f)S_2^*(f) \qquad (8\text{-}30)$$

By a similar procedure, we may also write

$$\mathcal{R}_{21}(\tau) \longleftrightarrow S_1^*(f)S_2(f) \qquad (8\text{-}31)$$

$$\mathcal{R}_{11}(\tau) \longleftrightarrow |S_1(f)|^2 \qquad (8\text{-}32)$$

Equation (8-32) shows that the transform of the autocorrelation function is the energy spectrum of $s_1(t)$. The autocorrelation function spectrum is therefore real and positive. If $s_1(t)$ is real, the autocorrelation function spectrum is also an even function of frequency.

8.3.2 Relationship between Correlation and Convolution

The similarity between the correlation integral and the convolution integral defined in Chapter 6 should be apparent. The convolution of two functions is defined as

$$s_3(t) = s_1(t) \otimes s_2(t) = \int_{-\infty}^{+\infty} s_1(\tau)s_2(t - \tau)\, d\tau \qquad (8\text{-}33)$$

We may interchange the roles of t and τ in (8-33) without changing the form of the operation.

$$s_3(\tau) = s_1(\tau) \otimes s_2(\tau) = \int_{-\infty}^{+\infty} s_1(t)s_2(\tau - t)\, dt \qquad (8\text{-}34)$$

Now write the cross-correlation function for $s_1(t)$ and $s_2(t)$ as

$$\mathcal{R}_{12}(\tau) = \int_{-\infty}^{+\infty} s_1(t)s_2^*(t - \tau)\, dt$$

and define a new signal related to $s_2(t)$ as follows:

$$s_4(\tau - t) = s_2^*(t - \tau) \tag{8-35}$$

Substitution of (8-35) in the expression for $\mathcal{R}_{12}(\tau)$ gives

$$\mathcal{R}_{12}(\tau) = \int_{-\infty}^{+\infty} s_1(t) s_4(\tau - t) \, dt \tag{8-36}$$

Comparing (8-34) and (8-36) shows that $\mathcal{R}_{12}(\tau)$ is obtained by the convolution of $s_1(t)$ with $s_4(t)$.

$$\mathcal{R}_{12}(\tau) = s_1(\tau) \otimes s_4(\tau)$$

Letting $t = 0$ in (8-35), we see that

$$s_4(\tau) = s_2^*(-\tau) \tag{8-37}$$

Therefore,

$$\mathcal{R}_{12}(\tau) = s_1(\tau) \otimes s_2^*(-\tau) \tag{8-38}$$

The cross-correlation function is now identified as the convolution integral with the appropriate function replaced by its conjugate image.

8.3.3 Effect of Translation on the Correlation Function

With the convolution operation, translations are additive. To determine the effect of translations on the correlation function, let

$$\mathcal{R}_{12}(\tau) = s_1(\tau) \otimes s_2^*(-\tau)$$

and

$$s_3(t) = s_1(t - t_1) = s_1(t) \otimes \delta(t - t_1)$$
$$s_4(t) = s_2(t - t_2) = s_2(t) \otimes \delta(t - t_2)$$

Then

$$\mathcal{R}_{34}(\tau) = s_3(\tau) \otimes s_4^*(-\tau) = [s_1(\tau) \otimes \delta(\tau - t_1)] \otimes [s_2^*(-\tau) \otimes \delta(t_2 - \tau)]$$
$$= [s_1(\tau) \otimes s_2^*(-\tau)] \otimes [\delta(\tau - t_1) \otimes \delta(t_2 - \tau)]$$
$$= \mathcal{R}_{12}(\tau) \otimes \delta[\tau - (t_1 - t_2)]$$
$$= \mathcal{R}_{12}[\tau - (t_1 - t_2)]$$

Thus translations are *subtractive* in the correlation operation. If t_1 and t_2 are thought of as representing the *location* of the signals in time, the cross-correlation function contains information concerning the time separation of the two signals. This property is demonstrated in Figure 8-4, which shows the cross-correlation function of two rectangular pulses.

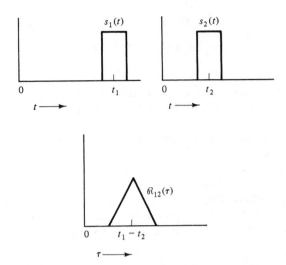

Figure 8-4 Cross-correlation of two rectangular pulses.

8.3.4 Frequency-Domain Correlation Functions

The correlation process is not restricted to time-domain functions. In signal analysis, we are frequently interested in the correlation of functions in the frequency domain. Thus if

$$s_1(t) \longleftrightarrow S_1(f)$$
$$s_2(t) \longleftrightarrow S_2(f)$$

then

$$\mathcal{K}_{12}(\nu) = \int_{-\infty}^{+\infty} S_1(f)S_2^*(f - \nu) \, df \qquad (8\text{-}39)$$

where

$\mathcal{K}_{12}(\nu) = $ cross-correlation function

$\nu = $ frequency variable of translation between $S_1(f)$ and $S_2(f)$

Also,

$$\mathcal{K}_{21}(\nu) = \int_{-\infty}^{+\infty} S_1^*(f - \nu)S_2(f) \, df \qquad (8\text{-}40)$$

and the spectral autocorrelation function is

$$\mathcal{K}_{11}(\nu) = \int_{-\infty}^{+\infty} S_1(f)S_1^*(f - \nu) \, df \qquad (8\text{-}41)$$

The inverse transform of the spectral correlation function can be found using the procedures outlined in Section 8.3.1. Thus

$$s_1(t)s_2^*(t) \longleftrightarrow \mathcal{K}_{12}(\nu)$$

$$s_1^*(t)s_2(t) \longleftrightarrow \mathcal{K}_{21}(\nu) \qquad (8\text{-}42)$$

$$|s_1(t)|^2 \longleftrightarrow \mathcal{K}_{11}(\nu)$$

If the time-domain functions are real, the corresponding frequency-domain correlation operations are identical to a convolution operation.

8.3.5 Correlation Functions for Power-Type Signals

So far, we have assumed that the integral of the squared magnitude of the functions being correlated is finite (energy-type functions). If the signals $s_1(t)$ and $s_2(t)$ contain finite average power but infinite total energy (power-type signals), the correlation function is defined as follows:

$$\begin{aligned} R_{12}(\tau) &= \lim_{T \to \infty} \frac{1}{T} \int_{-T/2}^{+T/2} s_1(t)s_2^*(t-\tau)\, dt \\[2mm] &= \langle s_1(t)s_2^*(t-\tau) \rangle \end{aligned} \qquad (8\text{-}43)$$

where the notation $\langle \cdot \rangle$ is used to denote the *time-average* integral operation.

8.3.6 Correlation Interval

Now assume that $s_1(t)$ has finite average power and that $s_2(t)$ has finite total energy. In this case, the integral of the product is finite. A useful correlation factor for this situation is called the *effective correlation interval*, defined as follows:

$$T_{12} = \frac{\left| \int_{-\infty}^{+\infty} s_1(t)s_2^*(t)\, dt \right|^2}{\langle |s_1(t)|^2 \rangle \int_{-\infty}^{+\infty} |s_2(t)|^2\, dt} \qquad (8\text{-}44)$$

For time-domain functions, T_{12} has units of seconds and represents the effective time interval over which the energy signal may be used to extract energy from the power signal.

As an example, let $s_1(t)$ have unit amplitude for all time and let $s_2(t)$ be one half-cycle of a cosine function, as shown in Figure 8-5. The average power in $s_1(t)$ is obviously unity.

The value of T_{12} for this example is

$$T_{12} = \frac{\left| \int_{-1/2}^{+1/2} \cos(\pi t)\, dt \right|^2}{(1)\int_{-1/2}^{+1/2} \cos^2(\pi t)\, dt} = \frac{8}{\pi^2} \qquad \text{sec}$$

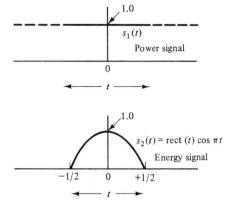

Figure 8-5 Examples of simple power signal and energy signal.

Notice that T_{12} is the product of c_{12} and c_{21}, as defined earlier, with $c_{12} = 4/\pi$ and $c_{21} = 2/\pi$. Now form the difference signal

$$s_d(t) = s_1(t) - c_{12}s_2(t) = 1.0 - \frac{4}{\pi} \cos (\pi t) \text{ rect } (t)$$

The energy *removed* from $s_1(t)$ by the extraction of the component $c_{12}s_2(t)$ is equal to:

$$\text{energy removed} = \int_{-\infty}^{+\infty} [s_1^2(t) - s_d^2(t)] \, dt$$

$$= \int_{-\infty}^{+\infty} 2c_{12}s_2(t) \, dt - \int_{-\infty}^{+\infty} c_{12}^2 s_2^2(t) \, dt$$

$$= \frac{8}{\pi} \int_{-1/2}^{+1/2} \cos (\pi t) \, dt - \frac{16}{\pi^2} \int_{-1/2}^{+1/2} \cos^2(\pi t) \, dt$$

$$= \frac{8}{\pi^2} \quad \text{v}^2\text{-sec}$$

For this example, the energy removed is the product of T_{12} and the average power in $s_1(t)$. That is,

$$\text{energy removed} = T_{12}\langle s_1^2(t) \rangle$$

In Figure 8-6, the squared-difference signal for this example is shown, together with a rectangular waveform of width T_{12} and amplitude equal to the root-mean-squared value of $s_1(t)$. Notice that the rectangular waveform is *equivalent* to $c_{12}s_2(t)$ in the sense that the same amount of energy is removed from $s_1(t)$ when the difference signal is formed. The factor T_{12} may be thought of as the *equivalent duration* of $s_2(t)$.

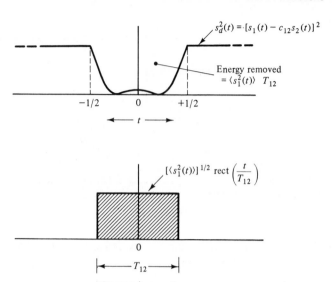

Figure 8-6 Use of correlation interval to define an equivalent signal duration.

8.4 RESOLUTION AND SIGNAL AMBIGUITY FUNCTIONS

The correlation function arises quite naturally in the performance analysis of acoustic systems [3]. Consider an active acoustic system used to determine the location of remote objects under water. Such a system typically generates a limited-duration acoustic waveform that radiates from the source. Objects to be detected present an impedance discontinuity in the water, causing a portion of the incident acoustic energy to be reflected back toward the source. Assuming that the reflecting objects can be considered as point reflectors, and are not moving relative to the source, the received signal at the source location is a delayed replica of the transmitted waveform, reduced in amplitude by the round-trip transmission loss and the reflecting characteristics of the target object. The envelope of a typical transmitted waveform, together with the received wave-forms from two point targets, is shown in Figure 8-7(a).

The design of the transmitted waveform and the receiving system is selected to enhance the possibility of recognizing that two targets are present in this example, and to measure accurately the range (or time location) of each target. In the absence of noise, these are trivial problems, and the target locations may be determined, in principle, with unlimited accuracy. The presence of noise complicates the total received waveform so that the actual shape and location of the reflected target signals are uncertain. The rigorous analysis of signals in the presence of noise is reserved for a later chapter. In this chapter, a heuristic

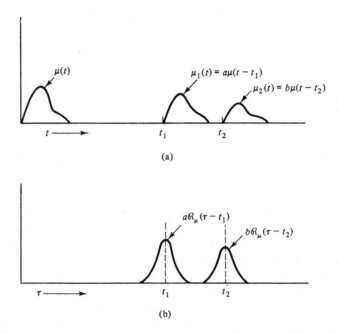

Figure 8-7 Transmitted and received waveforms, and cross-correlation of transmitted and received waveforms.

approach is used to identify the waveform characteristics that are important in optimizing system performance.

Intuitively, a reasonable approach to determining the presence and location of reflected target signals is to shift a replica of the transmitted waveform to various time locations, and determine the locations resulting in the minimum value for the *integrated squared difference* between this shifted reference waveform and the received signal. Thus let the received signal be

$$\mu_r(t) = \mu_1(t) + \mu_2(t)$$
$$= a\mu(t - t_1) + b\mu(t - t_2)$$

and form

$$\varepsilon^2(\tau) = \int_{-\infty}^{+\infty} |\mu(t - \tau) - \mu_r(t)|^2 \, dt$$

$$= \int_{-\infty}^{+\infty} |\mu(t - \tau)|^2 \, dt + \int_{-\infty}^{+\infty} |\mu_r(t)|^2 \, dt - 2 \operatorname{Re}\left[\int_{-\infty}^{+\infty} \mu_r(t)\mu^*(t - \tau)\right] dt$$

The first two terms represent the energy in the transmitted and received waveforms, and may be considered constant for our purposes. The final term is twice the real part of the cross-correlation function for the transmitted and received signal. Let W be the value of the constant terms and write

$$
\begin{aligned}
\varepsilon^2(\tau) &= W - 2\,\mathrm{Re}\left[a\int_{-\infty}^{+\infty} \mu(t - t_1)\mu^*(t - \tau)\,dt \right. \\
&\qquad\qquad\qquad\left. + b\int_{-\infty}^{+\infty} \mu(t - t_2)\mu^*(t - \tau)\,dt \right] \quad (8\text{-}45) \\
&= W - 2\,\mathrm{Re}\left[a\mathcal{R}_\mu(\tau - t_1) + b\mathcal{R}_\mu(\tau - t_2) \right]
\end{aligned}
$$

where $\mathcal{R}_\mu(\tau - t_n)$ is the autocorrelation function of $\mu(t)$ shifted to the location $\tau = t_n$.

The magnitude of $\varepsilon^2(\tau)$ is minimized at locations t_1 and t_2, where the shifted correlation functions reach their peak values. The correlation functions are shown in Figure 8-7(b). Intuitively, the accuracy with which t_1 or t_2 can be determined is inversely related to the *width* of the autocorrelation function of $\mu(t)$. Similarly, the ability to recognize the presence of two (or more) target signals is enhanced if the width of the correlation function is small compared with the separation in time between the targets. It may seem surprising that the ability to resolve two or more targets, or to determine range accurately, is determined by the width of the autocorrelation function of the transmitted waveform, rather than by the width of the waveform itself. For very simple waveforms, the waveform duration and the autocorrelation function duration are directly related. However, in general, it is possible for the signal duration to be much greater than the duration of the autocorrelation function.

8.4.1 Effective Bandwidth: Range Resolution

The ultimate shape for the autocorrelation function, to enhance performance in the areas of range measurement and target resolution, would be the impulse. As a measure of the waveform quality for these purposes, we may use the correlation interval, discussed in Section 8.3.6, to determine how "impulse-like" the autocorrelation function is. Equivalently, we may compare the *spectrum* of the autocorrelation function with the *spectrum* of an impulse. The result is a factor with units of $\sec^{-1}$, or bandwidth, and can be interpreted as the *equivalent* or *effective bandwidth* β_e of the waveform. This comparison is normally made using the complex envelope function for the waveform because the carrier frequency is not directly involved in the resolution process. For example, let the envelope function and its Fourier transform be

$$
\mu(t) \longleftrightarrow M(f)
$$

from which

$$
\mathcal{R}_\mu(\tau) \longleftrightarrow |M(f)|^2
$$

For the impulse, we have

$$\delta(\tau) \longleftrightarrow 1.0$$

The correlation interval, or effective bandwidth, for the frequency-domain functions, using the form in (8-44), is

$$\beta_e = \frac{\left[\int_{-\infty}^{+\infty}(1.0)|M(f)|^2\,df\right]^2}{\left[\lim_{B\to\infty}(1/B)\int_{-B/2}^{+B/2}(1.0)\,df\right]\int_{-\infty}^{+\infty}|M(f)|^4\,df} \tag{8-46}$$

Because the bracketed integral in the denominator in (8-46) has a value of unity, this becomes

$$\beta_e = \frac{\left[\int_{-\infty}^{+\infty}|M(f)|^2\,df\right]^2}{\int_{-\infty}^{+\infty}|M(f)|^4\,df} \tag{8-47}$$

An alternative expression for the effective bandwidth may be obtained using Parseval's theorem. Thus

$$\mathcal{R}_\mu(0) = \int_{-\infty}^{+\infty}|\mu(t)|^2\,dt = \int_{-\infty}^{+\infty}|M(f)|^2\,df \tag{8-48}$$

and

$$\int_{-\infty}^{+\infty}|\mathcal{R}_\mu(\tau)|^2\,d\tau = \int_{-\infty}^{+\infty}|M(f)|^4\,df \tag{8-49}$$

Substitution of (8-48) and (8-49) in (8-47) gives

$$\beta_e = \frac{\mathcal{R}_\mu^2(0)}{\int_{-\infty}^{+\infty}|\mathcal{R}_\mu(\tau)|^2\,d\tau} \tag{8-50}$$

The effective bandwidth defined in (8-47) and (8-50) increases as the autocorrelation function becomes more impulse-like.

As an example, assume that $\mu(t)$ is a sinc function, in which case the autocorrelation function is also a sinc function and the spectrum is rectangular in shape. That is, let

$$\mu(t) = \text{sinc}\left(\frac{t}{t_p}\right) \longleftrightarrow t_p\,\text{rect}\,(ft_p)$$

from which

$$\mathcal{R}_\mu(\tau) = t_p\,\text{sinc}\left(\frac{\tau}{t_p}\right) \longleftrightarrow t_p^2\,\text{rect}\,(ft_p)$$

The effective bandwidth, using (8-47), is

$$\beta_e = \frac{\left[t_p^2 \int_{-1/2t_p}^{+1/2t_p} df \right]^2}{t_p^4 \int_{-1/2t_p}^{+1/2t_p} df} = \frac{t_p^2}{t_p^3} = \frac{1}{t_p} \tag{8-51}$$

This gives the intuitively satisfying result that for a rectangular spectrum the effective bandwidth equals the width of the frequency-domain rectangle.

Now assume that

$$\mu(t) = \text{rect}\left(\frac{t}{t_p}\right) \longleftrightarrow t_p \, \text{sinc}\,(ft_p)$$

and

$$\mathcal{R}_\mu(\tau) = t_p\left(1 - \frac{|\tau|}{t_p}\right)\text{rect}\left(\frac{t}{2t_p}\right) \longleftrightarrow t_p^2 \, \text{sinc}^2(ft_p)$$

Using (8-47) or (8-50) the effective bandwidth is found to be

$$\beta_e = \frac{3}{2t_p} \qquad \text{(for rectangular pulse)} \tag{8-52}$$

For the simple sinc pulse and rectangular envelope functions, we see from (8-51) and (8-52) that the effective bandwidth is inversely proportional to the pulse width. Now consider a much more complicated envelope of the form

$$\mu(t) = \text{rect}\left(\frac{t - t_p/2}{t_p}\right)\exp(jkt^2) \tag{8-53}$$

This envelope is a complex function of time with a quadratic phase characteristic, kt^2. Defining frequency in terms of the rate of change of phase, we may identify the *instantaneous frequency* as

$$f_i = \frac{1}{2\pi}\frac{d(kt^2)}{dt} = \frac{kt}{\pi} \tag{8-54}$$

The instantaneous frequency is thus a linear function of time over the duration of the pulse. When applied to a high-frequency carrier, the envelope function in (8-53) results in a *linear frequency-modulated* (FM) *pulse*, also called a *chirp pulse*.

The maximum instantaneous frequency, f_m, is reached when $t = t_p$ in (8-54). Then

$$f_m = \frac{kt_p}{\pi} \tag{8-55}$$

If f_m is large compared with $1/t_p$, the effective bandwidth of the linear-FM pulse

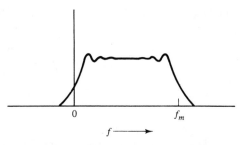

Figure 8-8 Spectrum of a linear-FM
envelope function.

is determined by the frequency modulation rather than by the shape and duration
of the rectangular function. For $f_m t_p \gg 1.0$, the shape of the spectrum of the
chirp envelope function is nearly rectangular, as shown in Figure 8-8. This
spectrum cannot be expressed in closed form, and must be computed using tables
of Fresnel integrals [7]. The effective bandwidth of this waveform approaches
f_m for large $f_m t_p$.

For large $f_m t_p$, the autocorrelation function for the chirp waveform is
approximately a sinc function with a width inversely proportional to f_m. Thus,

$$\mathcal{R}_\mu(\tau) \approx \text{sinc } (\tau f_m) \qquad (8\text{-}56)$$

The width of the autocorrelation in this case is significantly less than the time
duration of the original envelope function. It is reasonable to assume that
stationary targets may be separately identified with this waveform provided that
their separation is at least $1/f_m$.

8.4.2 Effective Time Duration: Velocity Resolution

Assume that the targets are in motion relative to the source. As a result of
the Doppler effect, the observed frequency of the signal reflected from a target
is affected by this relative motion, and by the location of the receiver. For
instance, for a target moving away from the source with a speed v, the signal
arriving at the target location will have a frequency

$$f_1 = f_0\left(1 - \frac{v}{c}\right) \qquad (8\text{-}57)$$

where f_0 = transmitted frequency from a stationary source
v = target radial speed away from source
c = sound speed in stationary medium

The signal reflected back to the source from the moving target undergoes an
additional Doppler shift, resulting in a received frequency at the source location
given by

$$f_2 = f_0\left(\frac{1 - v/c}{1 + v/c}\right) \qquad (8\text{-}58)$$

The Doppler shift, given by the difference between the received and transmitted frequencies, is

$$f_d = f_2 - f_0 = -\frac{2vf_0}{c}\left(1 + \frac{v}{c}\right)^{-1}$$

$$= -\frac{2v}{\lambda_0}\left(1 + \frac{v}{c}\right)^{-1}$$

(8-59)

If great accuracy is not required in calculating the Doppler frequency shift, it is convenient to simplify (8-59) by recognizing that v/c is typically much less than unity, so that

$$f_d \simeq \frac{-2v}{\lambda_0}$$

(8-60)

In the remainder of this discussion it is assumed that this approximation is permissible. Note that, as defined here, the Doppler frequency shift is negative for an opening target ($v > 0$), and positive for a closing target ($v < 0$).

Let the transmitted signal be of the form

$$s(t) = \mu(t) \exp (j2\pi f_0 t)$$

Assume that the transmitted signal is reflected by two targets that are in motion relative to the source. The target signals received at the source location will be

$$s_r(t) = a\mu(t - t_1) \exp \left[j2\pi(f_0 + f_{d_1})(t - t_1)\right]$$

$$+ b\mu(t - t_2) \exp \left[j2\pi(f_0 + f_{d_2})(t - t_2)\right]$$

(8-61)

where t_1, t_2 = time locations of targets 1 and 2

f_{d_1}, f_{d_2} = Doppler frequency shifts for targets 1 and 2

If the time difference, $t_2 - t_1$, is less than the reciprocal of the effective envelope bandwidth, we learned in the preceding section that it would be difficult to resolve the two targets based on time-domain cross-correlation with the transmitted waveform. As an extreme example, assume that $t_1 = t_2$. With no loss in generality, t_1 and t_2 can be set equal to zero, resulting in a received envelope function as follows:

$$\mu_r(t) = a\mu(t) \exp (j2\pi f_{d_1} t) + b\mu(t) \exp (j2\pi f_{d_2} t)$$

(8-62)

The two targets may still be resolved using the Doppler separation, $f_{d_2} - f_{d_1}$, by forming the frequency-domain cross-correlation function of the transmitted envelope spectrum with the received envelope spectrum. Thus let

$$\mu(t) \longleftrightarrow M(f)$$

$$\mu_r(t) \longleftrightarrow aM(f - f_{d_1}) + bM(f - f_{d_2})$$

The frequency-domain cross-correlation function is

$$\mathcal{K}_{\mu_r\mu}(\nu) = a\mathcal{K}_\mu(\nu - f_{d_1}) + b\mathcal{K}_\mu(\nu - f_{d_2}) \qquad (8\text{-}63)$$

where

$$\mathcal{K}_\mu(\nu) = \int_{-\infty}^{+\infty} M(f)M^*(f - \nu)\, df$$

The magnitudes of the received spectrum and the cross-correlation function are shown in Figure 8-9. In this figure, f_{d_1} and f_{d_2} are assumed positive, indicating that the targets are closing on the source. The ability to resolve the two targets based on this Doppler separation is inversely related to the width of the frequency-domain correlation function. The more "impulse-like" the correlation function is, the closer together in frequency may the targets be and still be resolved.

The correlation interval may be used to compare the spectral correlation function with a frequency-domain impulse. Alternatively, the transform of $\mathcal{K}_\mu(\nu)$ may be compared with the transform of $\delta(f)$. Thus

$$1.0 \longleftrightarrow \delta(f)$$

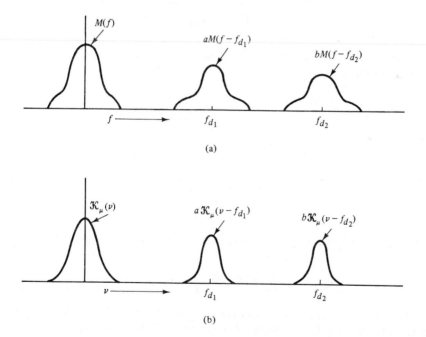

(a)

(b)

Figure 8-9   (a) Received spectrum; (b) cross-correlation of received and transmitted spectra, assuming closing targets.

and

$$|\mu(t)|^2 \longleftrightarrow M(\nu) \otimes M^*(-\nu) = \mathcal{H}_\mu(\nu)$$

from which

$$\tau_e = \frac{\left[\int_{-\infty}^{+\infty} (1.0)|\mu(t)|^2 \, dt\right]^2}{\left[\lim_{T\to\infty} (1/T) \int_{-T/2}^{+T/2} (1.0)^2 \, dt\right] \int_{-\infty}^{+\infty} |\mu(t)|^4 \, dt}$$

$$= \frac{\left[\int_{-\infty}^{+\infty} |\mu(t)|^2 \, dt\right]^2}{\int_{-\infty}^{+\infty} |\mu(t)|^4 \, dt} \qquad (8\text{-}64)$$

Equation (8-64) defines the *effective time duration* τ_e of the envelope function. The larger the value of τ_e, the more impulse-like will be $\mathcal{H}_\mu(\nu)$, and the greater will be the frequency-domain resolving capability of the waveform. Using Parseval's theorem, an equivalent expression for τ_e is

$$\tau_e = \frac{\mathcal{H}_\mu^2(0)}{\int_{-\infty}^{+\infty} |\mathcal{H}_\mu(\nu)|^2 \, d\nu} \qquad (8\text{-}65)$$

For the rectangular envelope function, the effective time duration is equal to the pulse width, t_p. Thus if

$$\mu(t) = \text{rect}\left(\frac{t}{t_p}\right)$$

then

$$\tau_e = \frac{\left[\int_{-t_p/2}^{+t_p/2} dt\right]^2}{\int_{-t_p/2}^{+t_p/2} dt} = \frac{t_p^2}{t_p} = t_p \qquad (8\text{-}66)$$

This is also true for the chirp waveform with a rectangular envelope since

$$\left|\text{rect}\left(\frac{t}{t_p}\right) \exp (jkt^2)\right| = \text{rect}\left(\frac{t}{t_p}\right)$$

For the chirp waveform, the product of effective bandwidth and effective time duration is typically large compared with unity. That is, because

$$\beta_e \simeq f_m \gg \frac{1}{t_p}$$

and

$$\tau_e = t_p$$

$$\beta_e \tau_e \simeq f_m t_p \gg 1.0 \qquad \text{for chirp waveform} \qquad (8\text{-}67)$$

8.4.3 Simultaneous Range and Velocity Resolution

Consider two received target signals separated in time by τ, and in Doppler frequency by an amount ν. The combined received envelope function is

$$\mu_r(t) = \mu(t) \exp(-j2\pi\nu t) + \mu(t - \tau)$$

where the amplitudes of the two received signals are assumed equal for simplicity. The envelope function, $\mu(t)$, should be selected to maximize the difference between the received target signals over as wide a range of τ and ν as possible. Proceeding as before, we form the integrated squared difference between the two received signals and attempt to maximize this quantity through proper waveform design. This leads to the cross-correlation function for the received signals, which can be expressed as follows:

$$\chi(\tau, \nu) = \int_{-\infty}^{+\infty} \mu(t)\mu^*(t - \tau) \exp(-j2\pi\nu t) \, dt \qquad (8\text{-}68)$$

Notice that in this case the correlation is a function of the two variables τ and ν. Intuitively, the resolution capability is improved by making this two-dimensional correlation function impulse-like in both the τ and ν domains. The combined resolution factor, by analogy with (8-50) and (8-65), is

$$\text{combined resolution factor} = \frac{\chi^2(0, 0)}{\displaystyle\iint_{-\infty}^{+\infty} |\chi(\tau, \nu)|^2 \, d\tau \, d\nu} \qquad (8\text{-}69)$$

This factor is also called the *signal ambiguity factor,* and $|\chi(\tau, \nu)|^2$ is the *signal ambiguity function.* This function describes the total ambiguity in resolving, or locating, targets accurately in range (time delay) and velocity (Doppler frequency).

In Section 8.4.1 we found that, with known target velocity, the range resolution capability is improved by minimizing the area under the squared magnitude of the time-domain envelope autocorrelation function (maximize effective bandwidth). Similarly, with known target range, in Section 8.4.2 the velocity resolution capability was shown to be inversely related to the area of the squared magnitude of the frequency-domain correlation function. By logical extension, with both range and velocity unknown, the ability to simultaneously resolve, or locate, targets in range and velocity is inversely proportional to the *volume* under the signal ambiguity function. However, we shall now show that this volume is invariant with respect to waveform design, and is actually equal to the square of the waveform energy, $|\chi(0, 0)|^2$. We first write the signal ambiguity function as a double integral. Thus

$$|\chi(\tau, \nu)|^2 = \int\!\!\int_{-\infty}^{+\infty} \mu(t)\mu^*(t - \tau)\mu^*(t')\mu(t' - \tau) \exp\left[-j2\pi\nu(t - t')\right] dt\, dt'$$

from which, with some rearranging, the integral over τ is

$$\int_{-\infty}^{+\infty} |\chi(\tau, \nu)|^2\, d\tau = \left\{ \int\!\!\int_{-\infty}^{+\infty} \mu(t)\mu^*(t') \left[\int_{-\infty}^{+\infty} \mu(t' - \tau)\mu^*(t - \tau)\, d\tau \right] \right.$$

$$\left. \times \exp\left[-j2\pi\nu(t - t')\right] dt\, dt' \right\} \qquad (8\text{-}70)$$

The bracketed term in (8-70) is $\mathcal{R}_\mu^*(t - t')$. Let

$$t' = t - T$$

$$dt' = -dT$$

and write

$$\int_{-\infty}^{+\infty} |\chi(\tau, \nu)|^2\, d\tau = \int\!\!\int_{-\infty}^{+\infty} \mu(t)\mu^*(t - T)\mathcal{R}_\mu^*(T) \exp\left(-j2\pi\nu T\right) dt\, dT$$

Integration with respect to t gives

$$\int_{-\infty}^{+\infty} |\chi(\tau, \nu)|^2\, d\tau = \int_{-\infty}^{+\infty} |\mathcal{R}_\mu(T)|^2 \exp\left(-j2\pi\nu T\right) dT \qquad (8\text{-}71)$$

Now form the integral of (8-71) over ν to obtain the final result.

$$\int\!\!\int_{-\infty}^{+\infty} |\chi(\tau, \nu)|^2\, d\tau\, d\nu = \int_{-\infty}^{+\infty} |\mathcal{R}_\mu(T)|^2 \left[\int_{-\infty}^{+\infty} \exp\left(-j2\pi\nu T\right) d\nu \right] dT$$

$$= \int_{-\infty}^{+\infty} |\mathcal{R}_\mu(T)|^2\, \delta(T)\, dT = \mathcal{R}_\mu^2(0) = |\chi(0, 0)|^2 \qquad (8\text{-}72)$$

Substitution of (8-72) in (8-69) produces the interesting result that the total signal ambiguity is unity, independent of waveform design. That is,

$$\frac{\chi^2(0, 0)}{\displaystyle\int\!\!\int_{-\infty}^{+\infty} |\chi(\tau, \nu)|^2\, d\tau\, d\nu} = 1.0 \qquad (8\text{-}73)$$

The significance of this result is that if the waveform is modified to improve the resolution in the range domain, the velocity resolution capability is inevitably degraded. Similarly, improving the velocity resolution capability degrades the range resolution.

8.4.4 Gaussian Signal Ambiguity Function

A mathematically convenient waveform to demonstrate the signal ambiguity concept is the Gaussian pulse. We choose the peak amplitude to give unity signal energy and write

$$\mu(t) = \frac{1}{\sigma^{1/2}\pi^{1/4}} \exp\left(\frac{-t^2}{2\sigma^2}\right) \longleftrightarrow \sigma^{1/2}\pi^{1/4}\sqrt{2}\,\exp\left(-2\pi^2 f^2 \sigma^2\right)$$

The two-dimensional autocorrelation function for this waveform is

$$\chi_\mu(\tau, \nu) = \exp\left[-\frac{\tau^2}{4\sigma^2} + j\pi\nu\tau - (\pi\sigma)^2\nu^2\right] \tag{8-74}$$

The time-delay autocorrelation function results if $\nu = 0$. Thus

$$\mathcal{R}_\mu(\tau) = \chi_\mu(\tau, 0) = \exp\left(\frac{-\tau^2}{4\sigma^2}\right) \tag{8-75}$$

Similarly, the Doppler-domain autocorrelation function is

$$\mathcal{K}_\mu(\nu) = \chi(0, \nu) = \exp\left[-(\pi\sigma\nu)^2\right] \tag{8-76}$$

The effective bandwidth and time duration for the Gaussian pulse are

$$\beta_e = \frac{1}{\sigma\sqrt{2\pi}}$$

$$\tau_e = \sigma\sqrt{2\pi} \tag{8-77}$$

The signal ambiguity function is the squared magnitude of (8-74). Thus

$$|\chi(\tau, \nu)|^2 = \exp\left[-\left(\frac{\tau^2}{2\sigma^2}\right) - 2(\pi\sigma\nu)^2\right] \tag{8-78}$$

Now substitute the relationships in (8-77) into (8-78) and obtain

$$|\chi(\tau, \nu)|^2 = \exp\left[-\pi(\beta_e^2\tau^2 + \tau_e^2\nu^2)\right] \tag{8-79}$$

A three-dimensional plot of the Gaussian ambiguity function is shown in Figure 8-10. For any fixed value of ν, this ambiguity function has a Gaussian shape when plotted as a function of τ. Similarly, for any fixed value of τ, it is a Gaussian function of ν. The area under the intersection of the function with a vertical plane at $\nu = 0$ is inversely related to the effective envelope bandwidth. Similarly, the cross-sectional area along the ν-axis with $\tau = 0$ is inversely related to the effective time duration.

The crosshatched area in Figure 8-10 is the intersection of the ambiguity function with a plane parallel to the τ-ν plane. This intersection can be shown to have an elliptical shape as follows. Let the intersection be defined by

$$|\chi(\tau, \nu)|^2 = K < |\chi(0, 0)|^2 \tag{8-80}$$

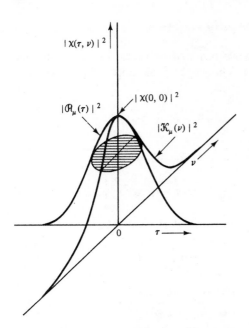

Figure 8-10 Three-dimensional Gaussian signal ambiguity diagram.

Now substitute (8-79) for $|\chi(\tau, \nu)|^2$, and take the natural logarithm of both sides.

$$\pi(\beta_e^2 \tau^2 + \tau_e^2 \nu^2) = -\ln K$$

or

$$\beta_e^2 \tau^2 + \tau_e^2 \nu^2 = -\frac{1}{\pi}\ln K \qquad (8\text{-}81)$$

The constant K may be chosen so that the right side of (8-81) is unity. This results in the equation for an ellipse with the semimajor and semiminor axes equal to $1/\beta_e$ and $1/\tau_e$. This ellipse is plotted in Figure 8-11 as a two-dimensional figure, portraying the pertinent information contained in the more complicated three-dimensional ambiguity diagram.

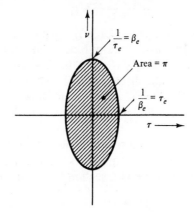

Figure 8-11 Two dimensional ambiguity diagram for a Gaussian pulse.

The width of the ambiguity ellipse in the τ-direction is a measure of the range resolution capability of the waveform, and the width in the ν-direction is related to the velocity resolution. For the Gaussian waveform, we note from (8-76) and (8-77) that the product $\beta_e \tau_e$ is unity. Furthermore, the area of the ellipse is given by:

$$\text{area of ambiguity ellipse} = \frac{\pi}{\beta_e \tau_e}$$

which is a constant. This is in keeping with the behavior of the three-dimensional ambiguity function, in that any attempt to improve the waveform capability in one domain results in an offsetting degradation in the other domain.

8.4.5 Gaussian Waveform with Linear Frequency Modulation

The product of effective bandwidth and effective duration is not in general equal to unity. However, a large waveform time–bandwidth product does not result in the ability to improve range and velocity resolution simultaneously, assuming that both target ranges and velocities are unknown and unbounded. If, on the other hand, either the range or velocity is known (or bounded within sufficiently close limits) for each target, the resolution in the other domain may be improved without limit, as we shall now see.

Assume a Gaussian waveform with linear frequency modulation, as follows:

$$\mu(t) = \frac{1}{\pi^{1/4} \sigma^{1/2}} \exp\left[-\left(\frac{t^2}{2\sigma^2} \right) - jkt^2 \right] \tag{8-82}$$

The effective duration of this waveform is not affected by the frequency modulation, and is therefore the same as for the simple Gaussian pulse. Thus

$$\tau_e = \sigma \sqrt{2\pi}$$

The effective bandwidth is dominated by the frequency-modulation term and can be put in the form

$$\beta_e = \frac{(1 + k^2 \tau_e^4 / \pi^2)^{1/2}}{\tau_e} \tag{8-83}$$

from which

$$\beta_e \tau_e = \left(\frac{1 + k^2 \tau_e^4}{\pi^2} \right)^{1/2} \tag{8-84}$$

The time–bandwidth product is proportional to k and to τ_e^2, and can be made arbitrarily large.

The signal ambiguity function for this waveform is

$$|\chi(\tau, \nu)|^2 = \exp\left[-\pi\left(\beta_e^2\tau^2 - \frac{2k}{\pi}\tau_e^2\nu\tau + \tau_e^2\nu^2\right)\right] \quad (8\text{-}85)$$

A cross section of this function, parallel to the τ-ν plane, again has an elliptical shape, obtained by setting the exponent in (8-85) equal to an appropriate constant. For instance, let

$$\beta_e^2\tau_e^2 - \frac{2k}{\pi}\tau_e^2\nu\tau + \tau_e^2\nu^2 = 1 \quad (8\text{-}86)$$

As a result of the frequency modulation, the principal axes of the ellipse in (8-86) are not parallel to the τ and ν axes, as they were with the simple Gaussian waveform. The two-dimensional ambiguity ellipse for the frequency-modulated waveform is shown in Figure 8-12. On this same figure, the ambiguity ellipse with $k = 0$ and with the same effective time duration is shown as the dashed curve. As k increases from zero, a little thought will show that the area of the ellipse remains constant.

With the waveform defined by (8-82), assume a target with zero relative velocity at a range corresponding to a time t_1. Correlation in the time domain with the transmitted envelope function results in the function

$$\mathcal{R}_\mu(\tau - t_1) = \chi[(\tau - t_1), 0]$$

The squared value of this function is the intersection of the signal ambiguity function (centered at t_1, 0) with a vertical plane through the τ axis. A second target, also with zero velocity, located at time (range) t_2, results in a correlation

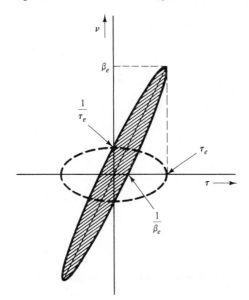

Figure 8-12 Two-dimensional ambiguity ellipse for a Gaussian pulse with frequency modulation.

function centered at t_2. The resulting correlation functions for these two targets are shown in Figure 8-13(a). These targets are easily resolved provided that $t_2 - t_1$ exceeds the inverse of the effective signal bandwidth.

Now assume that the first target relative velocity is not zero, resulting in a Doppler frequency shift ν_1. Time-domain correlation with the transmitted waveform is no longer centered at t_1, but is shifted in time by an amount proportional to the Doppler shift and to the frequency modulation constant, k. If the target velocity is unknown, the actual time location of the target is uncertain within the range $\pm\tau_e$ of the actual target location. Therefore, the two targets may not be resolved in range, without ambiguity, unless their separation exceeds the effective time duration. Figure 8-13(b) demonstrates the apparent shift in time location for the first target and the resulting possible overlap with the second target.

In a similar manner, targets at known ranges may be resolved in Doppler frequency provided that they are separated in frequency by at least $1/\tau_e$. If, however, the target ranges are unknown, the location of the correlation peaks in the frequency domain cannot unambiguously be associated with the true target locations in the frequency domain. Target frequency separations on the order of β_e are then required to guarantee resolution.

With these qualifications, it may appear that the large time–bandwidth

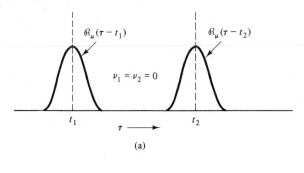

(a)

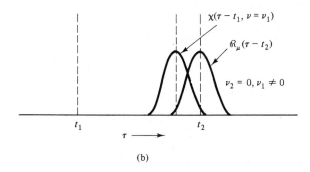

(b)

Figure 8-13   Target time-domain correlation functions with (a) equal target velocities; (b) unequal target velocities.

product obtained with the frequency-modulated waveform has no practical value. However, although target velocities and ranges may not be known a priori, they are seldom unbounded. Thus target velocities under water are generally small. Provided that the resulting Doppler shift is not large compared with $1/\tau_e$, resolution in range consistent with the inverse of the effective bandwidth may be approached.

For non-Gaussian waveforms, the total ambiguity, as defined by (8-69), is still unity. However, the intersection of the signal ambiguity function with a horizontal plane does not in general result in an ellipse as it does with the Gaussian waveform. Because of the simplicity of the two-dimensional elliptical ambiguity diagram, it is often used as a schematic representation of the ambiguity distribution in the τ-ν plane even for non-Gaussian envelope functions. For this purpose, the elliptical function in (8-86) is put in the following form, with the help of (8-84):

$$\beta_e^2 \tau^2 - 2(\beta_e^2 \tau_e^2 - 1)^{1/2} \tau \nu + \tau_e^2 \nu^2 = 1 \qquad (8\text{-}87)$$

This equation is rigorously valid only for the Gaussian envelope. However, using the actual effective bandwidth and duration for other envelope shapes, the ellipse of (8-87) demonstrates the pertinent behavior of the more complicated three-dimensional ambiguity function. The area is constant, and the τ and ν axes intersections are $1/\beta_e$ and $1/\tau_e$, respectively.

PROBLEMS

8.1. Determine the normalized correlation coefficient for the two waveforms

$$s_1(t) = \text{rect}\left(\frac{t}{t_{p1}}\right)$$

$$s_2(t) = \text{rect}\left(\frac{t}{t_{p2}}\right) \cos\left(\frac{\pi t}{t_{p2}}\right), \qquad t_{p2} \geq t_{p1}$$

Determine the relationship between t_{p2} and t_{p1} that maximizes the normalized correlation coefficient. Calculate the maximum possible correlation coefficient.

8.2. The cross-correlation function of $s_1(t)$ and $s_2(t)$ is defined as

$$\mathfrak{R}_{12}(\tau) = \int_{-\infty}^{+\infty} s_1(t) s_2^*(t - \tau)\, dt$$

Show that

$$\int_{-\infty}^{+\infty} \mathfrak{R}_{12}(\tau)\, d\tau = \int_{-\infty}^{+\infty} s_1(t)\, dt \int_{-\infty}^{+\infty} s_2^*(t)\, dt$$

8.3. Let $s(t)$ be a real function with a Fourier transform $S(f)$. If $\mathfrak{R}(\tau)$ is the auto-correlation function of $s(t)$, show that

$$\int_{-\infty}^{+\infty} [\mathfrak{R}'(\tau)]^2\, d\tau = (2\pi)^2 \int_{-\infty}^{+\infty} f^2 |S(f)|^4\, df$$

8.4. Using either (8-47) or (8-50), verify that the effective bandwidth of the Gaussian waveform $s(t) = \exp(-t^2/2\sigma^2)$ is $1/\sigma\sqrt{2\pi}$.

8.5. Let the Hilbert transform of $x(t)$ be designated $\hat{x}(t)$ and form the analytic signal $z(t)$ as follows:

$$z(t) = x(t) + j\hat{x}(t)$$

Show that

$$\mathcal{R}_z(\tau) = 2[\mathcal{R}_x(\tau) + j\hat{\mathcal{R}}(\tau)]$$

where $\hat{\mathcal{R}}_x(\tau)$ is the Hilbert transform of the autocorrelation function of $x(t)$.

8.6. Sketch the three-dimensional signal ambiguity function for a signal with a rectangular envelope function.

SUGGESTED READING

1. Mason, S. J., and H. J. Zimmerman, *Electronic Circuits, Signals, and Systems*. New York: John Wiley & Sons, Inc., 1960, Chap. 6.

2. Burdic, W. S., *Radar Signal Analysis*. Englewood Cliffs, N.J.: Prentice-Hall, Inc., 1968, Chaps. 3 and 5.

3. Woodward, P. M., *Probability and Information Theory with Applications to Radar*. Elmsford, N.Y.: Pergamon Press, Inc., 1955.

4. Helstrom, C. W., *Statistical Theory of Target Detection*. Elmsford, N.Y.: Pergamon Press, Inc., 1960.

5. Westerfield, E. C., R. H. Prager, and J. L. Stewart, "Processing Gains against Reverberation (Clutter) Using Matched Filters," *IRE Trans. Inf. Theory*, Vol. IT-6, pp. 342–348 (June 1960).

6. Rihaczek, A. W., *Principles of High-Resolution Radar*, New York: McGraw-Hill Book Company, 1969.

7. Abramowitz, M. and Stegun, I. A., *Handbook of Mathematical Functions*, Nat. Bureau of Standards, Applied Math Series (55), Wash., DC: U.S. Govt. Printing Office, 1965.

9

Random Processes

9.0 INTRODUCTION

System, signal, and environmental parameters required in the analysis of underwater acoustic systems are seldom known with great accuracy and may involve a great deal of uncertainty. In Chapter 5 we considered the complicated interaction of the ocean surface, ocean bottom, and the body of the ocean in determining the acoustic transmission loss between two points. An accurate prediction of transmission loss would require the detailed measurement of all pertinent physical characteristics of the medium as a function of range and time between the two points. This is not practical in most cases, so we must be satisfied with *average* characteristics of the medium. Acoustic signals of interest are often noise-like in character, and the ambient noise background in the ocean obviously results from nondeterministic phenomena. However, useful system analysis results may be obtained provided that there is an average pattern, or *statistical regularity,* in the involved phenomena.

   The methods of probability and statistics are important tools in the analysis of systems dealing with random, or noise-like, signals and phenomena. In this chapter the relevant fundamentals of these subjects are developed for use in subsequent chapters.

9.1 DEFINITION OF PROBABILITY

The concept of probability is conveniently presented by considering the *relative frequency of occurrence* of *chance events*. This approach, although lacking in mathematical rigor, is sufficient for our purposes.

A typical chance event is the result obtained by tossing a single die. Assuming a true die, the number of dots appearing on any one toss cannot be predicted in advance with certainty. However, with a large number of trials, there is a strong tendency for each number to appear a *predictable percentage* of the time. The possible outcomes of tossing a die are called *events*. If we label a particular die A, then the event of two dots appearing on this die is labeled A_2. In general, the possible events are labeled A_k, where k ranges from 1 to 6.

In a given experiment consisting of N tosses, let the number of times the event A_2 occurs be n. The *relative frequency of occurrence* of the event A_2 is $n(A_2)/N$. If the experiment is repeated, the number n may not be the same, but if N is allowed to approach infinity the relative frequency of occurrence tends toward a constant. The *probability of occurrence* or, more simply, the *probability* of the event A_2 is defined as the limiting value of the relative frequency of occurrence as the number of trials approaches infinity.

$$P(A_k) = \lim_{N \to \infty} \left[\frac{n(A_k)}{N} \right] \qquad (9\text{-}1)$$

In the case of a die-tossing experiment, only a single event can occur on each toss. The different possibilities in this case are said to be *mutually exclusive*. Thus the probability of getting a two *or* a three is

$$P(A_2 \text{ or } A_3) = \lim_{N \to \infty} \left[\frac{n(A_2) + n(A_3)}{N} \right]$$

$$= P(A_2) + P(A_3) \qquad (9\text{-}2)$$

or, in general,

$$P(A_1 \text{ or } A_2 \cdots A_L) = \sum_{k=1}^{L} P(A_k) \qquad (9\text{-}3)$$

For mutually exclusive events, the probability of any one of several possible events is the sum of the probabilities of each of these events. If all possible events are included, this sum is of course unity. Thus, for K possible outcomes,

$$\sum_{k=1}^{K} P(A_k) = 1 \qquad (9\text{-}4)$$

9.1.1 Joint Probability

Chance events are not limited to single results on each trial, as in the tossing of a single die. For instance, if we toss two dies, or a pair of dice, the result is characterized by two numbers for each trial. The trial result in this case is called a *compound event,* as compared with the *simple event* obtained with a single die. The probability associated with the simple event is called the *elemen-*

tary probability, and the probability associated with the compound event is the *joint probability.*

In the case of two die, let the compound event be labeled A_k, B_ℓ, where k and ℓ are the possible results on dies A and B, respectively. The joint probability $P(A_k, B_\ell)$ is the probability of obtaining the kth result on die A, and *at the same time* obtaining the ℓth result on die B. Because there are six possible outcomes on each die, there are a total of 36 possible compound events. The compound events are mutually exclusive and, therefore, the sum of the probabilities for all possible events is unity. For the two-die experiment,

$$\sum_{k=1}^{6} \sum_{\ell=1}^{6} P(A_k, B_\ell) = 1 \tag{9-5}$$

Consider the probability of obtaining a particular result, A_k, on one die, while allowing any one of the six possibilities on die B. The compound events that satisfy this requirement are

$$(A_k, B_1) \quad \text{or} \quad (A_k, B_2) \quad \text{or} \quad \cdots (A_k, B_6)$$

This is simply a tabulation of all possible ways of getting the simple event A_k, and the probability must therefore be $P(A_k)$. Because the associated compound events are mutually exclusive, this probability is also the sum of the joint probabilities of the compound events. Thus

$$P(A_k) = \sum_{\ell=1}^{6} P(A_k, B_\ell) \tag{9-6}$$

9.1.2 Conditional Probability

In an experiment involving compound events, (A_k, B_ℓ), suppose that we desire to calculate the probability of a particular A_k given that a particular B_ℓ has already occurred. This type of probability is called a *conditional probability* and is given the notation:

$$\text{probability of } A_k \text{ given } B_\ell \text{ has occurred} = P(A_k|B_\ell)$$

Assume that an experiment has been performed N times with the simple event B_ℓ occurring $n(B_\ell)$ times. Of the times B_ℓ occurs, we find that the desired compound event, A_k, B_ℓ has occurred $n(A_k, B_\ell)$ times. The *relative conditional frequency* of A_k, given that B_ℓ has occurred is, therefore,

$$\frac{n(A_k, B_\ell)}{n(B_\ell)}$$

Dividing numerator and denominator by N and passing to the limit as N approaches infinity, we arrive at the defining equation for conditional probability:

$$P(A_k \mid B_\ell) = \frac{P(A_k, B_\ell)}{P(B_\ell)} \qquad (9\text{-}7)$$

In a similar manner, the probability of a particular B_ℓ, given that A_k has occurred, is

$$P(B_\ell \mid A_k) = \frac{P(A_k, B_\ell)}{P(A_k)} \qquad (9\text{-}8)$$

Conditional probability is defined in (9-7) and (9-8) in terms of the joint probability and an elementary probability.

9.1.3 Statistical Independence

In an experiment involving a toss of a pair of dice, a particular result on one die in no way affects the result on the other die, assuming that we have an honest pair. That is, for any B_ℓ the conditional probability of A_k is equal to the elementary probability of A_k. When this is true, the events A_k and B_ℓ are said to be *statistically independent,* and we may write

$$P(A_k \mid B_\ell) = \frac{P(A_k, B_\ell)}{P(B_\ell)} = P(A_k)$$

$$P(B_\ell \mid A_k) = \frac{P(A_k, B_\ell)}{P(A_k)} = P(B_\ell)$$

from which

$$P(A_k, B_\ell) = P(A_k)P(B_\ell) \qquad (9\text{-}9)$$

Thus for statistically independent events the joint probability of the compound event is equal to the product of the elementary probabilities of the simple events.

9.2 DISCRETE RANDOM VARIABLES

The outcomes of experiments involving the tossing of a die can yield only certain discrete values. This characteristic is shared by many physical phenomena of practical interest. In general, let the outcome of a trial be designated x, limited to the discrete values x_k. The variable x is called a *discrete random variable,* and the set of probabilities $P(x_k)$ corresponding to the values x_k is the *probability distribution* for x when considered as a function of x.

A convenient representation of the distribution of discrete probabilities involves the use of the impulse function. Let the result x_k occur with probability

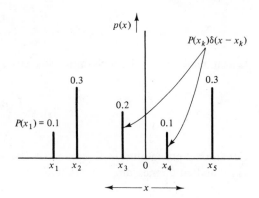

Figure 9-1   Discrete probability distribution, or probability density function.

$P(x_k)$. From the properties of an impulse, $P(x_k)$ can also be written as

$$P(x_k) = \int_{-\infty}^{+\infty} P(x_k)\delta(x - x_k)\, dx \qquad (9\text{-}10)$$

If the discrete variable has N possible outcomes, the distribution of probability, or *probability density,* as a function of x can be written as

$$p(x) = \sum_{k=1}^{N} P(x_k)\delta(x - x_k) \qquad (9\text{-}11)$$

As an example, assume that the values x_k are the discrete possible outputs of a
random voltage generator. A discrete probability distribution, or probability
density, is shown in Figure 9-1 for $N = 5$. Notice that the sum of discrete
probabilities is given by the integral of the probability density function, and
when evaluated over all possible values of x is unity. Thus,

$$\int_{-\infty}^{+\infty} p(x)\, dx = \int_{-\infty}^{+\infty} \sum_{k=1}^{5} P(x_k)\delta(x - x_k)\, dx = \sum_{k=1}^{5} P(x_k) \int_{-\infty}^{+\infty} \delta(x - x_k)\, dx$$

$$= \sum_{k=1}^{5} P(x_k) = 1.0 \qquad (9\text{-}12)$$

We may also be interested in the probability that the output voltage x is
equal to or less than some value X. For mutually exclusive events, this is given
by the sum of discrete probabilities for x_k equal to or less than X. In general
terms, we write

$$P(x_k \leq X) = \int_{-\infty}^{X} p(x)\, dx = \sum_{k=1}^{M} P(x_k) \qquad (9\text{-}13)$$

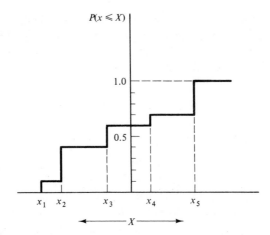

Figure 9-2   Cumulative probability function for discrete voltage generator.

where M is the greatest value for k for which $x_k \le X$. For instance, with the discrete random variable defined by Figure 9-1, the probability that x is equal to or less than x_4 is 0.7.

The function described by (9-13), when considered as a function of the upper limit X is the *cumulative probability function* for the variable x. For the discrete random voltage generator, the cumulative probability is shown in Figure 9-2. The cumulative probability is a monotonically increasing function of X. In the case of the discrete variable, the cumulative function increases in discrete steps at the locations of the allowed values of x. The maximum value of the cumulative probability function is unity.

Referring to (9-13), it is evident that the probability density function may be obtained by differentiating the cumulative probability function. The probability density function (p.d.f.) is a nonnegative function whose integral over all x must be unity.

If a discrete random variable is composed of the compound events (x_k, y_ℓ), we may define a *joint cumulative probability function* and a *joint probability density function* as follows:

$$P(x_k \le X, y_\ell \le Y) = \int_{-\infty}^{Y} \int_{-\infty}^{X} p(x, y) \, dx \, dy$$

$$= \int_{-\infty}^{Y} \int_{-\infty}^{X} \sum_k \sum_\ell P(x_k, y_\ell)\delta(x - x_k, y - y_\ell) \, dx \, dy$$

$$= \sum_{\substack{x_k \le X \\ y_\ell \le Y}} \sum P(x_k, y_\ell) \qquad\qquad (9\text{-}14)$$

Integration of the joint p.d.f. over the full range of both x and y must give unity because this includes all possible compound events. If the integration is carried out over all y, but over a limited range of x, the result is the cumulative probability function for x. That is,

$$P(x \leqq X) = \int_{-\infty}^{+\infty} \int_{-\infty}^{X} p(x, y) \, dx \, dy$$

$$= \int_{-\infty}^{X} p(x) \, dx \qquad (9\text{-}15)$$

From (9-15) we may also recognize that integration of the joint p.d.f. over the full range of one variable gives the elementary p.d.f. of the remaining variable.

$$p(x) = \int_{-\infty}^{+\infty} p(x, y) \, dy \qquad (9\text{-}16)$$

9.3 CONTINUOUS RANDOM VARIABLES

Let the number of possible discrete output voltages of a random voltage generator increase without bound. Because the sum of probabilities over the whole range must remain unity, the probability associated with each discrete event must approach zero as the number of possible events approaches infinity. However, the probability of the output voltage being in any finite range may be nonzero. The cumulative probability function therefore remains a nondecreasing function with a maximum value of unity.

9.3.1 Probability Density Functions

In the limit, the discrete variable becomes a continuous variable with a continuous cumulative probability function. In the absence of discontinuities in the cumulative probability function, the p.d.f. of the continuous variable will not contain impulsive elements.

Figure 9-3 presents a hypothetical cumulative probability function and probability density function for a continuous random voltage generator. Notice that although the p.d.f. is nonzero, the probability associated with any particular value is $p(x) \, dx$, and therefore infinitesimally small.

It is possible for a random variable to contain both discrete and continuous components. The p.d.f. for the combined variable can be expressed as the sum of the continuous p.d.f. and the impulsive p.d.f. for the discrete portion. Thus

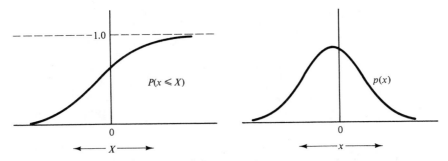

Figure 9-3 Cumulative probability and probability density function for a continuous random variable.

let

$$p_1(x) = \text{p.d.f. of continuous portion}$$

$$p_2(x) = \sum_{k=1}^{N} P(x_k)\delta(x - x_k) = \text{p.d.f. of discrete portion}$$

from which the p.d.f. of the complete variable is

$$p(x) = p_1(x) + \sum_{k=1}^{N} P(x_k)\delta(x - x_k) \tag{9-17}$$

9.3.2 Joint Probability Density Functions

For two continuous random variables, x and y, a two-dimensional cumulative distribution function and a joint p.d.f. may be defined in the same manner as for multiple discrete variables. Thus

$$P(x \leq X, y \leq Y) = \int_{-\infty}^{Y} \int_{-\infty}^{X} p(x, y) \, dx \, dy \tag{9-18}$$

The total volume under the joint p.d.f. is unity, and as with discrete variables, the simple probability functions for each continuous variable are obtained by integration of the joint density function over the appropriate ranges. Hence we may write

$$\int_{-\infty}^{+\infty} \int_{-\infty}^{+\infty} p(x, y) \, dx \, dy = 1.0 \tag{9-19}$$

$$\int_{-\infty}^{+\infty} \int_{-\infty}^{X} p(x, y) \, dx \, dy = P(x \leq X) \tag{9-20}$$

$$\int_{-\infty}^{Y} \int_{-\infty}^{+\infty} p(x, y) \, dx \, dy = P(y \le Y) \tag{9-21}$$

$$\int_{-\infty}^{+\infty} p(x, y) \, dy = p(x) \tag{9-22}$$

$$\int_{-\infty}^{+\infty} p(x, y) \, dx = p(y) \tag{9-23}$$

9.3.3 Conditional Probability Density Functions and Statistical Independence

A conditional probability density function can be defined and related to the joint p.d.f. and the elementary p.d.f.'s by analogy with the results for discrete variables.

$$p(x, y) = p(x|y)p(y) = p(y|x)p(x) \tag{9-24}$$

If x and y are statistically independent, the conditional density functions equal the elementary density functions. Thus

$$
\left.
\begin{aligned}
p(y|x) &= p(y) \\
p(x|y) &= p(x)
\end{aligned}
\right\} \quad \text{for } x, y \text{ independent} \tag{9-25}
$$

and

$$p(x, y) = p(x)p(y)$$

Also, for x and y independent,

$$P(x \le X, y \le Y) = P(x \le X)P(y \le Y) = \int_{-\infty}^{Y} p(y) \, dy \int_{-\infty}^{X} p(x) \, dx \tag{9-26}$$

9.4 MOMENTS

Analysis involving random variables requires the evaluation of *averages* or *expected values* of those variables. These averages may be related to the probability functions discussed in preceding sections.

Consider a discrete random variable, x, with K possible outcomes, x_k. If the defining experiment for x is repeated N times, the average value of the results is

$$x_{\text{av}} = \frac{x_1 n(1) + \cdots + x_k n(k) + \cdots + x_k n(K)}{N} \tag{9-27}$$

where N = number of trials
           $n(k)$ = number of times $x = x_k$

This is written more compactly as

$$x_{av} = \sum_{k=1}^{K} x_k \left[\frac{n(k)}{N} \right] \tag{9-28}$$

In the limit as N approaches infinity, the term inside the brackets becomes $P(x_k)$, and we write

$$E[x] = \bar{x} = \sum_{k=1}^{K} x_k P(x_k) \tag{9-29}$$

The term $E[x]$ is the *expected value* of the random variable x. This factor is also called the *mean, expectation, statistical average,* or the *ensemble average* of x. If x is a function of some other variable, such as time, it may be called a *stochastic variable,* and the associated averages are referred to as *stochastic averages.*

We are often interested in the expected value of various functions of a random variable. For instance, in the same manner used to arrive at (9-29), the expected value of the square of x can be obtained as

$$E[x^2] = \overline{x^2} = \sum_{k=1}^{K} x_k^2 P(x_k) \tag{9-30}$$

In the general case, if $f(x)$ is an arbitrary function of x,

$$E[f(x)] = \sum_{k=1}^{K} f(x) P(x_k) \tag{9-31}$$

By replacing the discrete probability with the probability density function, and the summation with an integration, the expected value for continuous random variables is obtained.

$$E[f(x)] = \int_{-\infty}^{+\infty} f(x) p(x) \, dx \tag{9-32}$$

The expected values of the various powers of x are of particular importance in system analysis problems. These are called the *moments* of the probability density function of x, defined as

$$n\text{th moment} = E[x^n] = \int_{-\infty}^{+\infty} x^n p(x) \, dx \tag{9-33}$$

The expected value of x is the *first moment* or *centroid* of $p(x)$. Likewise, the mean-squared value of x, $E[x^2]$, is the second moment of $p(x)$ with respect to the origin, and so on.

The *central moments* of $p(x)$ are defined as the nth-order expectations of the variable $(x - \bar{x})$. Thus

$$n\text{th central moment} = E[(x - \bar{x})^n] = \int_{-\infty}^{+\infty} (x - \bar{x})^n p(x) \, dx \qquad (9\text{-}34)$$

The first central moment is zero, by definition. The second central moment is the *variance* of x, often identified by special notation as follows:

$$\text{variance of } x = \text{Var}[x] = \sigma_x^2 \qquad (9\text{-}35)$$

Expanding the defining equation for the variance, we obtain

$$\sigma_x^2 = \int_{-\infty}^{+\infty} (x - \bar{x})^2 p(x) \, dx = \int_{-\infty}^{+\infty} x^2 p(x) \, dx - 2\bar{x} \int_{-\infty}^{+\infty} xp(x) \, dx$$

$$+ (\bar{x})^2 \int_{-\infty}^{+\infty} p(x) \, dx$$

$$= \int_{-\infty}^{+\infty} x^2 p(x) \, dx - (\bar{x})^2$$

$$= E[x^2] - E^2[x] \qquad (9\text{-}36)$$

Thus the variance is equal to the second moment minus the square of the first moment. The square root of the variance, σ_x, is the *standard deviation* of x.

If x is a random voltage as a function of time, the expected value, or average value, is recognized as the dc component of x. The varying, or ac component, is obtained as the difference between x and its expected value.

$$\text{ac component of } x = x - \bar{x}$$

The power contained in the ac component is proportional to the expected value of its square, which is also the variance of x.

$$P_{\text{ac}} = E[(x - \bar{x})^2] = \sigma_x^2$$

The total power in x is the sum of the ac power and dc power. The dc power is proportional to the square of the dc component. Thus

$$P_t = P_{\text{ac}} + P_{\text{dc}}$$

$$= \sigma_x^2 + (\bar{x})^2 = E[x^2] \qquad (9\text{-}37)$$

The concept of moments and expected values may be applied to multiple random variables. Let z be some function of x and y. Then

$$E[z] = E[f(x, y)] = \int_{-\infty}^{+\infty} \int_{-\infty}^{+\infty} f(x, y)p(x, y)\, dx\, dy \qquad (9\text{-}38)$$

The *joint moments* are

$$(n + m)\text{th moment} = E[x^n y^m] = \int_{-\infty}^{+\infty} \int_{-\infty}^{+\infty} x^n y^m p(x, y)\, dx\, dy \qquad (9\text{-}39)$$

and the joint central moments are

$$(n + m)\text{th central moment} = E[(x - \bar{x})^n (y - \bar{y})^m] \qquad (9\text{-}40)$$

Now let x and y be statistically independent, so that $p(x, y) = p(x)p(y)$. The $(n + m)$th moment now becomes

$$E[x^n y^m] = \int_{-\infty}^{+\infty} \int_{-\infty}^{+\infty} x^n y^m p(x)p(y)\, dx\, dy$$

$$= \int_{-\infty}^{+\infty} x^n p(x)\, dx \int_{-\infty}^{+\infty} y^m p(y)\, dy$$

$$= E[x^n]E[y^n] \qquad (9\text{-}41)$$

Thus, for statistically independent variables, the expected value of the product is the product of the expected values.

If z is the sum of random variables, x and y, the expected value of z is

$$E[z] = E[x + y] = \int_{-\infty}^{+\infty} \int_{-\infty}^{+\infty} (x + y)p(x, y)\, dx\, dy$$

$$= \int_{-\infty}^{+\infty} x \left[\int_{-\infty}^{+\infty} p(x, y)\, dy \right] dx$$

$$+ \int_{-\infty}^{+\infty} y \left[\int_{-\infty}^{+\infty} p(x, y)\, dx \right] dy \qquad (9\text{-}42)$$

Using the relationships in (9-22) and (9-23), (9-42) becomes

$$E[x + y] = \int_{-\infty}^{+\infty} xp(x)\, dx + \int_{-\infty}^{+\infty} yp(y)\, dy$$

$$= E[x] + E(y) \qquad (9\text{-}43)$$

Therefore, the expected value of the sum of random variables is equal to the sum of the expected values. Note that this result does not require that the random variables be statistically independent.

The variance of $z = x + y$ is, by definition,

$$\sigma_z^2 = E[(z - \bar{z})^2]$$

Because the expected value of z is the sum of the expected values of x and y, we may write

$$\sigma_z^2 = E\{[(x - \bar{x}) + (y - \bar{y})]^2\}$$
$$= E[(x - \bar{x})^2] + 2E[(x - \bar{x})(y - \bar{y})] + E[(y - \bar{y})^2]$$
$$= \sigma_x^2 + \sigma_y^2 + 2E[(x - \bar{x})(y - \bar{y})] \qquad (9\text{-}44)$$

If x and y are statistically independent, the expectation involving x and y in (9-44) can be separated into the product of expectations. That is, for x and y independent,

$$E[(x - \bar{x})(y - \bar{y})] = E[(x - \bar{x})]E[(y - \bar{y})] = 0$$

and the variance of z is the sum of the variances of x and y. It is useful to generalize this result for the sum of N independent random variables. Define, for all x_n independent,

$$z = \sum_{n=1}^{N} x_n$$

Then

$$\bar{z} = \sum_{n=1}^{N} \bar{x}_n \qquad (9\text{-}45)$$

and

$$\sigma_z^2 = \sum_{n=1}^{N} \sigma_{x_n}^2$$

Note from (9-44) that if random components of z are not statistically independent, the variance of z is *not* the sum of the variances of its component parts. In the general case, with all x_n not necessarily independent, we have

$$E[(z - \bar{z})^2] = E\left[\left(\sum_n x_n - \sum_n \bar{x}_n\right)^2\right] = E\left\{\left[\sum_n (x_n - \bar{x}_n)\right]^2\right\}$$
$$= E\left[\sum_m \sum_n (x_m - \bar{x}_m)(x_n - \bar{x}_n)\right]$$
$$= \sum_m \sum_n E[(x_m - \bar{x}_m)(x_n - \bar{x}_n)] \qquad (9\text{-}46)$$

The expectation involving x_m and x_n in (9-46) is called the *covariance* of x_m and x_n. For $m = n$, the covariance becomes the simple variance. For x_m and x_n statistically independent, (9-46) reduces to the result in (9-45).

9.5 EXAMPLES OF DENSITY FUNCTIONS

Before proceeding further, it is useful to consider several examples using the concepts developed to this point.

9.5.1 Uniform Density Function

Perhaps the simplest possible form for a probability density function is the *uniform* p.d.f., such as that shown in Figure 9-4. The continuous variable x has an amplitude that is equally likely in the region $b - a$ to $b + a$, with zero probability of being found outside this region. The probability density function for x is

$$p(x) = \frac{1}{2a} \, \text{rect}\left(\frac{x - b}{2a}\right) \tag{9-47}$$

By inspection, the area under $p(x)$ is unity. The mean value of x is

$$E[x] = \bar{x} = \int_{-\infty}^{+\infty} x p(x) \, dx = \frac{1}{2a} \int_{b-a}^{b+a} x \, dx$$

$$= \frac{x^2}{4a} \Bigg|_{b-a}^{b+a} = b$$

The variance of x is

$$E[(x - \bar{x})^2] = \sigma_x^2 = \int_{-\infty}^{+\infty} (x - \bar{x})^2 p(x) \, dx$$

$$= \frac{1}{2a} \int_{-\infty}^{+\infty} (x - b)^2 \, \text{rect}\left(\frac{x - b}{2a}\right) dx$$

$$= \frac{1}{2a} \int_{-\infty}^{+\infty} x^2 \, \text{rect}\left(\frac{x}{2a}\right) dx$$

$$\sigma_x^2 = \frac{x^3}{6a} \Bigg|_{-a}^{+a} = \frac{a^2}{3}$$

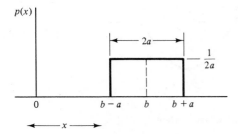

Figure 9-4   Uniform probability density function.

The standard deviation of x is the square root of the variance:

$$\sigma_x = \frac{a}{\sqrt{3}}$$

9.5.2 Probability Density Function for a Sine Wave

Let x be a sinusoidal function of θ, as shown in Figure 9-5. That is,

$$x = A \sin \theta$$

where θ is a random variable uniformly distributed over 2π radians, and A is a constant. We now wish to determine the p.d.f. of x.

The density function for x is defined by the equation

$$P(x \le X) = \int_{-\infty}^{X} p(x)\, dx$$

where $X \le A$. From Figure 9-5, the probability that $x \le X$ is equal to the probability that θ is *not* in the interval $[\theta_1, \pi - \theta_1]$. Because θ is equally likely anywhere over 2π, the required probability is

$$P(x \le X) = P(\theta \text{ not in } [\theta_1, \pi - \theta_1]) = \frac{\pi + 2\theta_1}{2\pi}$$

But

$$\theta_1 = \sin^{-1}\left(\frac{X}{A}\right)$$

Therefore,

$$P(x \le X) = \frac{1}{2} + \frac{1}{\pi} \sin^{-1}\left(\frac{X}{A}\right)$$

The required density function is obtained as the derivative of the cumulative

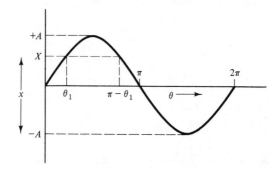

Figure 9-5 Sinusoidal function.

probability function:

$$p(X) = \begin{cases} \dfrac{dP(x \le X)}{dX} = \dfrac{1}{\pi(A^2 - X^2)^{1/2}} & \text{for } |X| \le A \\ 0 & \text{for } |X| > A \end{cases} \qquad (9\text{-}48)$$

This density function is shown in Figure 9-6. Notice that the p.d.f. approaches infinity as x approaches $\pm A$. This does not mean that the probability that $x = A$ or $-A$ is infinite. Remember that the probability that x is within any interval is given by the *area* under the p.d.f. within that interval. The total area must of course be unity.

The average value of x is zero because $p(x)$ is an even function of x, which requires that its first moment, or centroid, be zero. Thus

$$E[x] = \int_{-\infty}^{+\infty} xp(x)\, dx = 0$$

The variance of x is therefore equal to the second moment of the density function.

$$\sigma_x^2 = E[x^2] = \int_{-A}^{+A} \frac{x^2\, dx}{\pi(A^2 - x^2)^{1/2}}$$

Referring to a table of integrals, this is easily evaluated as

$$\sigma_x^2 = \frac{A^2}{2}$$

which gives the well-known result for a sine wave that the standard deviation, or rms value, is equal to the peak value divided by the square root of 2.

$$\sigma_x = \frac{A}{\sqrt{2}}$$

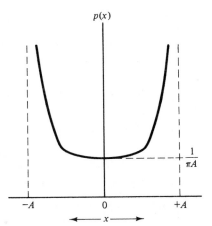

Figure 9-6 Probability density function for $x = A \sin \theta$.

9.5.3 Density Function at Output of a Perfect Rectifier

The output of a perfect half-wave rectifier is equal to the input when the input voltage is positive, and zero when the input is negative. Let the input voltage, $x(t)$, be a zero mean random function of time with a Gaussian p.d.f., defined by

$$p(x) = \frac{1}{\sigma_x \sqrt{2\pi}} \exp\left(\frac{-x^2}{2\sigma_x^2}\right) \tag{9-49}$$

The variance of $x(t)$ is σ_x^2, and the cumulative probability is obtained from

$$P(x \le X) = \frac{1}{\sigma_x \sqrt{2\pi}} \int_{-\infty}^{X} \exp\left(\frac{-x^2}{2\sigma_x^2}\right) dx \tag{9-50}$$

The integral in (9-50) cannot be solved in closed form, but tables of this function are available in normalized form. The Gaussian p.d.f. and cumulative probability are shown graphically in Figure 9-7.

The action of the perfect half-wave rectifier causes the output voltage to be zero half of the time—that is, whenever the input is negative. The p.d.f. for the output variable, $y(t)$, is therefore zero for $x(t)$ negative, and contains an impulse with a strength of 0.5 at the origin. For $x(t)$ positive, the output p.d.f. is identical to that for $x(t)$. The total output p.d.f. may be written as

$$p(y) = 0.5\delta(y) + \frac{u(y)}{\sigma_x \sqrt{2\pi}} \exp\left(\frac{-y^2}{2\sigma_x^2}\right) \tag{9-51}$$

where $u(y)$ is the unit step function. The output density function and cumulative probability are shown in Figure 9-8.

We now wish to determine the expected value and the variance of the output variable.

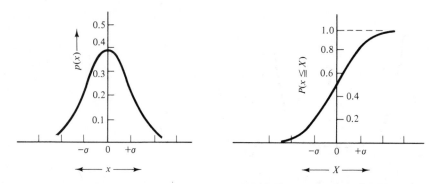

Figure 9-7 Gaussian p.d.f. and cumulative probability distribution function.

The expected value of y is, by definition,

$$E[y] = \frac{1}{2} \int_{-\infty}^{+\infty} y\delta(y)\, dy + \frac{1}{\sigma_x \sqrt{2\pi}} \int_{0}^{+\infty} y \exp\left(\frac{-y^2}{2\sigma_x^2}\right) dy$$

The first integral is obviously zero because the product $y\delta(y)$ is zero everywhere. The second integral is easily evaluated with the help of a table of definite integrals, with the result that

$$E[y] = \frac{\sigma_x}{\sqrt{2\pi}} \qquad (9\text{-}52)$$

The output variance is

$$\sigma_y^2 = E[y^2] - E^2[y]$$

Neglecting the integral involving the impulse function, this becomes

$$\sigma_y^2 = \frac{1}{\sigma_x \sqrt{2\pi}} \int_{0}^{+\infty} y^2 \exp\left(\frac{-y^2}{2\sigma_x^2}\right) dy - \frac{\sigma_x^2}{2\pi}$$

Again using a table of definite integrals, we obtain

$$\sigma_y^2 = \frac{\sigma_x^2}{2}\left(1 - \frac{1}{\pi}\right) \qquad (9\text{-}53)$$

From (9-52) and (9-53) we see that the rectifier action results in a dc component in the output, proportional to the magnitude of the input ac component, and an ac output component less than that of the input ac term.

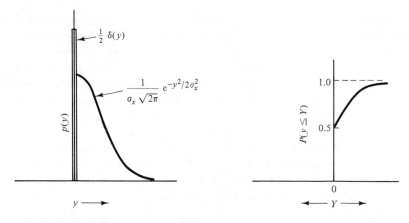

Figure 9-8 Output p.d.f. and cumulative probability function for a half-wave rectifier with a Gaussian input variable.

9.6 THE RANDOM PROCESS

A random voltage may be observed across the terminals of a resistor as a result of the thermal agitation of the electrons in the resistor material. This voltage is another example of a random variable, and its statistical properties may be determined by direct observation for a sufficient length of time.

We recognize that the voltage observed across a particular resistor obeys the same statistical and physical laws as that observed across any other resistor of the same type, under the same set of conditions. The voltages developed across the individual resistors are considered to be members, or *sample functions,* of a *random process:* in particular, a thermal noise random process.

A random process is defined by the set of all possible sample functions of that process, together with a complete set of probability statements for each sample function and for the relationships among sample functions. The complete set of sample functions is called the *ensemble* of sample functions.

It is not necessary that the sample functions of a random process be generated by the same physical process. For instance, imagine a switch that provides a connection, in a random manner, to one of several terminals, each of which provides a voltage generated by a different physical phenomenon. One terminal might be providing the thermal agitation voltage across a resistor, another voltage might be selected by the toss of a die, and so on. This collection of random variables is related through the probability statements necessary to describe the characteristics of the output voltage selected by the switch. In this case, the *process* probability functions are not identical to the corresponding sample function probability functions.

9.6.1 The Stationary Process

If the statistical properties of a random process do not change with time, the process is called a *stationary process.* Consider a large number of identical resistors at the same temperature. Let the set of voltages measured on each resistor at a particular time t_1, be labeled $\mathbf{x}(t_1)$, to distinguish it from the voltage on the kth resistor, $x_k(t_1)$. The *process* p.d.f. is then $p[\mathbf{x}(t_1)]$, or more simply $p(\mathbf{x}_1)$, and is the density function of the voltage readings across the ensemble of sample functions at the time t_1. If the process is stationary, a similar reading at a later time, t_2, will give an identical density function.

In a similar manner, we may compare the readings at times t_1 and t_2 and obtain the process joint p.d.f., $p(\mathbf{x}_1, \mathbf{x}_2)$. For the stationary process, the joint p.d.f. depends only on the time difference, $t_2 - t_1$, not on the absolute time location of the samples. That is, for the stationary process,

$$p(\mathbf{x}_1, \mathbf{x}_2) = p(\mathbf{x}_3, \mathbf{x}_4)$$

provided that $t_2 - t_1 = t_4 - t_3$. If this type of relationship holds for all higher-order joint density functions, the process is said to be *strict-sense stationary.* If

$p(\mathbf{x}_1)$ and $p(\mathbf{x}_1, \mathbf{x}_2)$ are invariant with respect to time shifts, but higher-order functions are not, the process is *wide-sense stationary*. If the process meets neither of these requirements, it is a *nonstationary* process. A thermal noise process at constant temperature with all elements in thermal equilibrium is a strict-sense stationary process. Obviously, if the temperature of the resistors changes slowly with time, the thermal agitation voltage will change and the process is nonstationary.

The concept of stationarity must be interpreted in the context of the time scale of importance to each problem under consideration. All physical phenomena are a part of the physical universe and, therefore, must change with time. However, it is a useful fiction to consider a process stationary if its statistical properties of interest do not change appreciably over the longest time scale of interest in a given problem.

9.6.2 The Ergodic Process

The time average of the voltage across a particular resistor in thermal equilibrium is defined by the equation

$$\langle x_k(t) \rangle = \lim_{T \to \infty} \frac{1}{T} \int_{-T/2}^{+T/2} x_k(t) \, dt \qquad (9\text{-}54)$$

where $x_k(t)$ is the voltage across the kth resistor of the ensemble. The statistical average, or ensemble average, is obtained by simultaneously measuring the voltage across each resistor in the ensemble at some instant of time. Thus

$$E[\mathbf{x}(t_1)] = \overline{\mathbf{x}(t_1)} = \ = \lim_{N \to \infty} \frac{1}{N} \sum_{k=1}^{N} x_k(t_1) = \int_{-\infty}^{+\infty} \mathbf{x}_1 p(\mathbf{x}_1) \, d\mathbf{x} \qquad (9\text{-}55)$$

If all resistors are equal and at the same constant temperature, the results obtained in (9-54) and (9-55) are equal. Hence for this process,

$$\langle x_k(t) \rangle = E[\mathbf{x}_1] = \overline{\mathbf{x}}_1 \qquad (9\text{-}56)$$

Furthermore, for this process we would find that

$$\langle x_k^2(t) \rangle = E[\mathbf{x}_1^2]$$

and

$$\langle x_k(t) x_k(t - \tau) \rangle = E[\mathbf{x}(t_1)\mathbf{x}(t_1 - \tau)] = E[\mathbf{x}_1 \mathbf{x}_2] \qquad (9\text{-}57)$$

where $\mathbf{x}_1 = \mathbf{x}(t_1)$ and $\mathbf{x}_2 = \mathbf{x}(t_1 - \tau)$. A process that behaves in this manner is called an *ergodic random process*. An ergodic random process is, of necessity, stationary. On the other hand, the fact that a process is stationary does not guarantee ergodicity. For instance, consider a thermal noise process with sample functions obtained from the thermal noise voltages developed across *unequal* resistors. The time averages obtained for the voltage across one particular resistor will not, in general, equal the corresponding ensemble averages obtained across the set of sample functions. The process is therefore nonergodic.

Much of our effort in system analysis is concerned with the detection and measurement of signals in combination with random noise derived from ergodic processes. This generally leads to tractable mathematical formulations, with uniquely defined optimization procedures. However, it is often necessary to include in the noise field the effects of discrete interfering signals, randomly distributed in both space and time. The noise field can no longer be considered ergodic in this case, and an exact analysis may be very difficult.

9.7 FUNCTIONS OF RANDOM VARIABLES

In system analysis, we must consider the effects of various mathematical operations on the statistical properties of random variables. In this section we show how this may be done for several commonly encountered operations.

9.7.1 Scaling of Random Variables

A common and very simple mathematical operation performed on signals and noise is that of amplitude scaling. Assume a random variable, x, with a p.d.f. $p(x)$. We now wish to determine the p.d.f., the mean value, and the variance of the variable $y = ax$, where a is a real constant.

The cumulative probability of y is related to the p.d.f. of x as follows:

$$P(y \le Y) = P\left(x \le \frac{Y}{a}\right) = \int_{-\infty}^{Y/a} p(x)\, dx \qquad (9\text{-}58)$$

The density function, $p(Y)$, is obtained as the partial derivative of (9-58) with respect to Y.

$$p(y) = \frac{\partial P(y \le Y)}{\partial Y} = \frac{\partial}{\partial Y}\left[\int_{-\infty}^{Y/a} p(x)\, dx\right] \qquad (9\text{-}59)$$

The partial derivative on the right side of (9-59) is evaluated using the general relationships:

if

$$h = \int_{f_1(z)}^{f_2(z)} g(x)\, dx$$

then

$$\frac{\partial h}{\partial z} = \frac{\partial f_2(z)}{\partial z}\, g[x = f_2(z)] - \frac{\partial f_1(z)}{\partial z}\, g[x = f_2(z)]$$

from which

$$p(Y) = \frac{1}{a}\, p\left(x = \frac{Y}{a}\right) \qquad (9\text{-}60)$$

As an example, let x be a Gaussian variable with mean value $\bar{x}$ and variance σ_x^2. Then

$$p(x) = \frac{1}{\sigma_x \sqrt{2\pi}} \exp\left[\frac{(x - \bar{x})^2}{2\sigma_x^2}\right]$$

and

$$p(Y) = \frac{1}{a\sigma_x \sqrt{2\pi}} \exp\left[\frac{[(Y/a) - \bar{x}]^2}{2\sigma_x^2}\right]$$

$$= \frac{1}{a\sigma_x \sqrt{2\pi}} \exp\left[\frac{(Y - a\bar{x})^2}{2a^2\sigma_x^2}\right] \tag{9-61}$$

Notice that $\bar{y} = a\bar{x}$ and $\sigma_y^2 = a^2\sigma_x^2$.

9.7.2 Sum of Random Variables

Let z be the sum of two random variables, x and y:

$$z = x + y \tag{9-62}$$

For some particular value of $z = Z$, solve (9-62) for y, and plot the straight-line relationship between x and y as shown in Figure 9-9. In region A of that figure, the sum of x and y is equal to or less than Z. The cumulative probability function for z may now be related to the properties of x and y. Thus

$$P(z \leq Z) = P[(x, y) \text{ are in } A]$$

From Figure 9-9, the region A is defined by the inequalities

$$-\infty \leq y \leq Z - x$$
$$-\infty \leq x \leq +\infty \tag{9-63}$$

The cumulative probability function for z is related to the joint density function

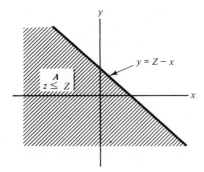

Figure 9-9 Graph defining region in x-y corresponding to $z \leq Z$, for $z = x + y$.

for x and y by

$$P(z \le Z) = \int_{-\infty}^{Z-x} \int_{-\infty}^{+\infty} p(x, y) \, dx \, dy \qquad (9\text{-}64)$$

Differentiation of (9-64) with respect to Z gives the probability density function for Z.

$$p(Z) = \frac{\partial}{\partial Z}\left[\int_{-\infty}^{Z-x} \int_{-\infty}^{+\infty} p(x, y) \, dx \, dy \right] \qquad (9\text{-}65)$$

Performing the indicated operations, we obtain

$$p(Z) = \int_{-\infty}^{+\infty} p(x, y = Z - x) \, dx \qquad (9\text{-}66)$$

To proceed further it is necessary to know the joint density function for x and y. If x and y are statistically independent, we obtain a particularly interesting result. Thus if

$$p(x, y) = p(x)p(y)$$

then

$$p(Z) = \int_{-\infty}^{+\infty} p(x)p(y = Z - x) \, dx \qquad (9\text{-}67)$$

For x and y independent, $p(Z)$ will be recognized as the *convolution* of the p.d.f.'s for x and y. This result may be extended to the case where z is the sum of a large number of statistically independent variables. For all x_n independent, let

$$z = x_1 + x_2 + \cdots + x_n + \cdots + x_N$$

Then   (9-68)

$$p(z) = p(x_1) \otimes p(x_2) \otimes \cdots \otimes p(x_N)$$

As a simple example, let all $p(x_n)$ be Gaussian functions with mean values $\bar{x}_n$ and variances $\sigma_{x_n}^2$. Because the convolution of two Gaussian functions results in a third Gaussian, $p(z)$ will also be Gaussian. For z defined by (9-68), let

$$p(x_n) = \frac{1}{\sigma_{x_n}\sqrt{2\pi}} \exp\left[\frac{-(x - \bar{x}_n)^2}{2\sigma_{x_n}^2} \right]$$

then

$$p(z) = \frac{1}{\sigma_z\sqrt{2\pi}} \exp\left[\frac{-(z - \bar{z})^2}{2\sigma_z^2} \right]$$

with

$$\bar{z} = \sum_{n=1}^{N} \bar{x}_n$$

$$\sigma_z^2 = \sum_{n=1}^{N} \sigma_{x_n}^2$$

As another example, let z be the sum of two statistically independent variables, x_1 and x_2, each with a zero-mean uniform density function of total width $2a$.

$$z = x_1 + x_2$$

$$p(x_1) = p(x_2) = \frac{1}{2a} \text{rect} \left(\frac{x}{2a} \right)$$

The density function for z is obtained from the self-convolution of the rect function p.d.f. This results in a triangular function for $p(z)$. Thus

$$p(z) = \frac{1}{4a^2} \text{rect} \left(\frac{x}{2a} \right) \otimes \text{rect} \left(\frac{x}{2a} \right)$$

$$= \frac{1}{4a^2} \left(1 - \frac{|z|}{2a} \right) \text{rect} \left(\frac{z}{4a} \right)$$

9.7.3 Characteristic Functions and the Central Limit Theorem

In Chapter 6 we learned that the convolution operation translates to a multiplication operation of the Fourier transforms of the original functions. This can be applied to the problem of determining the p.d.f. of the sum of independent random variables.

The transform of a probability density function is called the *characteristic function*, defined as

$$Q_x(s) = E[\exp(+j2\pi sx)] = \int_{-\infty}^{+\infty} p(x) \exp(j2\pi sx)\, dx \qquad (9\text{-}69)$$

The inverse transform is

$$p(x) = \int_{-\infty}^{+\infty} Q_x(s) \exp(-j2\pi sx)\, ds \qquad (9\text{-}70)$$

Now, with all x_n independent, let

$$z = \sum_{n=1}^{N} x_n$$

and

$$p(x_n) \longleftrightarrow Q_n(s)$$

The density function for z can be evaluated as the inverse transform of the product of characteristic functions. Thus

$$p(z) = p(x_1) \otimes \cdots \otimes p(x_N) \longleftrightarrow Q_1(s)Q_2(s) \cdots Q_N(s) \qquad (9\text{-}71)$$

Because a density function is nonnegative and has unit area, the maximum amplitude of the characteristic function is unity and occurs at $s = 0$. That is,

$$|Q_x(s)| = \left| \int_{-\infty}^{+\infty} p(x) \exp(j2\pi sx) \, dx \right| \le \int_{-\infty}^{+\infty} p(x) \, dx = 1.0 = Q_x(0) \qquad (9\text{-}72)$$

The moments of the density function may be evaluated with the help of the derivatives of the characteristic function. The nth derivative of the characteristic function is

$$\frac{d^n Q_x(s)}{ds^n} = (j2\pi)^n \int_{-\infty}^{+\infty} x^n p(x) \exp(j2\pi sx) \, dx$$

Evaluating both sides at $s = 0$, we obtain

$$\left. \frac{d^n Q_x(s)}{ds^n} \right|_{s=0} = (j2\pi)^n \int_{-\infty}^{+\infty} x^n p(x) \, dx$$

from which

$$E[x^n] = \frac{1}{(j2\pi)^n} \left. \frac{d^n Q_x(s)}{ds^n} \right|_{s=0} \qquad (9\text{-}73)$$

Using the result in (9-73), $Q_x(s)$ may be expanded as a power series about the point $s = 0$. For simplicity, assume that $p(x)$ is an even function of x, resulting in a real-valued characteristic function that is an even function of s. Therefore, the power series expansion of $Q_x(s)$ will contain only even powers of s. In this case, we write

$$Q_x(s) = 1 + a_2 s^2 + a_4 s^4 + \cdots \qquad (9\text{-}74)$$

where

$$a_k = \frac{1}{k!} \left. \frac{d^k Q_x(s)}{ds^k} \right|_{s=0} = \frac{(j2\pi)^k}{k!} E[x^k]$$

or

$$Q_x(s) = 1 - \frac{(2\pi s)^2}{2} E[x^2] + \frac{(2\pi s)^4}{24} E[x^4] \cdots \qquad (9\text{-}75)$$

The characteristic function for the sum of n statistically independent random variables with identical p.d.f.'s, $p(x)$, is

$$Q_z(s) = Q_x^n(s)$$

Now take the natural logarithm of both sides and write

$$\ln Q_z(s) = n \ln Q_x(s)$$

$$= n \ln \left[1 - \frac{(2\pi s)^2}{2} E[x^2] + \cdots \right]$$

For n large, the magnitude of $Q_z(s)$ will decrease rapidly as the magnitude of s increases from zero. In this case, we may restrict our attention to the region $|s| \ll 1.0$, and drop terms of higher order than s^2 in the expansion of $Q_x(s)$. For s very small

$$n \ln Q_x(s) \simeq n \ln \left[1 - \frac{(2\pi s)^2}{2} E[x^2] \right] \simeq -2(\pi s)^2 n E[x^2]$$

Therefore,

$$\ln Q_z(s) \simeq -2(\pi s)^2 n E[x^2]$$

Taking the antilog of both sides, and recognizing that with the assumption of an even p.d.f. $E[x^2] = \sigma_x^2$, we obtain

$$Q_z(s) = \exp\left[-2(\pi s)^2 n \sigma_x^2\right] \tag{9-76}$$

With the help of pair 6 in Table 6-1, the inverse transform of (9-76) is

$$p(z) = \frac{1}{\sigma_x \sqrt{2\pi n}} \exp\left(\frac{-z^2}{2n\sigma_x^2}\right) \tag{9-77}$$

Equation (9-77) is a Gaussian p.d.f. with zero mean and variance $\sigma_z^2 = n\sigma_x^2$, which is the sum of the variances of the random variables comprising z.

   Thus the sum of a large number of independent random variables with identical p.d.f.'s results in a variable z with a p.d.f. approaching the Gaussian shape. The restriction of identical p.d.f.'s that are even functions of their argument can be removed, and we find that the p.d.f. of the sum approaches a Gaussian shape provided that the means and variances of the individual variables are finite.

   This result is known as the *central limit theorem* and is one of the most useful theorems of probability theory. Many random variables encountered in nature are the resultant of a large number of independent influences. Hence, by the central limit theorem, the p.d.f. for such variables tends to approach the Gaussian shape. When the Gaussian assumption is valid, many analysis problems are simplified because of the convenient and well-documented properties of the Gaussian function.

   The central limit theorem should not be applied without some thought concerning the degree to which the required conditions for its validity are met. The number that may be considered a "large" number of independent variables may be hard to define. If one non-Gaussian variable with a variance large compared with the rest is added to the sum, it may cause the resultant density function to depart significantly from the Gaussian shape. In some situations, the resultant density function may approximate the Gaussian function quite closely in the vicinity of the mean, but depart significantly from the Gaussian shape for values far removed from the mean. In many problems, the shape of the "tail" of the density function is an important consideration. In these cases the validity of application of the central limit theorem should be carefully considered. Finally, if one or more of the variables in a sum has infinite mean or variance, the central limit theorem does not apply.

9.7.4 Square of a Random Variable

In Section 9.5.3 we determined the density function at the output of a half-wave linear rectifier. A more common type of rectifier encountered in system analysis is the full-wave square-law device. The input–output characteristic of such a device is shown in Figure 9-10.

Let the square-law device input be a random variable $x(t)$, resulting in an output $y(t)$. Thus

$$y(t) = x^2(t)$$

With the help of Figure 9-10, we see that the probability that the output is equal to or less than some value Y is equal to the probability that x is between the values $\pm\sqrt{Y}$.

$$P(y \leq Y) = \begin{cases} P(-\sqrt{Y} \leq x \leq +\sqrt{Y}) & \text{for } Y \geq 0 \\ 0 & \text{for } Y < 0 \end{cases}$$

Hence we may write

$$P(y \leq Y) = \int_{-\sqrt{Y}}^{+\sqrt{Y}} p(x)\, dx \tag{9-78}$$

The density function for the output is obtained by differentiation of (9-78) with respect to Y.

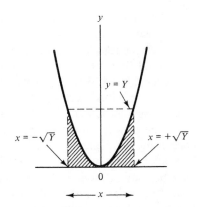

Figure 9-10 Input–output characteristics of perfect square-law device.

$$p(Y) = p(x = +\sqrt{Y}) \frac{d\sqrt{Y}}{dY} - p(x = -\sqrt{Y}) \frac{d(-\sqrt{Y})}{dY}$$

$$= \begin{cases} \dfrac{1}{2\sqrt{Y}} [p(x = +\sqrt{Y}) + p(x = -\sqrt{Y})] & \text{for } Y \geq 0 \\ 0 & \text{for } Y < 0 \end{cases} \qquad (9\text{-}79)$$

If the input p.d.f. is an even function of x, this simplifies to

$$p(Y) = \frac{1}{\sqrt{Y}} p(x = \sqrt{Y}) \qquad \text{for } Y \geq 0 \text{ and } p(x) = p(-x) \qquad (9\text{-}80)$$

As an example of considerable interest, let $x(t)$ be a Gaussian variable with

$$p(x) = \frac{1}{\sigma_x \sqrt{2\pi}} \exp\left(\frac{-x^2}{2\sigma_x^2}\right)$$

Application of (9-80) gives, for y positive,

$$p(y) = \frac{1}{\sigma_x \sqrt{2\pi y}} \exp\left(\frac{-y}{2\sigma_x^2}\right) \qquad (9\text{-}81)$$

This density function is shown in Figure 9-11, together with the Gaussian input density function.

The mean and variance of y may be calculated from the first and second moments of $p(y)$, or more directly from the moments of x. The even central moments for the Gaussian density function are easily shown to be provided by

$$E[(x - \bar{x})^n] = 1 \cdot 3 \cdot 5 \cdots (n - 1)\sigma_x^n \qquad (9\text{-}82)$$

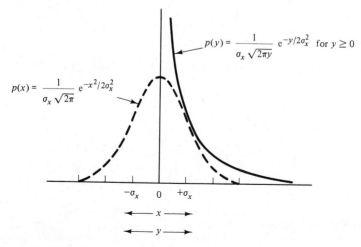

Figure 9-11 Input and output densities for a full-wave square-law device with Gaussian input variable.

Thus

$$E[(x - \bar{x})^2] = \sigma_x^2$$
$$E[(x - \bar{x})^4] = 3\sigma_x^4 \qquad\qquad (9\text{-}83)$$
$$E[(x - \bar{x})^6] = 15\sigma_x^6$$

For the output variable y we have

$$E[y] = E[x^2] = \sigma_x^2$$

and

$$\sigma_y^2 = E[y^2] - E^2[y] \qquad\qquad (9\text{-}84)$$
$$= E[x^4] - E^2[x^2]$$
$$= 3\sigma_x^4 - \sigma_x^4 = 2\sigma_x^4$$

Now consider a variable y, composed of the sum of the squares of two independent Gaussian variables x_1 and x_2 with zero mean and equal variance.

$$y = x_1^2 + x_2^2$$
$$p(x_1) = p(x_2) = \frac{1}{\sigma_x\sqrt{2\pi}} \exp\left(\frac{-x^2}{2\sigma_x^2}\right)$$

The density function for y in this case may be obtained by self-convolution of the density function in (9-81), or by the characteristic function method. This yields

$$p(y) = \frac{1}{2\sigma_x^2} \exp\left(\frac{-y}{2\sigma_x^2}\right) \qquad \text{for } y \geq 0 \qquad\qquad (9\text{-}85)$$

This is a simple exponential distribution, shown graphically in Figure 9-12. The

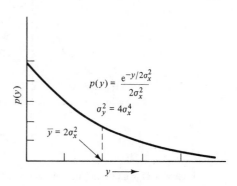

$$p(y) = \frac{e^{-y/2\sigma_x^2}}{2\sigma_x^2}$$
$$\sigma_y^2 = 4\sigma_x^4$$
$$\bar{y} = 2\sigma_x^2$$

Figure 9-12 Probability density function for $y = x_1^2 + x_2^2$, with x_1, x_2 zero-mean equal-variance Gaussian variables.

mean and variance of y in this case are

$$E[y] = E[x_1^2] + E[x_2^2] = 2\sigma_x^2$$
$$\sigma_y^2 = \text{Var}[x_1^2] + \text{Var}[x_2^2] \qquad (9\text{-}86)$$
$$= 2\sigma_{x_1}^4 + 2\sigma_{x_2}^4 = 4\sigma_x^4$$

Extending these results, if y is composed of the sum of the squares of N independent Gaussian variables with zero mean and equal variance, the resulting density function can be shown to be

$$p(y) = \frac{(y/\sigma_x^2)^{N/2-1}}{\sigma_x^2 2^{N/2}\Gamma(N/2)} \exp\left(\frac{-y}{2\sigma_x^2}\right) \qquad (9\text{-}87)$$

where $\Gamma(\cdot)$ is the gamma function. In normalized form with $\sigma_x^2 = 1.0$, this is called the *chi-squared* density function with N *degrees of freedom* and y is then an *N-degree-of-freedom chi-squared variable*. Often a variable of this type is referred to as chi-squared in form, even though σ_x^2 is not unity.

By direct extension of (9-86), the mean and variance of a variable with the chi-squared form is

$$E[y] = E\left[\sum_{n=1}^{N} x_n^2\right] = NE[x_n^2] = N\sigma_x^2$$
$$\text{Var}[y] = N\,\text{Var}[x^2] = 2N\sigma_x^4 \qquad (9\text{-}88)$$

The density function in (9-81) is chi-squared in form with a single degree

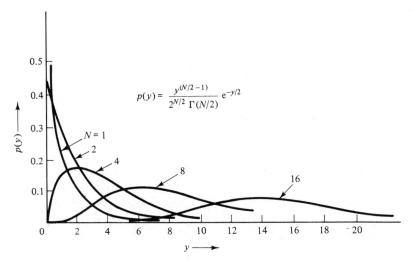

Figure 9-13   Probability density functions for chi-squared random variables.

of freedom, while that in (9-85) is a two-degree-of-freedom function. The chi-squared density functions for several different values of N are shown in Figure 9-13. Notice that for N large, the chi-squared density function approaches the Gaussian shape, as predicted by the central limit theorem.

9.7.5 Square Root of a Random Variable

Figure 9-14 shows the input–output relationship for a device that provides the positive square root of the input variable as an output. This relationship is

$$z = \sqrt{y} \quad \text{for } z \text{ and } y \geq 0$$

The p.d.f. for z is obtained from

$$P(0 \leq z \leq Z) = P(0 \leq y \leq Z^2) = \int_0^{Z^2} p(y)\, dy$$

Differentiation with respect to Z gives

$$p(Z) = \frac{dP(0 \leq z \leq Z)}{dZ} = \frac{d(Z^2)}{dZ} p(y = Z^2)$$

$$= 2Z p(y = Z^2) \quad \text{for } Z \geq 0 \quad (9\text{-}89)$$

As practical example, let y have the form of a two-degree-of-freedom chi-squared variable. That is, let

$$y = x_1^2 + x_2^2$$

where x_1 and x_2 are independent zero-mean Guassian variables with equal variance, from which

$$p(y) = \frac{1}{2\sigma_x^2} \exp\left(\frac{-y}{2\sigma_x^2}\right)$$

The variable z is the positive square root of the sum of the squares of x_1 and x_2. Direct substitution in (9-89) gives for the output density function

$$p(z) = \begin{cases} \dfrac{z}{\sigma_x^2} \exp\left(\dfrac{-z^2}{2\sigma_x^2}\right) & \text{for } z \geq 0 \\[2mm] 0 & \text{for } z < 0 \end{cases} \quad (9\text{-}90)$$

This density function is known as the *Rayleigh* density function, and z is a *Rayleigh variable*. The Rayleigh variable occurs quite naturally in system problems as the magnitude of the envelope function associated with a complex Gaussian noise waveform. For instance, consider the complex envelope function resulting from the summation of a large number of complex sinusoids as follows:

$$v = \sum_{n=1}^{N} a_n \exp(j\phi_n) \quad (9\text{-}91)$$

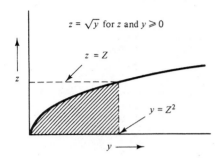

Figure 9-14 Input–output for a square-root device.

Let a_n be a zero-mean variable, statistically independent of a_m for $m \neq n$, and of all ϕ_n. The phase, ϕ_n, is assumed uniformly distributed over the interval $[-\pi, \pi]$, with ϕ_n independent of ϕ_m. Now define

$$x_1 = \sum_{n=1}^{N} a_n \cos \phi_n$$

$$x_2 = \sum_{n=1}^{N} a_n \sin \phi_n$$

and write (9-91) as

$$v = x_1 + jx_2 \qquad\qquad (9\text{-}92)$$

By the central limit theorem, x_1 and x_2 tend to become zero-mean Gaussian variables as N approaches infinity. Further, x_1 and x_2 become statistically independent with equal variances, σ_x^2. Now write (9-92) in the form

$$v = |v| \exp(j\psi)$$

where

$$|v| = (x_1^2 + x_2^2)^{1/2}$$

$$\psi = \tan^{-1}\left(\frac{x_2}{x_1}\right)$$

The magnitude of v is seen to be a Rayleigh variable. The phase term, ψ, can be shown to be uniformly distributed over the interval $[-\pi, +\pi]$. The variable v is called a *complex Gaussian* variable.

9.7.6 Product of Random Variables

Let z be the product of random variables x and y. To determine the density function for z, we must determine the limits on x and y that satisfy the condition that z is equal to or less than some value Z. The regions in the x-y plane that

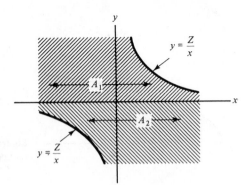

Figure 9-15 Graph for determining limits of integration in x and y for evaluating cumulative probability for $z = xy$.

satisfy this relationship are identified in Figure 9-15. From this figure, we see that $z \leq Z$ if x and y are in either A_1 or A_2.

$$P(z \leq Z) = P(x, y \text{ in } A_1) + P(x, y \text{ in } A_2)$$

$$= \iint\limits_{A_1} p(x, y) \, dx \, dy + \iint\limits_{A_2} p(x, y) \, dx \, dy \qquad (9\text{-}93)$$

Region A_1 is defined by the relationships

$$\left.\begin{array}{c} -\infty \leq x \leq \dfrac{Z}{y} \\[2mm] 0 \leq y = +\infty \end{array}\right\} A_1$$

Similarly, A_2 is defined by

$$\left.\begin{array}{c} \dfrac{Z}{y} \leq x \leq +\infty \\[2mm] -\infty \leq y \leq 0 \end{array}\right\} A_2$$

Applying these limits to the integrals in (9-93) and performing the differentiation with respect to Z, we obtain the density function for Z.

$$p(Z) = \frac{d}{dz}\left[\int_0^\infty \int_{-\infty}^{Z/y} p(x, y) \, dx \, dy + \int_{-\infty}^0 \int_{Z/y}^{+\infty} p(x, y) \, dx \, dy \right]$$

$$= \int_0^\infty \frac{1}{y} p\left(x = \frac{Z}{y}, y\right) dy - \int_{-\infty}^0 \frac{1}{y} p\left(x = \frac{Z}{y}, y\right) dy \qquad (9\text{-}94)$$

For x and y statistically independent, (9-94) simplifies somewhat to

$$p(Z) = \int_0^\infty \frac{1}{y} p\left(x = \frac{Z}{y}\right) p(y) \, dy - \int_{-\infty}^0 \frac{1}{y} p\left(x = \frac{Z}{y}\right) p(y) \, dy \qquad (9\text{-}95)$$

As an example of practical interest, let x and y be statistically independent zero-mean Gaussian random variables with density functions

$$p(x) = \frac{1}{\sigma_x \sqrt{2\pi}} \exp\left(\frac{-x^2}{2\sigma_x^2}\right)$$

$$p(y) = \frac{1}{\sigma_y \sqrt{2\pi}} \exp\left(\frac{-y^2}{2\sigma_y^2}\right)$$

Substitution in (9-95) gives the p.d.f. for $z = xy$.

$$p(z) = \frac{1}{\sigma_x \sigma_y 2\pi} \left\{ \int_0^\infty \frac{1}{y} \exp\left[-\left(\frac{z^2}{2\sigma_x^2 y^2} + \frac{y^2}{2\sigma_y^2} \right) \right] dy \right.$$

$$\left. - \int_{-\infty}^0 \frac{1}{y} \exp\left[-\left(\frac{z^2}{2\sigma_x^2 y^2} + \frac{y^2}{2\sigma_y^2} \right) \right] dy \right\} \tag{9-96}$$

The integrand in the integrals in (9-96) is an odd function of y, with the result that the second integral is the negative of the first. Therefore,

$$p(z) = \frac{1}{\sigma_x \sigma_y \pi} \int_0^\infty \frac{1}{y} \exp\left[-\left(\frac{z^2}{2\sigma_x^2 y^2} + \frac{y^2}{2\sigma_y^2} \right) \right] dy \tag{9-97}$$

The integral in (9-97) is similar to the defining integral for the zero-order Bessel function of the second kind. For instance, from [1] we find, for $u > 0$,

$$K_0(u) = \frac{1}{2} \int_0^\infty \frac{1}{t} \exp\left[-\left(\frac{u^2}{4t} + t \right) \right] dt \tag{9-98}$$

Now let $t = y^2/2\sigma_y^2$, $dt = y\, dy/\sigma_y^2$ and $u^2 = z^2/\sigma_x^2\sigma_y^2$. Substitution in (9-98) and (9-97) gives, for $z > 0$,

$$p(z) = \frac{1}{\sigma_x \sigma_y \pi} K_0\left(\frac{z}{\sigma_x \sigma_y} \right) \qquad \text{for } z > 0$$

Although (9-98) defines $K_0(u)$ only for positive values of u, $p(z)$ is obviously an even function in this example. Hence, we may write, for z either positive or negative,

$$p(z) = \frac{1}{\sigma_x \sigma_y \pi} K_0\left(\frac{|z|}{\sigma_x \sigma_y} \right) \tag{9-99}$$

This density function is shown in Figure 9-16.

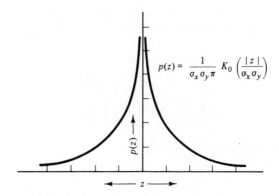

$$p(z) = \frac{1}{\sigma_x \sigma_y \pi} K_0 \left(\frac{|z|}{\sigma_x \sigma_y} \right)$$

Figure 9-16 Probability density function for product of independent, zero-mean Gaussian variables.

9.7.7 Ratio of Random Variables

Now let z be the ratio of the random variables x and y:

$$z = \frac{x}{y}$$

The regions in the x-y plane for which $z \leq Z$ are obtained by plotting the equation $x = yZ$, as shown in Figure 9-17. For y positive, region A_1 is defined by

$$-\infty \leq x \leq yZ$$

and for y negative, we have for A_2

$$Zy \leq x \leq +\infty$$

The cumulative probability for z may now be written as

$$P(z \leq Z) = \int_0^\infty \int_{-\infty}^{yZ} p(x, y) \, dx \, dy + \int_{-\infty}^0 \int_{yZ}^{+\infty} p(x, y) \, dx \, dy$$

Differentiation with respect to Z gives

$$p(Z) = \int_0^\infty yp(x = yZ, y) \, dy - \int_{-\infty}^0 yp(x = yZ, y) \, dy \qquad (9\text{-}100)$$

For x and y independent, this simplifies to

$$p(Z) = \int_0^\infty yp(x = yZ)p(y) \, dy - \int_{-\infty}^0 yp(x = yZ)p(y) \, dy \qquad (9\text{-}101)$$

As an example, assume x and y statistically independent, with x a Gaussian variable, and y a Rayleigh variable. Thus

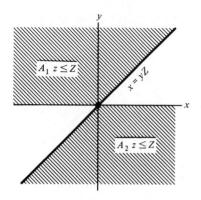

Figure 9-17 Regions in x-y for which $z \leq Z$, where $z = x/y$.

$$p(x) = \frac{1}{\sigma_x \sqrt{2\pi}} \exp\left(\frac{-x^2}{2\sigma_x^2}\right)$$

$$p(y) = \begin{cases} \dfrac{y}{\sigma_y^2} \exp\left(\dfrac{-y^2}{2\sigma_y^2}\right) & \text{for } y \geq 0 \\ 0 & \text{for } y < 0 \end{cases} \qquad (9\text{-}102)$$

and

$$p(x = yZ) = \frac{1}{\sigma_x \sqrt{2\pi}} \exp\left(\frac{-y^2 Z^2}{2\sigma_x^2}\right) \qquad (9\text{-}103)$$

Substitution of (9-102) and (9-103) in (9-101) gives

$$p(Z) = \frac{1}{\sigma_x \sigma_y^2 \sqrt{2\pi}} \int_0^\infty y^2 \exp(-ay^2)\, dy \qquad (9\text{-}104)$$

where

$$a = \frac{1}{2}\left(\frac{1}{\sigma_y^2} + \frac{Z^2}{\sigma_x^2}\right)$$

Only the integral over A_1 need be considered because $p(y)$ is zero for y negative. Evaluating the definite integral in (9-104) finally gives

$$p(Z) = \frac{\sigma_y}{2\sigma_x}\left[1 + \left(\frac{\sigma_y}{\sigma_x}\right)^2 Z^2\right]^{-3/2} \qquad (9\text{-}105)$$

This density function is shown in normalized form in Figure 9-18. The mean value of z is easily shown to be zero, and the variance is infinite. If a sum of independent random variables contains one or more variables of this type, we may not invoke the central limit theorem regardless of the form of the other variables.

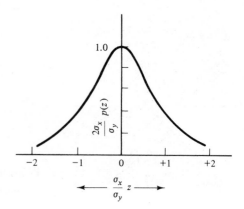

Figure 9-18 Probability density function for the ratio of Gaussian and Rayleigh variables.

9.7.8 Integrals of Random Variables

Assume a random process with sample functions $x_n(t)$. The integral of a particular sample function over the finite limits $[t_1, t_2]$ is itself a random variable, expressed as

$$y(n, t_1, t_2) = \int_{t_1}^{t_2} x_n(t)\, dt \tag{9-106}$$

The variable y will, in general, have a different value for different sample functions of x, and for different limits.

The expected value of y in (9-106) is obtained by averaging over all possible sample functions. Thus

$$E[y(t_1, t_2)] = \lim_{N\to\infty} \frac{1}{N} \sum_{n=1}^{N} \left[\int_{t_1}^{t_2} x_n(t)\, dt \right] \tag{9-107}$$

Now interchange the order of integration and summation and obtain

$$E[y(t_1, t_2)] = \int_{t_1}^{t_2} \left[\lim_{N\to\infty} \sum_{n=1}^{N} \frac{x_n(t)}{N} \right] dt$$

$$= \int_{t_1}^{t_2} E[x(t)]\, dt \tag{9-108}$$

Thus the expected value of the integral is the integral of the expected value. In general, $E[x(t)]$ may be a function of time. However, for a stationary process $E[x(t)]$ is not a function of time and may be removed from the integral. Equation (9-108) may then be written

$$E[y(t_2 - t_1)] = E[x] \int_{t_1}^{t_2} dt = (t_2 - t_1)E[x] \tag{9-109}$$

Let the variable $x_n(t)$ be multiplied by a deterministic function $h(t)$. The expected value of the integral of the product is

$$E[y(t_1, t_2)] = E\left[\int_{t_1}^{t_2} h(t)x_n(t)\, dt\right]$$

$$= \int_{t_1}^{t_2} h(t)E[x(t)]\, dt \qquad (9\text{-}110)$$

For a stationary process, this becomes

$$E[y(t_1, t_2)] = E[x]\int_{t_1}^{t_2} h(t)\, dt \qquad (9\text{-}111)$$

9.8 CORRELATION FUNCTIONS OF RANDOM PROCESSES

Let $x_n(t)$ be a sample function of a real random process. We may form the autocorrelation function of a limited time segment of $x_n(t)$ as follows:

$$\mathcal{R}_{x_{n_T}}(\tau) = \int_{-T/2}^{+T/2} x_{n_T}(t)x_{x_{n_T}}(t - \tau)\, dt \qquad (9\text{-}112)$$

where

$$x_{n_T} = x_n(t)\,\mathrm{rect}\left(\frac{t}{T}\right)$$

For a finite interval T, $\mathcal{R}_{x_{n_T}}(\tau)$ is itself a random variable.

Now let T approach infinity and form the time-average autocorrelation function

$$R_{x_n}(\tau) = \lim_{T\to\infty} \frac{1}{T}\int_{-T/2}^{+T/2} x_n(t)\, x_n(t - \tau)\, dt = \langle x_n(t)\, x_n(t - \tau)\rangle \qquad (9\text{-}113)$$

For an ergodic process, the time average in (9-113) is equal to the statistical average. Thus

$$R_{x_n}(\tau) = R_x(\tau) = \langle x_n(t)x_n(t - \tau)\rangle = E[\mathbf{x}_t\mathbf{x}_{t-\tau}]$$

$$= \int \mathbf{x}_t\mathbf{x}_{t-\tau}p(\mathbf{x}_t, \mathbf{x}_{t-\tau})\, d\mathbf{x} \qquad (9\text{-}114)$$

The expectation in (9-114) is the *process autocorrelation function,* and for the ergodic process is equal to the *time-average autocorrelation function* of a sample function of the process.

The properties of the process autocorrelation function of the ergodic process are the same as for the time autocorrelation function. Thus

$$R_\mathbf{x}(\tau) \le R_\mathbf{x}(0) = E[\mathbf{x}^2]$$

and

$$R_x(\tau) = R_x^*(-\tau)$$

If the process is purely real, the autocorrelation function is an even function.

For the ergodic process, the Fourier transform of the process auto-correlation function is of particular interest. To obtain this relationship, consider first the transform of a particular sample function over a finite time interval.

$$X_{n_T}(f) = \int_{-\infty}^{+\infty} \text{rect}\left(\frac{t}{T}\right) x_n(t) \exp(-j2\pi ft)\, dt \tag{9-115}$$

The energy spectrum for this signal is

$$|X_{n_T}(f)|^2$$

$$= \left| \int_{-\infty}^{+\infty} \text{rect}\left(\frac{t}{T}\right) x_n(t) \exp(-j2\pi ft)\, dt \right|^2$$

$$= \int_{-\infty}^{+\infty} \text{rect}\left(\frac{t}{T}\right) x_n(t) \exp(-j2\pi ft)\, dt \int_{-\infty}^{+\infty} \text{rect}\left(\frac{t'}{T}\right) x_n(t') \exp(+j2\pi ft')\, dt'$$

Now make the substitution $t' = t - \tau$ and write

$$|X_{n_T}(f)|^2 = \int_{-\infty}^{+\infty} \int_{-\infty}^{+\infty} x_n(t) x_n(t - \tau) \text{rect}\left(\frac{t}{T}\right) \text{rect}\left(\frac{t - \tau}{T}\right) \exp(-j2\pi f\tau)\, dt\, d\tau$$

The expected value of this expression is

$$E[|X_{n_T}(f)|^2]$$

$$= \int_{-\infty}^{+\infty} \int_{-\infty}^{+\infty} E[x_n(t) x_n(t - \tau)] \text{rect}\left(\frac{t}{T}\right) \text{rect}\left(\frac{t - \tau}{T}\right) \exp(-j2\pi f\tau)\, dt\, d\tau$$

The expectation inside the integral is the process autocorrelation function. Therefore,

$$E[|x_{n_T}(f)|^2] = \int_{-\infty}^{+\infty} R_x(\tau) \exp(-j2\pi f\tau) \left[\int_{-\infty}^{+\infty} \text{rect}\left(\frac{t}{T}\right) \text{rect}\left(\frac{t - \tau}{T}\right) dt \right] d\tau$$

The bracketed integral is the autocorrelation function of a rect function, so that finally

$$E[|X_{n_T}(f)|^2] = T \int_{-T}^{+T} \left(1 - \frac{|\tau|}{T}\right) R_x(\tau) \exp(-j2\pi f\tau)\, d\tau \tag{9-116}$$

Divide both sides of (9-116) by T and take the limit as T approaches infinity. This defines the expected value of the sample function power spectral density, which for the ergodic process is called the *process power spectral density*, and identified as the transform of the process autocorrelation function. Thus

$$\Psi_x(f) = \lim_{T \to \infty} E\left[\frac{|X_{n_T}(f)|^2}{T}\right] = \int_{-\infty}^{+\infty} R_x(\tau) \exp(-j2\pi f\tau)\, d\tau \tag{9-117}$$

9.8.1 Process Autocovariance Functions

The covariance of two random variables was defined in (9-46) by first subtracting the mean value from each variable, and then forming the expectation of the product of the resulting zero-mean variables. The *autocovariance function* of a *random process* is similarly defined by forming the autocorrelation function, after first removing the mean value.

$$C_x(\tau) = E\{[x(t) - \bar{x}][x(t - \tau) - \bar{x}]\}$$
$$= E[(x(t)x(t - \tau)] - (\bar{x})^2 \tag{9-118}$$
$$= R_x(\tau) - (\bar{x})^2$$

The transform of the autocovariance function is identical to the transform of the autocorrelation function, minus the term at $f = 0$.

$$C_x(\tau) \longleftrightarrow \Psi_x(f) - \Psi_x(0) \tag{9-119}$$

9.8.2 Process Cross-Correlation Functions

Let $z(t)$ be the sum of two stationary variables $x(t)$ and $y(t)$. The variable $z(t)$ is necessarily stationary in this case, so that the autocorrelation function of the z-process is

$$R_z(\tau) = R_x(\tau) + R_y(\tau) + R_{xy}(\tau) + R_{yx}(\tau)$$
$$= R_x(\tau) + R_y(\tau) + 2\text{Re}[R_{xy}(\tau)] \tag{9-120}$$

The Fourier transform of the cross-correlation function, $R_{xy}(\tau)$, is the *cross-spectral density* of the **x** and **y** processes.

$$R_{xy}(\tau) \longleftrightarrow \Psi_{xy}(f) \tag{9-121}$$

The z-process spectral density can now be written as

$$\Psi_z(f) = \Psi_x(f) + \Psi_y(f) + \Psi_{xy}(f) + \Psi_{yx}(f)$$
$$= \Psi_x(f) + \Psi_y(f) + 2\text{Re}[\Psi_{xy}(f)] \tag{9-122}$$

If the **x** and **y** processes are uncorrelated, their cross-spectral densities are zero, and the z-process spectral density is the sum of the **x** and **y** process spectral densities.

9.8.3 Filtering of Random Variables

The concept of the power spectral density of a random process is useful in determining the characteristics of the output of a linear filter stimulated at the input by a sample function of the random process.

Let $x(t)$ be a sample function of an ergodic process, with zero mean and variance σ_x^2. The output of a linear filter, with $x(t)$ at the input, is a random variable, $y(t)$. The output *process*, assuming all possible sample functions at the input, is also ergodic with a power spectral density defined as

$$\Psi_y(f) = \Psi_x(f) \, |H(f)|^2 \tag{9-123}$$

where $\Psi_x(f)$ = power spectral density of the x-process
$\Psi_y(f)$ = power spectral density of the y-process
$|H(f)|^2$ = squared-magnitude of the filter frequency response

Now assume that the power in the input process is essentially uniformly distributed over a bandwidth, B, large compared with the frequency extent of $H(f)$. Then

$$\Psi_x(f) = \frac{\sigma_x^2}{B} \, \text{rect}\left(\frac{f}{B}\right)$$

and

$$\Psi_y(f) = \frac{\sigma_x^2}{B} \, \text{rect}\left(\frac{f}{B}\right) |H(f)|^2 \simeq \frac{\sigma_x^2}{B} \, |H(f)|^2 \tag{9-124}$$

The autocorrelation function of the output process is obtained as the inverse transform of the process power spectral density. For this example, this gives

$$R_y(\tau) \longleftrightarrow \frac{\sigma_x^2}{B} \, |H(f)|^2 \tag{9-125}$$

Thus, for a flat input noise spectrum, the output autocorrelation function has a form determined by the characteristics of the linear filter.

For a zero-mean input process, the output process is also zero mean, with an output variance defined by

$$\sigma_y^2 = R_y(0) = \int_{-\infty}^{+\infty} \Psi_x(f) \, |H(f)|^2 \, df$$

For $\Psi_x(f)$ flat,

$$\sigma_y^2 = \frac{\sigma_x^2}{B} \int_{-\infty}^{+\infty} |H(f)|^2 \, df \tag{9-126}$$

For illustration, let the filter frequency response also be flat over some band $\beta \ll B$.

$$H(f) = \text{rect}\left(\frac{f}{\beta}\right) = |H(f)|^2$$

Then

$$R_y(\tau) \longleftrightarrow \frac{\sigma_x^2}{B} \, \text{rect}\left(\frac{f}{\beta}\right)$$

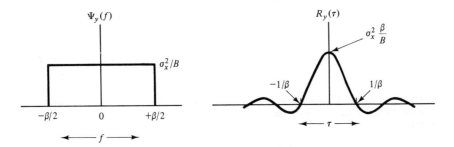

Figure 9-19   Output process spectral density and autocorrelation function for a flat input process spectral density and a rectangular filter.

from which

$$R_y(\tau) = \sigma_x^2 \left(\frac{\beta}{B}\right) \text{sinc}(\tau\beta) \qquad (9\text{-}127)$$

and

$$R_y(0) = \sigma_y^2 = \sigma_x^2 \left(\frac{\beta}{B}\right)$$

The output power spectral density and autocorrelation function for this example are shown in Figure 9-19. Notice that the output autocorrelation function is zero for a time separation of $1/\beta$. Time samples separated by at least $1/\beta$ will tend to be uncorrelated.

9.8.4 Process Correlation Time

For a rectangular process spectral density, time samples separated by exactly $1/\beta$ are uncorrelated. In this case, the separation is also equal to the reciprocal of the effective bandwidth, as defined in Section 8.4.1. In the general case, the *process correlation time* is defined as the reciprocal of the process effective bandwidth. That is,

$$T_{\text{cor}} = \frac{1}{\beta_e} = \frac{\int_{-\infty}^{+\infty} |R(\tau)|^2 \, d\tau}{R^2(0)} \qquad (9\text{-}128)$$

Time samples separated by less than T_{cor} are likely to be highly correlated, while separations greater than T_{cor} should result in essentially uncorrelated samples. Figure 9-20 shows the spectral density of a real narrowband ergodic process centered at a carrier frequency, f_0. The correlation function for this process can be expressed in terms of the *envelope correlation function* with the complex envelope defined as in 6.10. The real process spectral density is obtained by

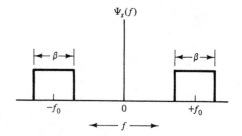

Figure 9-20 Process spectral density for
a narrowband random process centered at
$\pm f_0$.

translating the envelope spectral density to $\pm f_0$ and dividing by 4. Thus

$$R_s(\tau) \longleftrightarrow \Psi_s(f) = \tfrac{1}{4}[\Psi_\mu(f + f_0) + \Psi_\mu(f - f_0)]$$

where $\Psi_\mu(f)$ is the complex envelope spectral density. From the rule for translation, this gives

$$R_s(\tau) = \tfrac{1}{4}R_\mu(\tau)[\exp(j2\pi f_0\tau) + \exp(-j2\pi f_0\tau)]$$

$$= \tfrac{1}{2}R_\mu(\tau)\cos \omega_o\tau \tag{9-129}$$

The effective bandwidth of the envelope function is one-half the effective
bandwidth of the underlying real function. That is,

$$\frac{R_s^2(0)}{\int R_s^2(\tau)\, d\tau} = 2\left[\frac{R_\mu^2(0)}{\int R_\mu^2(\tau)\, d\tau}\right] \tag{9-130}$$

This results from the fact that the spectral density of the complex envelope is
determined by the positive frequency portion of the spectrum of the real function. In most system analysis problems, the effective bandwidth of the envelope
is the parameter of primary interest.

9.8.5 Autocorrelation Function of the Square of a
Random Variable

Let $x(t)$ be sample functions of a real ergodic process. A typical system
may form the square of $x(t)$. Define

$$x_1 = x(t)$$

$$x_2 = x(t - \tau)$$

$$y(t) = x^2(t)$$

$$E[y] = E[x^2] = E[x_1^2] = E[x_2^2]$$

Then

$$R_y(\tau) = R_{x^2}(\tau) = E[x_1^2 x_2^2]$$

In the general case, the autocorrelation function of the **y**-process may be difficult to evaluate. Fortunately, in the frequently encountered case where the **x**-process is zero-mean Gaussian, $R_y(\tau)$ bears a simple relationship to $R_x(\tau)$. To show this, we require an important characteristic of multivariate Gaussian variables.

Assume that the **x**-process is zero-mean Gaussian with variance σ_x^2. For τ sufficiently large, x_1 and x_2 are essentially statistically independent. The joint density function then becomes

$$p(x_1, x_2) = p(x_1)p(x_2) = \frac{1}{2\pi\sigma_x^2} \exp\left(\frac{-x_1^2}{2\sigma_x^2}\right) \exp\left(\frac{-x_2^2}{2\sigma_x^2}\right)$$

$$= \frac{1}{2\pi\sigma_x^2} \exp\left[\frac{-(x_1^2 + x_2^2)}{2\sigma_x^2}\right] \qquad (9\text{-}131)$$

If x_1 and x_2 are not statistically independent, the joint density function can be shown to be [2]

$$p(x_1, x_2) = \frac{1}{2\pi\sigma_x^2\sqrt{1 - \rho_{12}^2}} \exp\left[\frac{-(x_1^2 - 2\rho_{12}x_1x_2 + x_2^2)}{2\sigma_x^2(1 - \rho_{12}^2)}\right] \qquad (9\text{-}132)$$

where $\rho_{12} = E[x_1x_2]/\sigma_x^2$.

Notice that if the normalized correlation ρ_{12} is zero, the joint Gaussian density function of (9-132) reduces to that for statistically independent variables in (9-131). That is, for joint-Gaussian random variables, zero correlation guarantees statistical independence. This is a property that is not generally true for non-Gaussian variables. Non-Gaussian variables for which the normalized correlation is zero are said to be *linearly independent*. For Gaussian variables, *linear independence* implies *statistical independence*.

We now return to the problem of evaluating the autocorrelation function for the square of a zero-mean Gaussian variable. Using the correlation coefficients developed in Section 8.2, separate x_1 into two orthogonal components as follows:

$$x_1 = c_{12}x_2 + x_d$$

where

$$c_{12} = \frac{E[x_1x_2]}{E[x_2^2]} = \rho_{12}$$

Recall that x_d and x_2 are orthogonal. Being Gaussian, this also guarantees statistical independence.

Now write

$$x_1^2x_2^2 = (\rho_{12}x_2 + x_d)^2x_2^2$$

$$= \rho_{12}^2x_2^4 + 2\rho_{12}x_2^3x_d + x_2^2x_d^2 \qquad (9\text{-}133)$$

Because of the independence of x_2 and x_d, the expected value of (9-133) reduces to

$$E[x_1^2 x_2^2] = \rho_{12}^2 E[x_2^4] + E[x_2^2]E[x_d^2] \qquad (9\text{-}134)$$

From (9-83),

$$E[x_2^4] = E[x^4] = 3E^2[x^2] \qquad (9\text{-}135)$$

Also

$$\begin{aligned} E[x_d^2] &= E[x_1^2] - 2\rho_{12}E[x_1 x_2] + \rho_{12}^2 E[x_2^2] \\ &= E[x^2](1 - \rho_{12}^2) \end{aligned} \qquad (9\text{-}136)$$

Substitution of (9-135) and (9-136) in (9-134) gives

$$\begin{aligned} E[x_1^2 x_2^2] &= 3\rho_{12}^2 E^2[x^2] + (1 - \rho_{12}^2)E^2[x^2] \\ &= (1 + 2\rho_{12}^2)E^2[x^2] \end{aligned}$$

or

$$E[x_1^2 x_2^2] = E^2[x^2] + 2E^2[x_1 x_2]$$

This may also be written

$$R_y(\tau) = R_{x^2}(\tau) = R_x^2(0) + 2R_x^2(\tau) \qquad (9\text{-}137)$$

Thus, for the Gaussian variable the autocorrelation function at the output of a square-law device can be found directly from a knowledge of the autocorrelation function at the input.

The power spectrum of y is found by direct transformation of (9-137):

$$\Psi_y(f) = R_x^2(0)\delta(f) + 2\mathcal{H}_{\Psi_x}(f) \qquad (9\text{-}138)$$

where

$$\mathcal{H}_{\Psi_x}(f) = \Psi_x(f) \otimes \Psi_x(f) \longleftrightarrow R_x^2(\tau)$$

The Gaussian variable is a rare example of a random variable for which both the p.d.f. and the power spectrum of the square of the variable may be readily obtained.

In analysis involving complex random variables, the squared magnitude of the complex variable is often required. For the complex Gaussian variable, the output autocorrelation function is again simply related to the input autocorrelation function. Let z be defined as

$$z = x + jy$$

and

$$R_z(\tau) = E[z(t)z^*(t - \tau)] = E[z_1 z_2^*]$$

where x and y are statistically independent, zero-mean, equal-variance Gaussian variables. It is left to the reader to verify that for $|z|^2 = x^2 + y^2$,

$$R_{|z|^2}(\tau) = E[|z_1|^2|z_2|^2] = E^2[|z|^2] + E^2[z_1 z_2^*]$$
$$= R_z^2(0) + R_z^2(\tau) \tag{9-139}$$

Comparison of (9-137) and (9-139) reveals a factor-of-2 difference on the term associated with the autocorrelation function of the input variable.

The results given in (9-137) and (9-139) are special cases of a more general relationship involving the zero-mean jointly Gaussian variables x_1, x_2, x_3, and x_4. For these variables we may write

$$E[x_1 x_2 x_3 x_4] = E[x_1 x_2]E[x_3 x_4] + E[x_1 x_3]E[x_2 x_4] + E[x_1 x_4]E[x_2 x_3] \tag{9-140}$$

It is easily verified that if $x_1 = x_2$ and $x_3 = x_4$, the result obtained is identical to (9-137).

9.9 EXAMPLE: MULTIPATH TRANSMISSION CHANNEL

Acoustic transmission between two points in the ocean often involves multiple paths, each with its own propagation delay and loss factor. We now consider a simplified model of such a transmission channel, as shown in Figure 9-21, for the purpose of demonstrating some aspects of random processes developed in preceding sections.

The input signal $x(t)$ is a sample function of an ergodic zero-mean-Gaussian (ZMG) process with variance σ_x^2. The coefficients a_n and time delays t_n contained in the channel impulse response $h(t)$ are random parameters that are assumed fixed for any particular realization of $h(t)$. Thus, let $h_k(t)$ be a particular realization of the impulse response expressed as

$$h_k(t) = \sum_{n=1}^{N} a_{n_k} \delta(t - t_{n_k})$$

The channel output conditioned on this particular impulse response is $z_k(t)$.

The parameters a_n and t_n are assumed independent of frequency and independent of each other. Also, a_n is zero mean and independent of a_m for all $n \neq m$. The range of time delays, $t_N - t_1$, is assumed small compared to the total time period of interest.

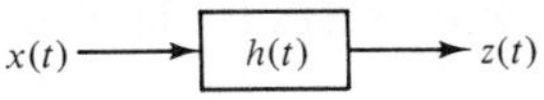

$$x(t) \longrightarrow \boxed{h(t)} \longrightarrow z(t)$$

$$h(t) = \sum_{n=1}^{N} a_n \delta(t - t_n)$$

$$z(t) = x(t) \otimes h(t) = \sum_n a_n x(t - t_n)$$
$$= \sum_n y_n(t - t_n)$$

Figure 9-21 Random multipath transmission channel.

Because the impulse response $h_k(t)$ is time invariant, the output variable $z_k(t)$ is a stationary random variable. That the z_k process is also ergodic can be verified by showing that the time average and statistical average of $z_k^2(t)$ are equal. Thus the statistical average over the ensemble of sample functions of $x(t)$ is

$$E[z_k^2(t)] = E[z^2|h_k] = E\left\{\left[\sum_n a_{n_k} x(t - t_{n_k})\right]^2\right\}$$

$$= E\left[\sum_m \sum_n a_{m_k} a_{n_k} x(t - t_{m_k}) x(t - t_{n_k})\right]$$

$$= \sum_m \sum_n a_{m_k} a_{n_k} E[x(t - t_{m_k}) x(t - t_{n_k})]$$

$$= \sum_m \sum_n a_{m_k} a_{n_k} R_x(t_{m_k} - t_{n_k}) \tag{9-141}$$

The time average of $z_k^2(t)$ is

$$\langle z_k^2(t) \rangle = \lim_{T \to \infty} \frac{1}{T} \int_T z_k^2(t) \, dt$$

$$= \sum_m \sum_n a_{m_k} a_{n_k} \left\{ \lim_{T \to \infty} \frac{1}{T} \int_T x(t - t_{m_k}) x(t - t_{n_k}) \, dt \right\} \tag{9-142}$$

Because the x-process is ergodic, the time average in (9-142) is equal to the statistical average in (9-141). Thus

$$\langle z_k^2(t) \rangle = \sum_m \sum_n a_{m_k} a_{n_k} R_x(t_{m_k} - t_{n_k}) = E[z_k^2(t)] \tag{9-143}$$

The *unconditioned* z-process includes all possible sample functions $z_k(t)$ obtained by allowing all possible realizations of $h(t)$. The statistical average of the unconditioned variable z^2 is

$$E[z^2] = E\left[\sum_m \sum_n a_m a_n x(t - t_m) x(t - t_n)\right]$$

$$= \sum_m \sum_n E[a_m a_n] E[x(t - t_m) x(t - t_n)] \tag{9-144}$$

Because of the independence of a_m and a_n, (9-144) is zero except for $m = n$. Therefore,

$$E[z^2] = \sum_n E[a_n^2] E[x^2] = N\sigma_a^2 \sigma_x^2 \tag{9-145}$$

Equation (9-145) is obviously not equal to the time average of a particular sample function $z_k(t)$, leading to the conclusion that the unconditioned z-process is stationary but not ergodic.

In a similar manner we could show that $y_n(t) = a_n x(t - t_n)$ is stationary and ergodic when conditioned on a particular $h_k(t)$. Thus

$$E[y_n^2|h] = a_{n_k}^2 E[x^2(t - t_n)|h] = a_{n_k}^2 \sigma_x^2$$

$$\langle y_n^2|h \rangle = a_{n_k}^2 \lim_{T \to \infty} \frac{1}{T} \int_T x^2(t - t_n)\, dt = a_{n_k}^2 \sigma_x^2$$

(9-146)

On the other hand, the unconditioned statistical average of $y_n^2(t)$ is

$$E[y_n^2] = E[a_n^2 x^2(t - t_n)] = E[a_n^2]E[x^2] = \sigma_a^2 \sigma_x^2 \qquad (9\text{-}147)$$

indicating that the unconditioned y-process is not ergodic.

Because the y-process is not ergodic, the *temporal* p.d.f. for a particular sample function $y_n(t)$, conditioned on $h_k(t)$, is not constrained to have the same form as the *ensemble* p.d.f. of the unconditioned process. For a particular sample function the temporal p.d.f. is ZMG with variance $a_{n_k}^2 \sigma_x^2$. Considered across the ensemble, the value of y is the product of the random variables a_n and x. As a particular example, if a_n is a ZMG variable, y is the product of ZMG variables, resulting in an ensemble p.d.f. in the form of a zero-order Bessel function of the second kind as derived in Section 9.7.6. Thus

$$p(y) = \frac{1}{\sigma_a \sigma_x \pi} K_0 \left(\frac{|y|}{\sigma_a \sigma_x} \right) \qquad (9\text{-}148)$$

The temporal p.d.f. of $z_k(t)$ is obviously ZMG because it is the sum of the ZMG variables $y_{n_k}(t)$. The ensemble p.d.f. of the unconditioned variable z is not in general Gaussian because it is the sum of N variables, not necessarily independent, with density functions given by (9-148). If the elements of z are independent, the ensemble p.d.f. may approach the Gaussian form by the central limit theorem as N becomes large.

The spectral properties of $z_k(t)$ will now be examined. Because $z_k(t)$ is a continuous random variable, it is necessary to estimate the spectral properties from the Fourier transform of finite-duration segments of $z_k(t)$. Define such a limited segment of duration T centered at ℓT as

$$z_{k_T}(t, \ell) = z_k(t)\, \text{rect}\left(\frac{t - \ell T}{T} \right) \qquad (9\text{-}149)$$

The duration T is assumed long compared with the total range of delays in $h_k(t)$; that is, $T \gg (t_N - t_1)$.

The Fourier transform of $z_{k_T}(t, \ell)$ is

$$Z_{k_T}(f, \ell) = \int_{-\infty}^{+\infty} z_{k_T}(t, \ell) \exp(-j2\pi ft)\, dt$$

$$= \sum_n a_{n_k} \int_{-\infty}^{+\infty} x(t - t_{n_k}) \text{rect}\left(\frac{t - \ell T}{T} \right) \exp(-j2\pi ft)\, dt$$

$$= \sum_n a_{n_k} X_{k_T}(f, \ell, n) \qquad (9\text{-}150)$$

where

$$x(t - t_{n_k}) \operatorname{rect}\left(\frac{t - \ell T}{T}\right) \longleftrightarrow X_{k_T}(f, \ell, n)$$

For the range of delays small compared with T we have approximately

$$X_{k_T}(f, \ell, n) \simeq X_T(f, \ell, 0) \exp(-j2\pi f t_{n_k}) \qquad (9\text{-}151)$$

from which

$$Z_{k_T}(f, \ell) \simeq X_T(f, \ell, 0)\left[\sum_n a_{n_k} \exp(-j2\pi f t_{n_k})\right] \qquad (9\text{-}152)$$

But the bracketed term in (9-152) is the Fourier transform of $h_k(t)$. Thus

$$Z_{k_T}(f, \ell) \simeq X_T(f, \ell, 0) H_k(f) \qquad (9\text{-}153)$$

Consider now the random variable in (9-153) with a fixed value of $f = f_0$. Let the x-process be a wideband process with a constant spectral density equal to σ_x^2/β. The variance of $Z_{k_T}(f_0, \ell)$ is then

$$\operatorname{Var}[Z_{k_T}(f_0, \ell)] = \frac{T\sigma_x^2}{\beta} |H_k(f_0)|^2 = \sigma_{Z_T, f_0}^2 \qquad (9\text{-}154)$$

Conditioned on a particular $h_k(t)$, $Z_{k_T}(f_0, \ell)$ is zero-mean complex Gaussian variable. The squared magnitude of $Z_{k_T}(f_0, \ell)$ is therefore chi-squared in form with two degrees of freedom. Let $\Psi_{f_0} = |Z_{k_T}(f_0, \ell)|^2$ and write

$$p(\Psi_{f_0}|h) = \frac{1}{\overline{\Psi_{f_0}}} \exp\left(\frac{-\Psi_{f_0}}{\overline{\Psi_{f_0}}}\right) \qquad (9\text{-}155)$$

The variance of Ψ_{f_0} is

$$\operatorname{Var}[\Psi_{f_0}|h] = \sigma_{Z_T, f_0}^4 = \frac{T^2 \sigma_x^4}{\beta^2} |H_k(f_0)|^4 \qquad (9\text{-}156)$$

Because $H_k(f)$ is not constant over the frequency band, the variance of Ψ_{f_0} is a function of f_0.

Let the time index ℓ be fixed at $\ell = \ell_0$, and consider the variable $Z_{k_T}(f, \ell_0) = X_T(f, \ell_0) H_k(f)$. If the lowest frequency of interest is limited such that the phase terms $2\pi f t_n$ in $H_k(f)$ are uniformly distributed over 2π radians, then $H_k(f)$ is a zero-mean complex Gaussian variable. The variable $Z_{k_T}(f, \ell_0)$ is therefore the product of two independent complex Gaussian variables, and the power spectral density is the product of two independent chi-squared variables, each with two degrees of freedom. Let $\Psi_{\ell_0} = |Z_{k_T}(f, \ell_0)|^2$ and write

$$\Psi_{\ell_0} = |X_T(f, \ell_0)|^2 |H_k(f)|^2$$

$$E[\Psi_{\ell_0}] = \sigma_{X_T}^2 \sigma_H^2 = \frac{T\sigma_x^2 \sigma_H^2}{\beta} \qquad (9\text{-}157)$$

and

$$\text{Var}[\Psi_{\ell_0}] = E[\Psi_{\ell_0}^2] - E^2[\Psi_{\ell_0}]$$

but

$$E[\Psi_{\ell_0}^2] = E[|X_T|^4] E[|H_k|^4]$$
$$= (2\sigma_{X_T}^4)(2\sigma_H^4)$$

hence

$$\text{Var}[\Psi_{\ell_0}] = 4\sigma_{X_T}^4 \sigma_H^4 - \sigma_{X_T}^4 \sigma_H^4 = 3\sigma_{X_T}^4 \sigma_H^4 \tag{9-158}$$

Notice that, conditioned on $h_k(t)$, the variance with fixed ℓ_0 is different from the variance with fixed f_0.

$$\text{Var}[\Psi_{\ell_0}] = \frac{3 \, \text{Var}[\Psi_{f_0}]\sigma_H^4}{|H(f_0)|^4} \tag{9-159}$$

The p.d.f. for Ψ_{ℓ_0} is obtained by the method discussed in Section 9.7.6 for the product of statistically independent random variables. Substitution of the chi-squared-form density functions for $|X_T|^2$ and $|H_k|^2$ into (9-95) leads, with some manipulation, to a Bessel function solution. The final result is

$$p(\Psi_{\ell_0}|h) = \frac{2}{\sigma_{X_T}^2 \sigma_H^2} K_0 \left(\frac{2\sqrt{\Psi_{\ell_0}}}{\sigma_{X_T}\sigma_H} \right) \tag{9-160}$$

It is instructive to compare the density functions of (9-155) and (9-160). This comparison is facilitated by defining a mean-shifted and standard-deviation-normalized variable u as

$$u = \frac{\Psi - \overline{\Psi}}{[\text{Var}(\Psi)]^{1/2}}$$

The density functions for u corresponding to (9-155) and (9-160) are shown graphically in Figure 9-22. Perhaps the most significant difference between the two p.d.f.'s occurs for large values of the variable u. The modified Bessel function p.d.f. approaches zero much more slowly than does the chi-squared p.d.f. for large values of u. This difference is important in those applications where system performance is affected by the "tails" of the probability density function. The importance of the p.d.f. tail is discussed in more detail in Chapter 13 relative to false-alarm performance and detection threshold setting for a receiving system.

As a final variation on this example, assume that the impulse response $h(t)$ is replaced with a new and independent realization $h(t, \ell)$ at the beginning of each interval T. Thus

$$h(t, \ell) = \sum_n a_{n_\ell} \delta(t - t_{n_\ell})$$

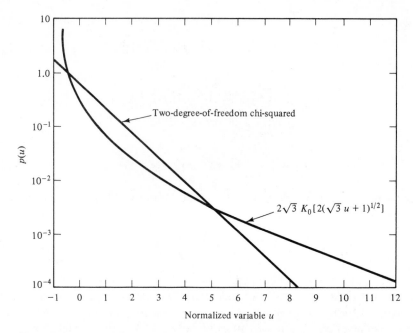

Figure 9-22 Comparison of normalized 2-DOF chi-squared and zero order modified Bessel function p.d.f.'s.

The random variable $|Z_T(f, \ell)|^2$ in this case becomes the product of independent chi-squared variables for either $f = f_0$ or $\ell = \ell_0$. The p.d.f. in either case will be of the form given in (9-160).

This example illustrates the fact that care must be exercised in determining the properties of random variables as they proceed through a series of system operations.

PROBLEMS

9.1. A random variable z is related to the zero-mean statistically independent random variables x_1, x_2, x_3, and x_4 as follows:

$$z = (x_1 + x_2)(x_3 + x_4)$$

Write the expression for $E[z^2]$.

9.2. A random variable x has a p.d.f. of the form shown in Figure P9-1. Write the properly normalized expression for this p.d.f. Calculate the mean value and the variance of x.

9.3. Let z be the sum of n statistically independent random variables, each of the form $y_i = \sin \theta_i$, where θ_i is uniformly distributed over the interval $[0, 2\pi]$. Using the characteristic function method, show that the p.d.f. of z approaches the Gaussian

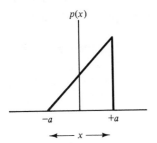

Figure P9-1

form as n approaches infinity. As a hint, the characteristic function for the p.d.f. of y_i will be zero-order Bessel function of the first kind.

9.4. Let $z = xy$, with x and y statistically independent. Determine the p.d.f. of z given that

$$p(x) = \frac{1}{2a} \text{rect}\left(\frac{x}{2a}\right)$$

$$p(y) = \frac{1}{2b} \text{rect}\left(\frac{y}{2b}\right)$$

9.5. Let x be a random variable with a p.d.f. given by

$$p(x) = \frac{1}{2a} \text{rect}\left(\frac{x}{2a}\right)$$

Use (9-73) to confirm that the variance of x is $a^2/3$.

9.6. Let $z(t)$ be a complex Gaussian variable as described in Section 9.7.5. Prove the relationship given in (9-139). That is, show that for $y(t) = |z(t)|^2$,

$$R_y(\tau) = R_z^2(0) + R_z^2(\tau)$$

9.7. Two statistically independent variables x and y have the density functions

$$p(x) = \frac{1}{\pi(1 - x^2)^{1/2}} \text{rect}\left(\frac{x}{2}\right)$$

$$p(y) = y \exp\left(\frac{-y^2}{2}\right) \qquad \text{for } y \geq 0$$

Show that the product xy has a Gaussian p.d.f.

SUGGESTED READING

1. Gradshteyn, A. A., and A. H. Ryzhik, *Tables of Integrals, Series, and Products.* New York: Academic Press, Inc., 1979, p. 959.
2. Davenport, W. B., and W. L. Root, *Random Signals and Noise.* New York: McGraw-Hill Book Company, 1958.

3. Parzen, E., *Modern Probability Theory and Its Applications*. New York: John Wiley & Sons, Inc., 1960.

4. Miller, K. S., *Engineering Mathematics*. New York: Dover Publications, Inc., 1963.

5. Abramowitz, M., and I. A. Stegun, *Handbook of Mathematical Functions,* Natl. Bur. Stand., Appl. Math. Ser. 55. Washington, D.C.: U.S. Government Printing Office, 1964, Chs. 9 and 26.

10 Ambient Noise in the Ocean

10.0 INTRODUCTION

A hydrophone placed in the ocean provides an output signal that for significant periods of time is best described as "noise." That is, the amplitude fluctuates randomly, permitting only a statistical description based on observation over extended intervals. The observed output voltage results from a combination of electrical noise generated in the electronic portions of the system and acoustic noise produced by random pressure fluctuations in the ocean at the hydrophone location. As discussed in Chapter 3, proper system design should result in the acoustic noise being the predominant component. In this chapter we discuss the sources and characteristics of the background acoustic noise, or *ambient noise*. This noise establishes a lower limit on the intensity of useful acoustic signals in an underwater acoustic system.

Many useful acoustic signals in the ocean are noise-like in character. The term "ambient noise" refers to the noise that remains after all easily identifiable sound sources are eliminated. For instance, the presence of many ships randomly distributed over the ocean surface results in a component of ambient noise ascribed to "distant shipping" or "ship traffic." However, the noise produced by a single nearby ship is easily identified and localized, and is therefore treated as an acoustic signal rather than as a part of ambient noise.

For the purposes of system analysis, both the spatial and temporal correlation properties of the ambient noise field are of importance. The relationship between spatial correlation and the angular distribution of ambient noise is developed in this chapter for several simple noise models of practical interest.

Excellent detailed descriptions of the physical sources of ambient noise are available in the literature. The reader is referred especially to Wenz [1] and Urick [2]. A review of this material is presented here as an aid to understanding the mechanisms that lead to the temporal and spatial noise properties.

10.1 SOURCES OF AMBIENT NOISE

The sources of ambient noise are both natural and human-made, with different sources exhibiting different directional and spectral characteristics. Among the natural sources of noise are seismic disturbances, agitation of the sea surface by wind, and thermal activity of the water molecules. In addition, biological sources such as snapping shrimp, croakers, dolphins, and various other fishes and ocean mammals add significantly to the noise background, especially in harbors and coastal waters. The principal human-made component of ambient noise is the sound generated by distant shipping. The term "distant" in this case includes the range from perhaps several hundred to 1000 nautical miles or more.

It is evident that ambient noise levels are highly dependent on geographic location, acoustic transmission characteristics, season of the year, and weather. Distant shipping noise is dependent on the level and type of commercial activity on the high seas, and in recent years this component of noise has been steadily increasing. Even political events, such as the closing of the Suez Canal, have a measurable effect on ambient noise through their influence on global shipping patterns.

Consideration of the characteristics of ambient noise is facilitated by dividing the acoustic spectrum into three bands. The low band covers the range from 1 Hz or lower to several hundred hertz. The midfrequency band extends from several hundred hertz to about 50 kHz, and the high band includes everything above 50 kHz. In each of these bands, a dominant source of ambient noise can be identified, although there may be considerable overlap of effects at the band edges.

The ambient noise in the low-frequency band is dominated by the sounds of distant shipping. In the deep ocean, distant shipping noise has a broad peak in the vicinity of 30 Hz and falls off rapidly above 100 Hz. Below 10 Hz, noise resulting from ocean turbulent pressure fluctuations and seismic activity exceed that caused by distant shipping, but this very low frequency region is generally not of great interest in the design of underwater acoustic systems.

The radiated acoustic noise from a ship measured at close range covers a frequency band far exceeding that attributed to distant shipping. The reduction in the high-frequency portion of the radiated ship spectrum is the result of absorption in the ocean at the long ranges associated with distant shipping. Figure 10-1 is a typical plot of spectral intensity at close range for a merchant ship, together with the modification of the spectral shape resulting from measurement at a range of 500 nm. Notice the very rapid drop in the spectrum above 100 Hz when measured at 500 nm.

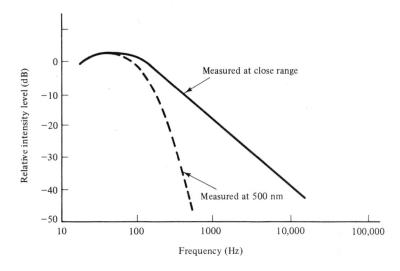

Figure 10-1   Effect of absorption loss on shape of radiated noise spectrum for a surface ship.

The observed noise level in the band 10 to 100 Hz is a function of shipping density and of long-range propagation characteristics of the ocean in this frequency region. Low-frequency sound arriving from a great distance must depend primarily on refracted transmission paths of the convergence zone or sound channel type. Paths requiring multiple bottom reflections introduce large attenuation and therefore do not contribute significantly to the ambient noise field. Hence high ambient levels in the low-frequency region generally occur with water depths and sound speed profiles that support convergence zone transmission. Figure 10-2 gives a range of levels to be expected for distant shipping noise, including the effects of shipping density and propagation characteristics. Notice that the ordinate is the intensity level in $dB//\mu Pa$ as measured in a 1.0-Hz band (see Section 2.5.3).

From the fact that distant shipping noise is restricted primarily to refractive paths, we may draw at least qualitative conclusions concerning the angular distribution of ambient noise in the vertical plane in the low-frequency band. Using typical sound speed profiles for the deep ocean, it is easy to verify that the vertical angle of arrival at a receiving hydrophone, for acoustic rays that do not intersect either the ocean surface or bottom, is limited approximately to the range $\pm 15°$ relative to the horizontal. Thus the noise intensity for distant shipping, as measured by a directional sensor, is significantly higher for angles close to the horizontal plane than for angles near the vertical. In Figure 10-3 a typical angular noise distribution in the vertical plane is sketched for the low-frequency band. The intensity level for the figure is in $dB//\mu Pa/Hz$ per steradian solid angle.

In the midfrequency band, ambient noise is dominated by the effects of

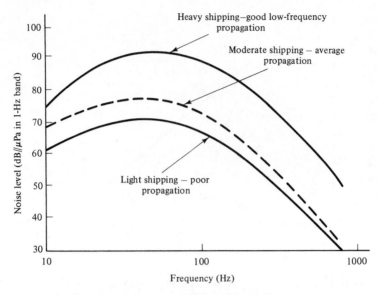

Figure 10-2 Distant shipping ambient noise levels.

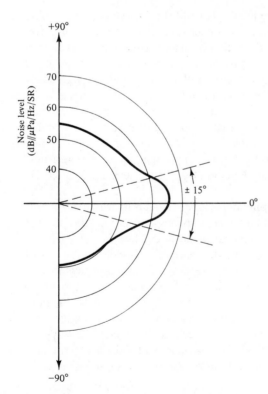

Figure 10-3 Typical deep-ocean vertical-plane noise distribution at low frequency.

wind acting on the sea surface. Knudsen et al. [3], as a result of studies during World War II, demonstrated a correlation between ambient noise and wind force or sea state. Curves of ambient noise in the midfrequency range with wind force, sea state, or wind speed as the family parameter are often referred to as "Knudsen curves." In very approximate terms, the ambient noise level in this band increases about 5 dB for every doubling of wind speed [1].

Wind-related noise has a broad spectral peak around 500 Hz and decreases at a rate of -5 to -6 dB per octave at higher frequencies. Among the processes that contribute noise in this region are breaking whitecaps, spray, bubble formation and collapse, and possibly turbulent pressures in the air coupling directly into the water. Acoustic noise generated in this manner does not propagate to great distances because of absorption. For this reason, and because of the geometric properties of the surface as a radiator, ambient noise in the midfrequency band tends to be more intense in the vertical direction than in the horizontal direction (see Section 10.2.4).

In the frequency band above 50 kHz, ambient noise is dominated by the thermal agitation of the water molecules. This component of noise increases at a rate of $+6$ dB per octave as frequency increases [4]. Thermal noise in the ocean is spatially isotropic.

Typical ambient acoustic noise spectra for the three bands discussed are shown in Figure 10-4 as measured with an omnidirectional hydrophone [2]. This figure represents average characteristics for the deep ocean. Significant departures from the levels indicated may be experienced related to geography,

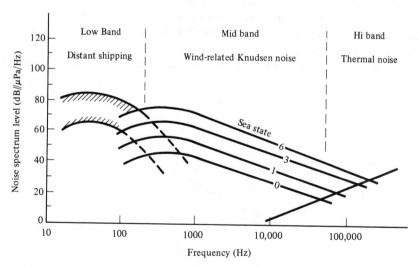

Figure 10-4   Typical deep-water ambient noise spectra as measured with an omni-directional hydrophone. (After Urick [2]; used with permission of McGraw-Hill Book Company.)

season, and weather. Figure 10-4 provides a guide as to the range of values to be expected, and should be supplemented with measured on-site data where possible.

10.2 CORRELATION PROPERTIES OF AMBIENT NOISE

Consider the noise signal output of two omnidirectional hydrophones separated in space as shown in Figure 10-5. Because the ambient noise is a random function, the exact differences between n_1 and n_2 are not predictable, so we must rely on a statistical description of the variation of the noise field in space and time. This leads to the formulation of the correlation function for ambient noise samples such as n_1 and n_2.

It should be evident from Section 10.1 that the ambient noise process is not stationary either in space or time. Spatial stationarity is affected in a gross sense by geographic location and in a finer sense by the proximity and state of the surface and bottom boundaries. Long-term temporal variations occur on a seasonal basis with short-term variations occurring over intervals ranging from daily down to a few minutes [5, 6]. However, many useful results are obtained by assuming stationarity in both space and time. This permits a tractable mathematical formulation and, together with the ergodic assumption, the use of transform techniques to relate time-frequency or linear space-angular space properties of the noise field. The degree to which stationarity assumptions are valid depends on the application. A common practice is to assume stationarity and qualify the results of the analysis as required to account for the effects of nonstationarity. With the stationarity assumption, the cross-correlation between n_1 and n_2 is a function only of the vector difference in position between $\mathbf{p}_1$ and $\mathbf{p}_2$ and of the time delay, τ.

To simplify the discussion further, the ambient noise field is assumed

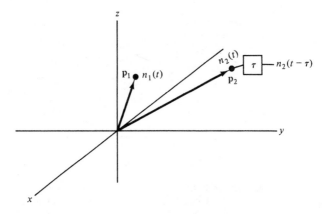

Figure 10-5 Geometry for defining noise field correlation properties.

isotropic in a horizontal plane with $\mathbf{p}_1$ and $\mathbf{p}_2$ located in a horizontal plane. With this assumption the cross-correlation function is dependent only on the magnitude of $\mathbf{p}_1 - \mathbf{p}_2$. Specific examples that deviate from this assumption are treated in Section 10.2.7. More general developments may be found in Cron and Sherman [7] and Cox [8].

10.2.1 Ambient Noise Field Spatial Model

The sources of ambient noise are assumed to originate at a distance such that the noise at a hydrophone can be considered as the summation of a large number of statistically independent plane waves, each with random orientation in space. This effect can be simulated by many random noise sources distributed on the surface of a large sphere with the omnidirectional hydrophone located at the center of the sphere.

To describe the correlation properties of the noise field we establish a coordinate system as shown in Figure 10-6. Direction is defined by the polar angles θ, measured in the z-y plane relative to the positive z-axis, and ψ measured from the z-y plane to the unit vector. Alternatively, azimuth and elevation angles, γ and ϕ, may be defined as shown in the figure.

Omnidirectional sensors are located at points 1 and 2 on the x-axis with separation d much less than the radius of the hypothetical noise sphere. Thus, noise arriving at points 1 and 2 from a small element of the spherical surface differ only in relative arrival time.

Let the noise signal arriving at the origin in the elemental solid angle $d\Omega$ and from the direction θ, ψ be defined as $n_0(\theta, \psi, t)\, d\Omega$. Assume that the

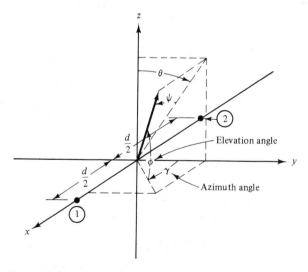

Figure 10-6 Polar coordinate system used in noise field definition.

frequency-domain Fourier transform of this signal exists so that we may also write

$$n_0(\theta, \psi, t)\, d\Omega \;=\; \left[\int_{-\infty}^{+\infty} N_0(\theta, \psi, f) \exp\left(j2\pi f t\right) df \right] d\Omega \qquad (10\text{-}1)$$

The total signal at the origin is obtained by integration over the 4π steradians of the spherical surface.

$$n_0(t) \;=\; \int_{4\pi} \int_{-\infty}^{+\infty} N_0(\theta, \psi, f) \exp\left(j2\pi f t\right) df\, d\Omega \qquad (10\text{-}2)$$

Because the noise generated by the elements of the spherical surface are assumed statistically independent, and assuming that the noises at different frequencies are also independent, the average noise intensity at the origin is

$$I_n \;=\; E[|n_0(t)|^2] \;=\; \int_{4\pi} \int_{-\infty}^{+\infty} \overline{|N_o(\theta, \psi, f)|^2}\, df\, d\Omega \qquad (10\text{-}3)$$

This average intensity is assumed constant over the spatial region of interest in keeping with the assumption of spatial stationarity.

The noise signals at the two sensors in Figure 10-6 arriving from a direction ψ, θ differ from the signal at the origin by a time delay factor that is a function of ψ but not of θ. In particular, the delay at point 1 relative to the signal at the origin is $-(d \sin \psi)/2c$, and at point 2 is $+(d \sin \psi)/2c$, where c is the sound speed. Using the rule developed in Chapter 6 for the effect of a time delay on the Fourier transform, the Fourier transforms of the signals at points 1 and 2 are related to the Fourier transform of the signal at the origin as follows:

$$N_1(\theta, \psi, f) = N_0(\theta, \psi, f) \exp\left(\frac{j\pi f d \sin \psi}{c} \right)$$
$$N_2(\theta, \psi, f) = N_0(\theta, \psi, f) \exp\left(\frac{-j\pi f d \sin \psi}{c} \right) \qquad (10\text{-}4)$$

from which

$$n_1(t) = \int_{4\pi} \int_{-\infty}^{+\infty} N_0(\theta, \psi, f) \exp\left[j2\pi f\left(t + \frac{d \sin \psi}{2c} \right) \right] df\, d\Omega$$
$$n_2(t) = \int_{4\pi} \int_{-\infty}^{+\infty} N_0(\theta, \psi, f) \exp\left[j2\pi f\left(t - \frac{d \sin \psi}{2c} \right) \right] df\, d\Omega \qquad (10\text{-}5)$$

Now form the temporal cross-correlation function for n_1 and n_2 by adding a time delay τ to the signal at the output of sensor 2.

$$R_{12}(d, \tau) = E[n_1(t)n_2^*(t - \tau)] \qquad (10\text{-}6)$$

Taking advantage of the statistical independence with respect to frequency and direction, (10-6) in integral form becomes

$$R_{12}(d, \tau) = \int_{-\infty}^{+\infty} \exp(j2\pi f\tau)\left\{\int_{4\pi} \overline{|N_0(\theta, \psi, f)|^2} \exp\left(j\frac{2\pi fd \sin \psi}{c}\right) d\Omega\right\} df$$

(10-7)

If $d = 0$, (10-7) becomes the usual expression relating the temporal auto-correlation and the frequency-domain power spectral density. Thus

$$R_{12}(d = 0, \tau) = \int_{-\infty}^{+\infty} \overline{|N_0(f)|^2} \exp(j2\pi f\tau) df$$

(10-8)

where

$$\overline{|N_0(f)|^2} = \int_{4\pi} \overline{|N_0(\theta, \psi, f)|^2} \, d\Omega = I_n(f)$$

The inner integral in (10-7) is in the form of a Fourier transform of the noise angular spectral density at frequency f to the linear domain of separation d along the x-axis. Thus, we define the single-frequency spatial cross-correlation for the signals at points 1 and 2 as

$$Q_{12}(d, f) = \int_{4\pi} \overline{|N_0(\theta, \psi, f)|^2} \exp\left(j\frac{2\pi fd \sin \psi}{c}\right) d\Omega$$

Converting the integral to polar coordinates, this becomes

$$Q_{12}(d, f) = \int_{-\pi}^{+\pi} \int_{-\pi/2}^{+\pi/2} \overline{|N_0(\theta, \psi, f)|^2} \exp\left(j\frac{2\pi fd \sin \psi}{c}\right) \cos \psi \, d\psi \, d\theta$$

(10-9)

The function $Q_{12}(d, f)$ is also called the *cross-spectral density* of the signals at points 1 and 2. The *space-time correlation function* of (10-7) is therefore identified as the Fourier transform of the *cross-spectral density* defined by (10-9).

$$R_{12}(d, \tau) = \int_{-\infty}^{+\infty} Q_{12}(d, f) \exp(j2\pi f\tau) df$$

(10-10)

For $d = 0$, (10-10) reduces to the temporal autocorrelation function of (10-8). For $\tau = 0$, we have

$$R_{12}(d, \tau = 0) = \int_{-\infty}^{+\infty} Q_{12}(d, f) df$$

(10-11)

The transform relationship in (10-9) indicates that the correlation function at frequency f for sensors located along a line such as the x-axis can be determined from a knowledge of the three-dimensional distribution of the noise field at frequency f. In this simple form, (10-9) is not a reversible operation because the three-dimensional noise distribution is not recoverable from a knowledge of the correlation function for sensors along a single line. A knowledge of the correlation function for sensors separated by the vector difference $\mathbf{p}_1 - \mathbf{p}_2$, where the direction of the difference can take on all possible values, is required to permit recovery of noise-field angular distribution in the general case.

For simplicity, it is convenient to consider the spatial correlation properties of the noise field at a single frequency. Hence we are interested in the cross-spectral density, $Q_{12}(d, f)$, including possibly the effect of a time delay τ applied to one of the sensor signals. Thus we define

$$Q_{12}(d, f, \tau) = Q_{12}(d, f) \exp(j2\pi f\tau) \tag{10-12}$$

It is also convenient to normalize the result with respect to the average noise intensity at frequency f. Thus, define

$$q_{12}(d, f, \tau) = \frac{Q_{12}(d, f, \tau)}{Q_{12}(d = 0, \tau = 0, f)} \tag{10-13}$$

Finally, we recognize that for real-valued noise waveforms, the normalized correlation is the real part of (10-13).

$$q_{12}'(d, f, \tau) = \text{Re}[q_{12}(d, f, \tau)] \tag{10-14}$$

We shall derive the cross-spectral densities, or single-frequency spatial correlation functions, for several angular noise distributions of practical interest. The geometries of the noise models to be considered are shown in Figure 10-7.

10.2.2 Isotropic Noise

The isotropic model for noise as shown in Figure 10-7(a) is by far the least complex mathematically, yet is of considerable practical interest. The transition frequency region between distant shipping-dominated and surface-effects-dominated ambient noise often has a nearly isotropic angular noise distribution. The noise field above 50 kHz is also essentially isotropic, and at other frequencies the noise field usually contains an isotropic component.

The total noise field intensity at f for the isotropic model is obtained from (10-8). Assuming a constant angular intensity K, this gives

$$I_n(f) = Q_{12}(0, f, 0) = \int_{4\pi} K \, d\Omega = 4\pi K \tag{10-15}$$

To obtain the cross-spectral density, we use (10-9) to write

$$Q_{12}(d, f) = K \int_{-\pi}^{+\pi} \int_{-\pi/2}^{+\pi/2} \exp\left(j\frac{2\pi f d}{c}\sin\psi\right)\cos\psi \, d\psi \, d\theta \tag{10-16}$$

Now let $u = (\sin\psi)/\lambda$, where $\lambda = c/f$, integrate over θ, and obtain

$$Q(d, \lambda) = 2\pi K\lambda \int_{-1/\lambda}^{+1/\lambda} \exp(j2\pi ud) \, du$$

$$= 2\pi K\lambda \int_{-\infty}^{+\infty} \text{rect}\left(\frac{u\lambda}{2}\right)\exp(j2\pi ud) \, du$$

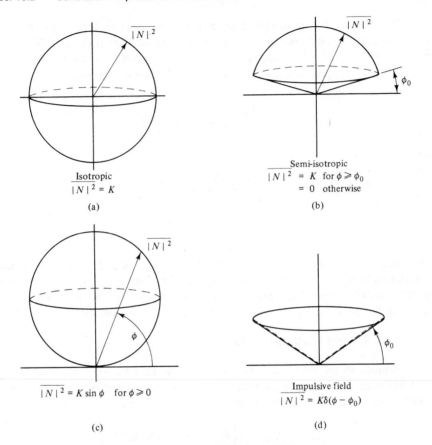

Figure 10-7   Simplified noise models.

which is recognized as a sinc function.

$$Q(d, \lambda) = 4\pi K \, \text{sinc}\left(\frac{2d}{\lambda}\right)$$

$$= 4\pi K \left[\frac{\sin\left(2\pi d / \lambda\right)}{2\pi d / \lambda}\right] \qquad (10\text{-}17)$$

From (10-12), (10-13), and (10-15), the normalized cross-spectral density, including the delay factor τ, is

$$q_{12}(d, \lambda, \tau) = \text{sinc}\left(\frac{2d}{\lambda}\right) \exp\left(\frac{j 2\pi c \tau}{\lambda}\right)$$

or for real-valued noise

$$q'_{12}(d, \lambda, \tau) = \text{sinc}\left(\frac{2d}{\lambda}\right) \cos\left(\frac{2\pi c \tau}{\lambda}\right) \qquad (10\text{-}18)$$

For $\tau = 0$, the spatial correlation is a sinc function with zeros at $d = n\lambda/2$, where n is any integer. Assuming that the ambient noise is a Gaussian random process, zero correlation also guarantees statistical independence.

Now let τ equal the acoustic travel time between sensors separated by a distance d. Then

$$\cos\left(\frac{2\pi c \tau}{\lambda}\right) = \cos\left(\frac{2\pi d}{\lambda}\right)$$

and

$$q'_{12}\left(d, \lambda, \tau = \frac{d}{c}\right) = \frac{\sin(2\pi d/\lambda)\cos(2\pi d/\lambda)}{2\pi d/\lambda}$$

$$= \frac{\sin(4\pi d/\lambda)}{4\pi d/\lambda} \tag{10-19}$$

The effect of the added delay in this case is to introduce additional zeros in the correlation function such that sensors separated by multiples of $\lambda/4$ will result in statistically independent noise signals. The spatial correlation functions for the isotropic noise field with $\tau = 0$ and with $\tau = d/c$ are shown in Figure 10-8.

10.2.3 Semi-isotropic Noise

Semi-isotropic noise, as shown in Figure 10-7(b), is defined as being uniform over some symmetric portion of the spherical surface and zero elsewhere. For instance, using the azimuth-elevation coordinate system, let

$$\overline{|N_0(\phi, \gamma, f)|^2} = \begin{cases} K & \text{for } \phi_0 \le \phi \le \dfrac{\pi}{2} \\ 0 & \text{elsewhere} \end{cases} \tag{10-20}$$

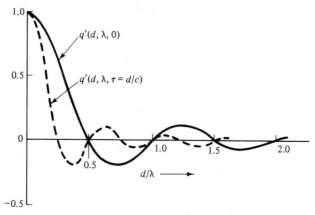

Figure 10-8 Spatial correlation functions for isotropic noise field.

For ϕ so limited, ψ and θ will be limited in a similar manner. Thus

$$\overline{|N_0(\psi, \theta, f)|^2} = \begin{cases} K & \text{for } \left(-\dfrac{\pi}{2} + \phi_0\right) \le \psi, \theta \le \left(\dfrac{\pi}{2} - \phi_0\right) \\ 0 & \text{elsewhere} \end{cases} \tag{10-21}$$

The cross-correlation function for this noise field is

$$Q_{12}(d, \lambda, \tau) = K \exp(j2\pi f\tau) \int\!\!\!\int_{-(\pi/2-\phi_0)}^{+(\pi/2-\phi_0)} \exp\left(\frac{j2\pi d \sin \psi}{\lambda}\right) \cos \psi \, d\psi \, d\theta$$

$$= K(\pi - 2\phi_0) \exp(j2\pi f\tau) \int_{-(\pi/2-\phi_0)}^{+(\pi/2-\phi_0)} \exp\left(\frac{j2\pi d \sin \psi}{\lambda}\right) \cos \psi \, d\psi$$

Performing the indicated integration and normalizing, we obtain for real-valued noise

$$q'_{12}(d, \lambda, \tau) = \text{sinc}\left(\frac{2d \cos \phi_0}{\lambda}\right) \cos\left(\frac{2\pi c\tau}{\lambda}\right) \tag{10-22}$$

In (10-22) consider first the case $\phi_0 = 0$. This corresponds to a noise field that is isotropic in the upper hemisphere and zero in the lower hemisphere. Although not strictly realizable, this is a crude approximation to an isotropic noise field as seen by a sensor located close to a highly absorptive bottom. Notice that with $\phi_0 = 0$, (10-22) is identical in form to (10-17) for the isotropic case. This interesting result can be generalized by considering the semi-isotropic field to be composed of a part that is an even function of ϕ and a part that is an odd function of ϕ. Thus let

$$\overline{|N(\phi, \gamma)|^2} = \overline{|N(\phi)|_e^2} + \overline{|N(\phi)|_0^2}$$

A little thought will show that, assuming that the noise field is uniform in the horizontal plane, the contribution of the odd portion of the noise field to the correlation function is zero. For the semi-isotropic noise field with $\phi_0 = 0$, the even portion is isotropic over the whole sphere while the odd portion has a constant magnitude but with a sign reversal between the upper and lower hemispheres. Thus we could describe this noise field as

$$\overline{|N(\phi)|^2} = \frac{K}{2}[1 + \text{sgn}(\phi)] \tag{10-23}$$

where

$$\text{sgn}(\phi) = \begin{cases} +1 & \text{for } \phi \text{ positive} \\ -1 & \text{for } \phi \text{ negative} \end{cases}$$

For any noise field, uniform in azimuth, for which the elevation pattern can be separated into even and odd components, the correlation function as measured in the horizontal plane may be obtained by considering only the even component. As an example, consider a noise field with a cardioid elevation pattern as follows:

$$\overline{|N(\phi, \gamma)|^2} = \frac{K}{2}(1 + \sin \phi) \tag{10-24}$$

Because $\sin \phi$ is an odd function of ϕ, the correlation function is determined entirely by the even component of (10-24), which is an isotropic field with amplitude $K/2$. Thus the cardioid noise field has the same correlation function as the isotropic field when evaluated in a plane perpendicular to the axis of the cardioid.

As ϕ_0 increases from zero in a positive direction, a larger spacing between sensors is required to obtain zero correlation. In the limit, as ϕ_0 approaches $\pi/2$, the noise field collapses to a point corresponding to a plane wave arriving from the vertical direction. In this case the signals at the two sensors have a normalized correlation of unity for all spacings, assuming that the delay τ is zero.

10.2.4 Surface-Generated Noise Field

As a result of the action of the wind on the ocean surface, assume that each elemental area of the surface radiates independent acoustic noise downward into the water with an angular radiation pattern $g(\phi)$ for $-\pi/2 \le \phi \le 0$. The noise intensity in a direction ϕ, referred to a unit distance from an area element dA, is

$$dI(\phi) = I_0|g(\phi)|^2 \, dA \tag{10-25}$$

where $I(\phi)$ has units of power per unit area and I_0 is the source intensity per unit area of surface.

Consider now the geometry of Figure 10-9. A sensor, H, located at depth d receives noise at an elevation angle ϕ from an annular ring of radius r, as shown. The received intensity from this annular ring is

$$dI_H = \frac{I_0|g(\phi)|^2 \, dA}{R^2} \tag{10-26}$$

Setting dA equal to the area of the annular ring, we obtain

$$dI_H = \frac{I_0|g(\phi)|^2 2\pi r \, dr}{R^2} \tag{10-27}$$

Equation (10-27) gives the noise intensity in units of power per unit area at range R, per element of surface. To convert to units of power per unit area per steradian of solid angle, we determine the projected area of the annular ring on a surface normal to the direction ϕ and divide by the area of a sphere of radius R. Thus

$$\text{steradians subtended by } dA \text{ from } H = \frac{2\pi r \sin \phi \, dr}{4\pi R^2} \tag{10-28}$$

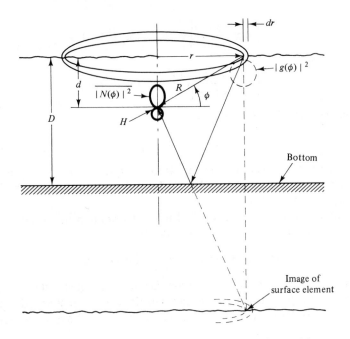

Figure 10-9   Geometry for surface-generated noise model.

Dividing (10-27) by (10-28) gives the angular noise field measured at H resulting from noise received directly from the surface.

$$\overline{|N(\phi)|^2} = \frac{I_0|g(\phi)|^2 2\pi r\, dr}{R^2} \left(\frac{4\pi R^2}{2\pi r \sin \phi\, dr} \right)$$

$$= \frac{4\pi I_0|g(\phi)|^2}{\sin \phi} \qquad \text{for } 0 \le \phi \le \frac{\pi}{2} \qquad (10\text{-}29)$$

Notice that the noise field described by (10-29) is independent of the sensor depth. Actually, the effect of absorption loss will cause the noise level to decrease with increasing depth [2]. This is especially true at the higher frequencies. The effects of refraction have also been neglected in deriving (10-29).

As shown in Figure 10-9, the sensor at H will also receive a noise signal by way of a bottom reflection of the surface-generated noise. The bottom reflected noise field at H may be derived by assuming its source is the image of the actual noise sources at the surface. Again neglecting absorption and refraction, this will result in a noise pattern in the lower hemisphere identical in form to (10-29), but reduced in intensity by a factor L_b. For simplicity, the bottom loss is assumed independent of angle. The total angular noise distribution in the elevation plane may now be written as

$$\overline{|N(\phi)|^2} = \begin{cases} 4\pi I_o \dfrac{|g(|\phi|)|^2}{\sin|\phi|} & \text{for } \phi \geq 0 \\[4mm] \dfrac{4\pi I_0 L_b |g(|\phi|)|^2}{\sin|\phi|} & \text{for } \phi < 0 \end{cases} \qquad (10\text{-}30)$$

A common assumption for $|g(\phi)|^2$ is

$$|g(\phi)|^2 = \sin^m \phi$$

where m is an arbitrary positive constant. At low frequencies $m = 0$ gives results consistent with experimental data, while in the midfrequency range $m = 2$ is the value mostly commonly used.

Assuming that $m = 2$, the vertical noise distribution in the upper hemisphere at H obtained by substitution in (10-30) is

$$\overline{|N(\phi)|^2} = 4\pi I_0 \sin \phi \qquad \text{for } \phi \geq 0 \qquad (10\text{-}31)$$

To determine the spatial correlation function, we make the coordinate transformation $\sin \phi = \cos \psi \cos \theta$ and substitute the transformed equation (10-31) into (10-9). Recognizing that the imaginary part of the result is zero, this gives

$$Q_{12}(d, \lambda) = 4\pi I_0 \int_{-\pi/2}^{+\pi/2} \int_{-\pi/2}^{+\pi/2} \cos \theta \cos^2 \psi \cos\left(\frac{2\pi d}{\lambda} \sin \psi\right) d\theta \, d\psi$$

$$= 8\pi I_0 \int_{-\pi/2}^{+\pi/2} \cos^2 \psi \cos\left(\frac{2\pi d}{\lambda} \sin \psi\right) d\psi$$

$$= (8\pi I_0)(\pi)\left[\frac{J_1(2\pi d/\lambda)}{2\pi d/\lambda}\right] \qquad (10\text{-}32)$$

where $J_1(\cdot)$ is the Bessel function of the first kind and order 1. The normalized correlation function is obtained by dividing (10-32) by the intensity received by an omnidirectional sensor at H. The result is

$$q_{12}(d, \lambda) = \frac{2J_1(2\pi d/\lambda)}{2\pi d/\lambda} \qquad (10\text{-}33)$$

The spatial correlation in the horizontal plane for a noise field with a $\sin|\phi|$ distribution in the vertical plane is shown in Figure 10-10 as a function of d/λ. Notice that this correlation function is zero for a sensor separation of approximately $d = 0.61\lambda$. This is slightly greater than the separation required in an isotropic noise field to achieve zero correlation ($d = 0.5\lambda$ for isotropic noise). Also notice that in Figure 10-10 the spacing between successive zeros is not constant.

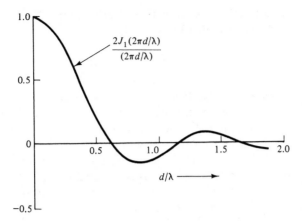

Figure 10-10 Spatial correlation function in the horizontal plane for $|N(\phi)|^2 = K$ $\sin |\phi|$.

10.2.5 Impulsive Noise

The impulsive noise model depicted in Figure 10-7(d) assumes that the noise field is uniformly distributed in azimuth but is zero everywhere in elevation except in the direction ϕ_0. Thus let

$$|N(\phi)|^2 = K\delta(\phi - \phi_0) \qquad (10\text{-}34)$$

where $\delta(\phi - \phi_0)$ is the unit impulse centered on ϕ_0.

With $\phi_0 = 0$, this model restricts the noise to the horizontal plane. For a sensor located on the axis of the deep sound channel (see Chapter 5), the noise signals from convergence zones arrive at predictable angles a few degrees above and below the horizontal plane. The impulsive noise model provides a reasonable representation for this component of the total noise field.

To obtain the spatial correlation function for the impulsive field, it is convenient to use the elevation–azimuth coordinate system. Recognizing that $\sin \psi = \cos \phi \sin \gamma$, the spatial correlation function becomes

$$Q_{12}(d, \lambda) = K \int_{-\pi}^{+\pi} \int_{-\pi/2}^{+\pi/2} \delta(\phi - \phi_0) \exp\left(j\frac{2\pi d}{\lambda} \cos \phi \sin \gamma\right) \cos \phi \, d\phi \, d\gamma$$

$$(10\text{-}35)$$

Upon integration we obtain

$$Q_{12}(d, \lambda) = 2K \cos \phi_0 \int_{0}^{+\pi} \exp\left(j\frac{2\pi d}{\lambda} \cos \phi_0 \sin \gamma\right) d\gamma \qquad (10\text{-}36)$$

$$= 2\pi K \cos \phi_0 J_0\left(\frac{2\pi d}{\lambda} \cos \phi_0\right) \qquad (10\text{-}37)$$

where $J_0(\bullet)$ is the zero-order Bessel function of the first kind. Normalization with respect to the total intensity finally gives

$$q_{12}(d, \lambda) = J_0\left(\frac{2\pi d}{\lambda} \cos \phi_0\right) \tag{10-38}$$

A plot of this zero-order Bessel function with $\phi_0 = 0$ is compared in Figure 10-11 with correlation functions for isotropic noise and for noise with a $\sin |\phi|$ distribution. The required sensor separation for zero correlation is less for the horizontal impulsive field than for either of the other noise field distributions. Note, however, that the correlation function for impulsive noise has the largest correlation excursions, or "side lobes," as d/λ increases.

As ϕ_0 increases from zero, the correlation function spreads out, requiring increased sensor separation for zero correlation. At $\phi_0 = \pi/2$, the noise field collapses about the z-axis and the noise signals at the two sensors are perfectly correlated for any separation (assuming that $\tau = 0$).

10.2.6 Effect of Noise Field Rotation

The examples considered thus far are of noise fields with principal axes perpendicular to the horizontal plane containing the two sensors, with the noise field uniform in the horizontal plane. Examination of Figure 10-11 suggests that for these conditions the spatial correlation in the horizontal plane is not very sensitive to the angular distribution of noise in the vertical plane. Vertical distributions ranging from isotropic to impulsive in the horizontal plane result in spatial correlation functions that are remarkably similar.

A little thought would show that for sensors located on the x-axis, as

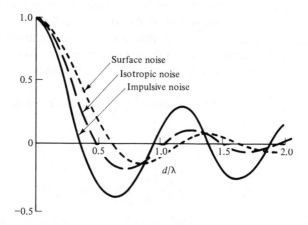

Figure 10-11 Comparison of spatial correlation functions for horizontal impulsive noise, isotropic noise, and surface-generated noise.

shown in Figure 10-6, the principal axis of the noise field may be rotated to any position in the z-y plane without affecting the shape of the spatial correlation function.

Although an arbitrary orientation of the noise field axis is beyond the intended scope of this discussion, it is of interest to consider the effect of rotating the noise field by 90° to align with the x-axis rather than with the z-axis. This is equivalent to locating the sensors on the z-axis instead of the x-axis.

Assume an impulsive noise field symmetrical about the x-axis. That is, let

$$|N(\psi, \theta)|^2 = K\delta(\psi - \psi_1) \tag{10-39}$$

This describes a conical shell about the x-axis with the surface at an angle ψ_1 relative to the z-y plane. The spatial correlation function is obtained by direct substitution in (10-9).

$$Q_{12}(d, \lambda, \tau)$$

$$= K \exp\left(\frac{j2\pi c\tau}{\lambda}\right) \int_{-\pi}^{+\pi} \int_{-\pi/2}^{+\pi/2} \delta(\psi - \psi_1) \cos \psi \exp\left(\frac{j2\pi d \sin \psi}{\lambda}\right) d\psi\, d\theta$$

$$\tag{10-40}$$

Performing the indicated integration, this gives

$$Q_{12}(d, \lambda, \tau) = 2\pi K \cos \psi_1 \exp\left(j\frac{2\pi d \sin \psi_1}{\lambda} + j\frac{2\pi c\tau}{\lambda}\right) \tag{10-41}$$

For real-valued noise, we require the real part of (10-42) which, upon normalizing, results in

$$q'_{12}(d, f, \tau) = \cos\left(\frac{2\pi d \sin \psi_1}{\lambda} + \frac{2\pi c\tau}{\lambda}\right) \tag{10-42}$$

With this orientation of the impulsive noise field, a value of τ can always be selected to give any value of correlation between plus and minus unity.

As an additional example of noise field rotation, let the sinusoidal noise distribution of Section 10.2.4 be rotated so that the principal axis is aligned with the x-axis. Thus

$$|N(\psi, \theta)|^2 = \begin{cases} K \sin \psi & \text{for } 0 \le \psi \le \dfrac{\pi}{2} \\ 0 & \text{elsewhere} \end{cases}$$

from which

$$Q_{12}(d, \lambda, \tau) = K \int_{-\pi}^{+\pi} \int_{0}^{+\pi/2} \sin \psi \cos \psi \exp\left[j\left(\frac{2\pi d \sin \psi}{\lambda} + \frac{2\pi c\tau}{\lambda}\right)\right] d\psi\, d\theta$$

$$\tag{10-43}$$

The real part of the normalized result is

$$q'_{12}(d, \lambda, \tau) = 2 \cos\left(\frac{2\pi c\tau}{\lambda}\right) \left[\frac{\sin(2\pi d/\lambda)}{2\pi d/\lambda} + \frac{\cos(2\pi d/\lambda) - 1}{(2\pi d/\lambda)^2}\right]$$

$$- 2 \sin\left(\frac{2\pi c\tau}{\lambda}\right) \left[\frac{\sin(2\pi d/\lambda)}{(2\pi d/\lambda)^2} - \frac{\cos(2\pi d/\lambda)}{2\pi d/\lambda}\right] \qquad (10\text{-}44)$$

This spatial correlation function is shown in Figure 10-12 for $\tau = 0$, $\tau = -d/c$ and $\tau = +d/c$. For $\tau = -d/c$, the two sensor signals are in phase for a plane wave signal arriving from the direction $\psi = \pi/2$; that is, from the direction coinciding with the noise field maximum. For $\tau = +d/c$, the sensor signals are in phase for a plane wave arriving from the direction $\psi = -\pi/2$, corresponding to the direction of the noise field minimum. The normalized spatial correlations for these three conditions simplify to

$$q'_{12}(d, f, \tau = 0) = 2\left[\frac{\sin(2\pi d/\lambda)}{2\pi d/\lambda} - \frac{\sin^2(\pi d/\lambda)}{2(\pi d/\lambda)^2}\right]$$

$$q'_{12}(d, f, \tau = -d/c) = \frac{\sin^2(\pi d/\lambda)}{(\pi d/\lambda)^2} \qquad (10\text{-}45)$$

$$q'_{12}(d, f, \tau = +d/c) = 2\left[\frac{\sin(4\pi d/\lambda)}{2(\pi d/\lambda)} + \frac{\cos(4\pi d/\lambda) - \cos(2\pi d/\lambda)}{(2\pi d/\lambda)^2}\right]$$

The zeros in the correlation function occur at multiples of approximately $\lambda/4$ with the sensors phased, or "steered," in the direction of the noise field minimum. With $\tau = 0$ the separation between correlation zeros is approximately $\lambda/2$, and with the sensors steered in the direction of the noise field maximum, the zeros are separated by λ.

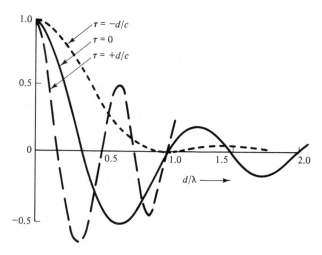

Figure 10-12 Spatial correlation functions for sensors on the x-axis with $|N(\psi, \theta)|^2 = K \sin \psi$ and with three values of relative electrical delay.

10.2.7 Spatial Correlation for More Complex Noise Fields

The spatial correlation function for a noise field composed of the sum of several statistically independent components is the sum of the correlation functions for each of the components considered separately. This provides a method for synthesizing the correlation function for more complex noise fields, using the results from the simple noise fields considered in preceding sections.

In the midfrequency band, the ambient noise field often consists of three components: a surface-generated component, together with its reflection from the bottom; an impulsive component oriented near the horizontal plane caused by distant shipping; and a small isotropic component caused by miscellaneous effects. Assuming that all components are uniform in the horizontal plane, assume an angular noise distribution in the vertical plane of the form

$$\overline{|N(\phi)|^2} = \begin{cases} K_0 + K_1 \sin|\phi| & \text{for } 0 \le \phi \le \dfrac{\pi}{2} \\[2mm] K_0 + L_b K_1 \sin|\phi| + K_2 \delta(\phi - \phi_0) & \text{for } \phi < 0 \end{cases}$$

This distribution is sketched in Figure 10-13 assuming that the noise intensity in the impulsive component, the isotropic component, and the bottom reflected component are each 10 dB below the intensity in the direct-path part of the surface-generated noise.

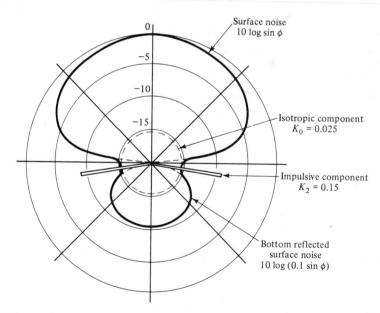

Figure 10-13   Vertical noise distribution for combination of surface-generated noise, isotropic noise, and impulsive noise.

With $\tau = 0$, the horizontal-plane spatial correlation function for this combination of the several noise sources is

$$Q_{12}(d, \lambda) = 2\pi K_1(1 + L_b)\frac{J_1(2\pi d/\lambda)}{2\pi d/\lambda} + 4\pi K_0 \, \text{sinc}\left(\frac{2d}{\lambda}\right)$$

$$+ 2\pi K_2 \cos \phi_0 J_0\left(\frac{2\pi d}{\lambda} \cos \phi_0\right) \quad (10\text{-}46)$$

With the noise field dominated by the surface-generated component, as in the example shown in Figure 10-13, the principal effect of the other components is to modify slightly the location of the zeros of the composite correlation function.

10.3 USE OF THE SPATIAL CORRELATION FUNCTION IN SYSTEM NOISE CALCULATIONS

In preceding sections the ambient noise signal was assumed to be sensed by hydrophones of zero dimensions located at discrete points in space. To produce a finite output signal, a sensor with zero dimensions must have an impulsive spatial response. Thus a proper formulation of the signal sensed by a point hydrophone on the x-axis would be

$$n(x_1, t) = \int_{-\infty}^{+\infty} n(x, t)\delta(x - x_1) \, dx \quad (10\text{-}47)$$

where $n(x, t)$ is the noise signal per unit length along the x-axis and $\delta(x - x_1)$ is the spatial response of a point hydrophone located at $x = x_1$.

Now consider a hydrophone that extends continuously from $-L/2$ to $+L/2$ along the x-axis. The spatial response of this "line hydrophone" can be described in terms of a spatial rect function. It is convenient to normalize the spatial response to have unit area so that the output signal resulting from a plane wave normal to the x-axis is independent of the hydrophone length. Thus define the hydrophone spatial response $g(x)$ as

$$g(x) = \frac{1}{L} \, \text{rect}\left(\frac{x}{L}\right) \quad (10\text{-}48)$$

from which the hydrophone output signal in the presence of the ambient noise field is

$$n_0(t) = \frac{1}{L} \int_{-\infty}^{+\infty} n(x, t) \, \text{rect}\left(\frac{x}{L}\right) dx \quad (10\text{-}49)$$

Because the noise field is assumed stationary in time, the mean-squared value of $n_0(t)$ is independent of time and we can write

$$E[|n_0|^2] = \frac{1}{L^2} \int_{-\infty}^{+\infty} \int_{-\infty}^{+\infty} \text{rect}\left(\frac{x}{L}\right) \text{rect}\left(\frac{x'}{L}\right) E[n(x)n^*(x')] \, dx \, dx' \quad (10\text{-}50)$$

The noise field is also assumed stationary in space, so that the expectation inside the double integral is a function only of the difference in location. Thus let $x - x' = \Delta$ and write

$$E[n(x)n*(x')] = R_n(x - x') = R_n(\Delta) \tag{10-51}$$

Substitution in (10-50) gives

$$E[|n_0|^2] = \frac{1}{L^2} \int_{-\infty}^{+\infty} \int_{-\infty}^{+\infty} \text{rect}\left(\frac{x}{L}\right) \text{rect}\left(\frac{x - \Delta}{L}\right) R_n(\Delta) \, dx \, d\Delta \tag{10-52}$$

Now integrate with respect to x to obtain

$$E[|n_0|^2] = \frac{1}{L} \int_{-L}^{+L} \left(1 - \frac{|\Delta|}{L}\right) R_n(\Delta) \, d\Delta \tag{10-53}$$

The triangular weighting function inside the integral in (10-53) is the autocorrelation function of the line-hydrophone spatial response $g(x)$. Therefore, in the general case of a sensor distributed along the x-axis, the mean-squared noise output is

$$E[|n_0|^2] = \int_{-\infty}^{+\infty} R_g(\Delta) R_n(\Delta) \, d\Delta \tag{10-54}$$

where

$$R_g(\Delta) = g(\Delta) \otimes g*(-\Delta)$$

The mean-squared noise output is shown by (10-54) to be the integral of the noise field spatial correlation function as weighted by the spatial autocorrelation function for the sensor.

As an example, let the noise field be isotropic with a cross-spectral density (or single-frequency spatial correlation function) at a wavelength λ given by

$$Q_n(\Delta, \lambda) = 4\pi K \, \text{sinc}\left(\frac{2\Delta}{\lambda}\right)$$

The mean-squared value of the sensor output at wavelength λ, assuming a sensor with a rectangular spatial response, is

$$\begin{aligned}
E[|n_0(\lambda)|^2] &= \frac{4\pi K}{L} \int_{-L}^{+L} \left(1 - \frac{|\Delta|}{L}\right) \text{sinc}\left(\frac{2\Delta}{\lambda}\right) d\Delta \\
&= \frac{8\pi K}{L} \left[\int_0^L \frac{\sin(2\pi\Delta/\lambda)}{2\pi\Delta/\lambda} \, d\Delta - \frac{\lambda}{2\pi L} \int_0^L \sin\left(\frac{2\pi\Delta}{\lambda}\right) d\Delta \right] \\
&= 8\pi K \left\{ \frac{\lambda}{2\pi L} \text{Si}\left(\frac{2\pi L}{\lambda}\right) + \frac{\lambda^2}{4\pi^2 L^2} \left[\cos\left(\frac{2\pi L}{\lambda}\right) - 1 \right] \right\} \tag{10-55}
\end{aligned}$$

where $\text{Si}(\cdot)$ is the sine-integral function defined as

$$\text{Si}\,(u) = \int_0^u \frac{\sin x}{x}\, dx$$

The sine-integral function is available in tabular form in most handbooks of mathematical functions [9].

For $L \ll \lambda$ it is easy to verify that (10-55) reduces to the result for a point hydrophone. That is,

$$E[|n_0(\lambda)|^2] = 4\pi K \qquad \text{for } L \ll \lambda \qquad\qquad (10\text{-}56)$$

For $L \gg \lambda$ we obtain

$$E[|n_0(\lambda)|^2] = 4\pi K\left(\frac{\lambda}{2L}\right) \qquad \text{for } L \gg \lambda \qquad\qquad (10\text{-}57)$$

For a fixed length L, (10-57) shows that the noise output signal, relative to the response to a constant plane wave, decreases with decreasing wavelength. The significance of this is considered further in Chapter 11.

The mean-squared output signal over a finite frequency band is obtained by integration of (10-55) over the frequency band of interest. Thus for $f = c/\lambda$ and $\beta = f_2 - f_1$,

$$
\begin{aligned}
E[|n_0(\beta)|^2] &= \int_{f_1}^{f_2} E[|n_0 f)|^2]\, df \\
&= \int_{f_1}^{f_2} \left[\int_{-\infty}^{+\infty} R_g(\Delta) Q_n(\Delta, f)\, d\Delta \right] df \\
&= \int_{-\infty}^{+\infty} R_g(\Delta) \left[\int_{f_1}^{f_2} Q_n(\Delta, f)\, df \right] d\Delta \qquad\qquad (10\text{-}58)
\end{aligned}
$$

If the range f_2 to f_1 covers the entire frequency region occupied by ambient noise, the inner integral in (10-58) becomes $R_n(\Delta)$ and (10-58) reduces to (10-54).

PROBLEMS

10.1. Using the ambient noise levels in Figure 10-4 for heavy shipping and sea state 3, calculate the approximate noise band levels for the frequency bands 10 to 100 Hz, 100 to 1000 Hz, and 1000 to 10,000 Hz.

10.2. The intensity level of broadband noise is sometimes given using a bandwidth of one-third of an octave as the reference bandwidth rather than a 1-Hz bandwidth. The bandwidth of a third octave is $0.23f_g$, where f_g is the geometric mean frequency of the band. Assume noise spectrum levels in 1-Hz bands at 100 Hz, 1000 Hz, and 10,000 Hz of $+80$ dB$//\mu$Pa, $+60$ dB$//\mu$Pa, and $+50$ dB$//\mu$Pa, respectively. Convert these levels to third-octave band levels and plot the resulting third-octave spectrum.

10.3. Assume an angular noise intensity given by

$$\overline{|N(\phi,\gamma)|^2} = \begin{cases} 4\pi I_0 \sin \phi & \text{for } 0 < \phi \le \dfrac{\pi}{2} \\ 0 & \text{for } \phi < 0 \end{cases}$$

where ϕ and γ are the elevation and azimuth angles, respectively. Calculate the noise intensity as measured by an omnidirectional sensor.

10.4. A line hydrophone of length L is aligned with the x-axis in the presence of a noise field with an impulsive angular response as described by (10-39). Derive an expression for the mean-squared output of the line hydrophone in response to this noise field.

SUGGESTED READING

1. Wenz, G. M., "Acoustic Ambient Noise in the Ocean: Spectra and Sources," *Acoust. Soc. Am.,* Vol. 34, No. 12, pp. 1936–1956 (Dec. 1962).

2. Urick, R. J., *Principles of Underwater Sound for Engineers,* 2nd ed. New York: McGraw-Hill Book Company, 1975, Chap. 7.

3. Knudsen, V. O., R. S. Alford, and J. W. Emling, "Underwater Ambient Noise," *J. Mar. Res.,* Vol. 7, p. 410 (1948).

4. Mellen, R. H., "Thermal Noise Limit in the Detection of Underwater Acoustic Signals," *J. Acoust. Soc. Am.,* Vol. 24, p. 478 (1952).

5. Hodgkiss, W. S., and V. C. Anderson, "Detection of Sinusoids in Ocean Acoustic Background Noise." *J. Acoust. Soc. Am.,* Vol. 67, No. 1, p. 214 (Jan. 1980).

6. Anderson, V. C., "Nonstationary and Nonuniform Oceanic Background in a High Gain Acoustic Array," Marine Physical Laboratory of Scripps Institute of Oceanography, MPL-U-80/78, Jan. 1979.

7. Cron, B. F., and C. H. Sherman, "Spatial-Correlation Functions for Various Noise Models," *J. Acoust. Soc. Am.,* Vol. 34, p. 1732 (1962).

8. Cox, H., "Spatial Correlation in Arbitrary Noise Fields with Application to Ambient Sea Noise," *J. Acoust. Soc. Am.,* Vol. 54, No. 5, p. 1289 (1973).

9. Abramowitz, M., and I. Stegun, Eds., *Handbook of Mathematical Functions,* Natl. Bur. Stand., Appl. Math. Ser. 55. Washington, D.C.: U. S. Government Printing Office, 1964, Chap. 5.

11

Spatial Filtering: Beamforming

~~~~~~

11.0 INTRODUCTION

The proper processing of acoustic signals begins with the design of the sensor system. In addition to converting the acoustic signal to an electrical signal, the sensor system is often required to accomplish the following tasks:

   1. Reduce the ambient noise with which a plane wave signal must compete

   2. Permit the separate detection, or resolution, of plane wave signals arriving from different directions

   3. Permit the measurement of the direction of arrival of plane wave signals

By replacing the words "plane wave signal" with "sine-wave signal," "ambient noise" with "broadband noise," and "direction" with "frequency," we see that these objectives are the same as those of a frequency-domain filter used to process narrowband time-domain waveforms. That is, narrowband filters reduce the noise with which a sine-wave signal must compete, permit the resolution of signals of different frequency, and aid in the measurement of the frequency of individual sine-wave signals. By analogy the sensor system performs the function of a *spatial filter* to enhance the detectability, resolution, and directional measurement of plane wave signals.

The analogy between the frequency-domain representation of a narrowband signal in broadband noise and a plane wave signal in ambient noise is demonstrated in Figure 11-1. For simplicity, the plot in Figure 11-1(b) represents the angular distribution of acoustic energy in one plane only.

The signal-to-noise power ratio, SNR, at the output of either the temporal or spatial filter is a matter of considerable interest. In Figure 11-1(a) define for the temporal case

$$a^2 \delta(f - f_0) = \text{signal power spectral density}$$

$$|N(f)|^2 = \text{noise power spectral density}$$

$$|H(f)|^2 = \text{filter power response}$$

The signal and noise powers at the filter output are

$$P_s = a^2 \int_{-\infty}^{+\infty} \delta(f - f_0)|H(f)|^2 \, df = a^2|H(f_0)|^2$$

$$P_n = \int_{-\infty}^{+\infty} |N(f)|^2|H(f)|^2 \, df$$

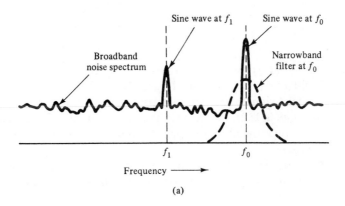

(a)

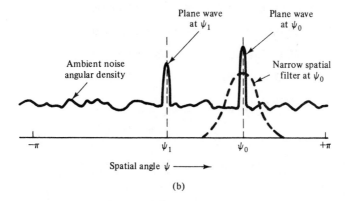

(b)

Figure 11-1 Comparison of temporal and spatial filter functions: (a) power spectral density of sine-wave signals in broadband noise; (b) angular density of acoustic plane wave signal in ambient noise.

from which

$$\text{SNR} = \frac{P_s}{P_n} = \frac{a^2 |H(f_0)|^2}{\int_{-\infty}^{+\infty} |N(f)|^2 |H(f)|^2 \, df}$$

As an example, let the noise spectral density be a constant N_0, and the filter response be a rect function centered at f_0 such that

$$|H(f)|^2 = \text{rect}\left(\frac{f - f_0}{\beta}\right)$$

$$|H(f_0)|^2 = 1.0$$

then

$$\text{SNR} = \frac{a^2}{N_0 \int_{-\infty}^{+\infty} \text{rect}\left[(f - f_0)/\beta\right] \, df}$$

$$= \frac{a^2}{\beta N_0} \tag{11-1}$$

Notice that in (11-1) the signal-to-noise ratio is inversely proportional to the width of the filter response as defined by the factor β. This is in general true provided that the total signal spectrum is narrow compared with the filter width. It should also be evident from Figure 11-1 that the capability to estimate signal frequency and to resolve signals closely spaced in frequency is inversely proportional to the filter width.

In a similar manner we may define the signal-to-noise ratio at the output of the spatial filter. For the two-dimensional signal and noise of figure 11-1(b), let

$$a^2 \delta(\psi - \psi_0) = \text{signal angular intensity density}$$

$$|N(\psi)|^2 = \text{noise angular intensity density}$$

$$|G(\psi)|^2 = \text{spatial filter angular power response}$$

The output signal-to-noise ratio is

$$\text{SNR} = \frac{a^2 |G(\psi_0)|^2}{\int_{-\pi}^{+\pi} |N(\psi)|^2 |G(\psi)|^2 \, d\psi} \tag{11-2}$$

With a constant angular noise field density K we may write

$$\text{SNR} = \frac{a^2}{K \psi_B} \tag{11-3}$$

where ψ_B is a measure of the width of the spatial filter response, defined by

$$\psi_B = \int_{-\pi}^{+\pi} \frac{|G(\psi)|^2}{|G(\psi_0)|^2} \, d\psi$$

As in the temporal case, the signal-to-noise ratio is inversely proportional to the filter width.

The spatial filter average output power for an arbitrary filter position ψ_k is given by

$$P_0(\psi_k) = \int_{-\pi}^{+\pi} \left[|N(\psi)|^2 + |S(\psi)|^2 \right] |G(\psi - \psi_k)|^2 \, d\psi \qquad (11\text{-}4)$$

where $|S(\psi)|^2$ is the angular intensity distribution associated with plane wave signals. Equation (11-4) is the angle-domain convolution of the combined signal-plus-noise angular density with the power response of the spatial filter. The result is shown in Figure 11-2 for the noise and signal fields of Figure 11-1.

It is evident from Figure 11-2 that the ability to resolve closely spaced plane wave targets is inversely related to the spatial filter width. Also, the accuracy of estimating target direction is improved as ψ_B is decreased. Thus the spatial filter performance with respect to its intended functions is improved by minimizing the width of its response in the angular domain. Although other system considerations may modify this conclusion, the minimization of ψ_B is often an important goal in the design of the spatial filter system.

A commonly used factor to judge the quality of a spatial filter is the degree of noise power reduction it provides relative to a device with uniform response at all angles (omnidirectional sensor). For the two-dimensional noise model with isotropic noise, this factor was shown to be inversely related to the filter angular width, where the angular width is defined as the integral of the normalized power response of the filter.

Now consider a three-dimensional noise field and spatial filter defined by

$|N(\psi, \theta)|^2$ = noise field angular intensity distribution

$$b(\psi, \theta) = \frac{|G(\psi, \theta)|^2}{|G_{\max}|^2} = \text{normalized power response of spatial filter}$$

For the purpose of detecting a plane wave signal the *array gain* of the spatial

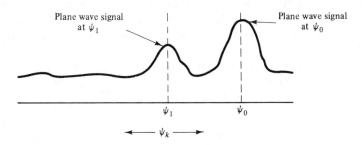

Figure 11-2 Spatial filter output with isotropic noise and two plane wave signals.

filter is defined as the ratio of the noise power out of an omnidirectional device to the noise power out of the spatial filter. Expressed in decibel notation, array gain is

$$\text{array gain} = \text{AG} = 10 \log \left[\frac{\int_{4\pi} |N(\psi, \theta)|^2 \, d\Omega}{\int_{4\pi} |N(\psi, \theta)|^2 b(\psi, \theta) \, d\Omega} \right] \qquad (11\text{-}5)$$

If the noise field is isotropic, this factor is called the *directivity index* and (11-5) simplifies to

$$\text{directivity index} = \text{DI} = 10 \log \left[\frac{4\pi}{\int_{4\pi} b(\psi, \theta) \, d\Omega} \right] \qquad (11\text{-}6)$$

The integral in the denominator of (11-6) may be interpreted as a solid angle measure, in steradians, of the "width" or extent of the three-dimensional response of the spatial filter.

11.1 ONE-DIMENSIONAL SPATIAL FILTERS

A one-dimensional device such as a line hydrophone satisfies the basic requirements of a spatial filter. It provides direction discrimination, at least in a limited sense, and a signal-to-noise improvement relative to an omni-directional sensor. Because of the simplified mathematics, relative to multidimensional devices, our detailed consideration of spatial filters will be limited primarily to the one-dimensional case. Extension of the basic principles to more complex configurations is straightforward, although often mathematically tedious. Multidimensional devices are covered briefly in Section 11.8.

11.1.1 Angular Response of a Line Hydrophone

Consider the angular response of a line hydrophone to a plane wave signal as a function of the direction of arrival (ψ, θ). Let the line hydrophone be aligned with the x-axis so that angle ψ is the angle between the direction of arrival and a plane perpendicular to the axis of the line hydrophone. With this choice, the signal received at any point on the hydrophone is a function only of ψ and not of θ.

Referring to Figure 11-3, let the signal at the origin be $s(t)$. The signal at any other point on the x-axis is delayed or advanced relative to the signal at the origin by an amount $(x \sin \psi)/c$, so that we may write

$$s(t, x) = s\left(t + \frac{x \sin \psi}{c} \right)$$

where c is the sound speed.

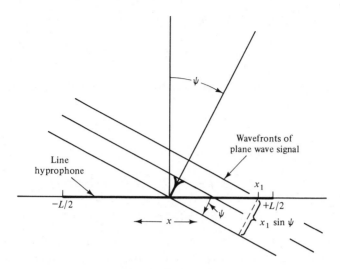

Figure 11-3 Plane wave signal incident on a line hydrophone.

The output of the line hydrophone is obtained by integration of its response to the signal along its length. Thus if $g(x) \, dx$ is the hydrophone response to a unit signal at x, the total output resulting from a plane wave at angle ψ is

$$s_0(t, \psi) = \int_{-\infty}^{+\infty} g(x) s\left(t + \frac{x \sin \psi}{c} \right) dx \qquad (11\text{-}7)$$

The hydrophone response $g(x)$ is commonly called the *aperture function*.

Assume that $s(t)$ has a frequency-domain Fourier transform, $S(f)$, so that

$$s\left(t + \frac{x \sin \psi}{c} \right) \longleftrightarrow S(f) \exp\left(\frac{j2\pi f x \sin \psi}{c} \right)$$

The line hydrophone output may now be expressed in the alternative form

$$s_0(t, \psi) = \int_{-\infty}^{+\infty} \left[\int_{-\infty}^{+\infty} S(f) \exp\left(\frac{j2\pi f x \sin \psi}{c} + j2\pi f t \right) df \right] g(x) \, dx$$

Rearranging, we have

$$s_0(t, \psi) = \int_{-\infty}^{+\infty} S(f) \left[\int_{-\infty}^{+\infty} g(x) \exp\left(\frac{j2\pi f x \sin \psi}{c} \right) dx \right] \exp(j2\pi f t) \, df \qquad (11\text{-}8)$$

The inner integral in (11-8) has the form of the Fourier transform of the aperture function $g(x)$ from the x-domain to a domain represented by the variable $u = (\sin \psi)/\lambda$, where $\lambda = c/f$. The transform of the aperture function is called the *pattern function,* defined by

$$G(u) = G(f, \psi) = \int_{-\infty}^{+\infty} g(x) \exp(j2\pi x u) \, dx \qquad (11\text{-}9)$$

Substitution of (11-9) in (11-8) gives

$$s_o(t, \psi) = \int_{-\infty}^{+\infty} S(f) G(f, \psi) \exp(j2\pi ft) \, df \qquad (11\text{-}10)$$

Let the incident plane wave be a single-frequency complex sinusoid of unit amplitude. Then

$$s(t) = \exp(j2\pi f_0 t) \longleftrightarrow S(f) = \delta(f - f_0)$$

and

$$s_o(t, \psi) = \int_{-\infty}^{+\infty} \delta(f - f_0) G(f, \psi) \exp(j2\pi ft) \, df$$

$$= G(f_0, \psi) \exp(j2\pi f_0 t) \qquad (11\text{-}11)$$

The function $G(f_0, \psi)$ is the single-frequency pattern function defining the shape of the spatial filter response provided by the line hydrophone at frequency f_0.

Let $g(x)$ be a constant over the length L and zero elsewhere. In normalized form we may write

$$g(x) = \frac{1}{L} \text{rect}\left(\frac{x}{L}\right)$$

The corresponding pattern function is

$$G(u) = \frac{1}{L} \int_{-\infty}^{+\infty} \text{rect}\left(\frac{x}{L}\right) \exp(j2\pi xu) \, dx = \text{sinc}(uL) \qquad (11\text{-}12)$$

The rectangular aperture function and the resulting pattern function are shown in Figure 11-4. Notice that the variable u is confined to the interval $\pm 1/\lambda$ as $\sin\psi$ varies over the range ± 1. As ψ varies over the range $[0, 2\pi]$, the pattern repeats the form shown in Figure 11-4 such that $G(\psi) = G(\pi - \psi)$. The angular response of the device is therefore ambiguous with respect to the angle ψ, responding equally to targets at angle ψ and $\pi - \psi$ as well as being independent of the spatial angle θ.

A three-dimensional plot of the line-hydrophone pattern function is sketched in Figure 11-5. For simplicity, only the central lobe of the pattern is shown in this figure. The pattern is roughly a disk in the z-y plane formed by the revolution of the main-lobe pattern about the x-axis.

A measure of the angular width of the main lobe is given by the angular distance from the peak of the pattern to the first zero. From Figure 11-4 the first zero occurs at $u = 1/L$. Converting to an angle, we obtain

$$u = \frac{1}{L} = \frac{\sin\psi_B}{\lambda}$$

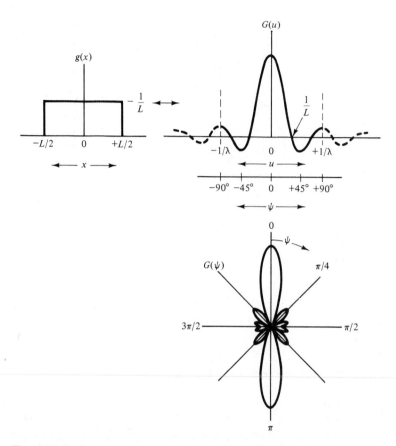

Figure 11-4   Rectangular aperture function and resulting pattern function.

$$\psi_B = \text{arc sin}\left(\frac{\lambda}{L}\right) \qquad \text{rad} \tag{11-13}$$

For $L \gg \lambda$, this gives

$$\psi_B \simeq \frac{\lambda}{L} \tag{11-14}$$

The main-lobe width, or beamwidth, defined by (11-14) is the angular width of the sinc function pattern between points approximately 4 dB down from the peak. The more commonly used 3-dB beamwidth, in degrees, is given by

$$\psi_{3dB} \simeq 50\frac{\lambda}{L} \qquad \text{deg} \tag{11-15}$$

The width of the main lobe is a measure of the angular resolution provided by the line hydrophone for signals separated in the ψ-direction. Resolution capability is obviously improved as L increases or λ decreases.

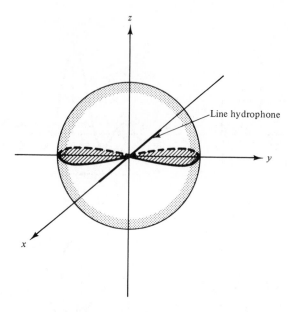

Figure 11-5 Three-dimensional main-lobe response of a line hydrophone oriented along the x-axis.

11.1.2 Directivity Index for Line Hydrophone

Determination of the directivity index of a device requires the calculation of the output noise power in the presence of an isotropic noise field. This may be accomplished using the normalized power pattern function in combination with the noise angular density function, or equivalently the device aperture function together with the noise spatial correlation function.

For the line hydrophone, the normalized power pattern function may be written as

$$b(\psi, \theta, \lambda) = \text{rect}\left(\frac{\theta}{2\pi}\right)\left[\frac{\sin^2(\pi L \sin \psi/\lambda)}{(\pi L \sin \psi/\lambda)^2}\right] \tag{11-16}$$

from which the denominator in (11-6) becomes

$$\int_{4\pi} b(\psi, \theta) \, d\Omega = \int_{-\pi}^{+\pi} \int_{-\pi/2}^{+\pi/2} \frac{\sin^2(\pi L \sin \psi/\lambda)}{(\pi L \sin \psi/\lambda)^2} \cos \psi \, d\psi \, d\theta \tag{11-17}$$

or, letting $u = (\sin \psi)/\lambda$ and integrating over θ, we have

$$\int_{4\pi} b(\psi, \theta) \, d\Omega = 2\pi\lambda \int_{-1/\lambda}^{+1/\lambda} \text{sinc}^2(Lu) \, du \tag{11-18}$$

With the help of Parseval's theorem, this result can be shown equivalent to that obtained using the device aperture function. Thus for the isotropic noise field, the single-frequency correlation function, or cross-spectral-density, and its u-domain transform are

$$Q_n(x, \lambda) = 4\pi \text{ sinc} \left(\frac{2x}{\lambda} \right) \longleftrightarrow 2\pi\lambda \text{ rect} \left(\frac{u\lambda}{2} \right) = |N(u)|^2 \qquad (11\text{-}19)$$

where the noise density in the (ψ, θ)-domain has been normalized to unity. For the line hydrophone with a rectangular aperture function, we have

$$g(x) = \frac{1}{L} \text{ rect} \left(\frac{x}{L} \right) \longleftrightarrow \text{ sinc} (uL)$$

$$R_g(x) = \frac{1}{L} \left(1 - \frac{|x|}{L} \right) \text{rect} \left(\frac{x}{2L} \right) \longleftrightarrow \text{ sinc}^2 (uL) = |G(u)|^2 \qquad (11\text{-}20)$$

From (10-54) and from Parseval's theorem, the single-frequency noise power output for the line hydrophone is

$$\int_{-\infty}^{+\infty} R_g(x) Q_n(x, \lambda) \, dx = \int_{-1/\lambda}^{+1/\lambda} |N(u)|^2 |G(u)|^2 \, du$$

$$= 2\pi\lambda \int_{-1/\lambda}^{+1/\lambda} \text{rect} (u\lambda/2) \text{ sinc}^2(uL) \, du$$

$$\frac{4\pi}{L} \int_{-L}^{+L} \left(1 - \frac{|x|}{L} \right) \text{ sinc} \left(\frac{2x}{\lambda} \right) dx = 2\pi\lambda \int_{-1/\lambda}^{+1/\lambda} \text{ sinc}^2(uL) \, du \qquad (11\text{-}21)$$

which is identical to (11-18).

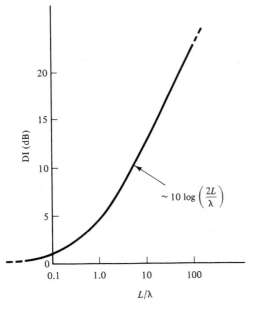

$$\sim 10 \log \left(\frac{2L}{\lambda} \right)$$

Figure 11-6   Directivity index for a continuous-line hydrophone.

Whether integration is performed in the x-domain or the u-domain, the result is given by (10-55). Using that result, the DI for the line hydrophone at a wavelength λ is given by

$$\text{DI} = 10 \log \left(\frac{2L}{\lambda} \left\{ \frac{2}{\pi} \text{ Si} \left(\frac{2\pi L}{\lambda} \right) + \frac{\lambda}{\pi^2 L} \left[\cos \left(\frac{2\pi L}{\lambda} \right) - 1 \right] \right\}^{-1} \right)$$

$$(11\text{-}22)$$

This expression is plotted as a function of L/λ in Figure 11-6. For $L \gg \lambda$, the DI is given approximately by

$$\text{DI} = 10 \log \left(\frac{2L}{\lambda} \right) \qquad\qquad (11\text{-}23)$$

and as L/λ approaches zero, the DI approaches 0 dB.

11.2 DISCRETE SPATIAL ARRAYS

Spatial filters may be formed using arrays of omnidirectional hydrophones. A simple and yet very practical discrete array consists of a number of hydrophone elements spaced at equal increments along a straight line. The characteristics of such a line array will now be developed by extension of the results for the continuous-line hydrophone.

Consider again the aperture function and pattern function for the continuous-line hydrophone as shown in Figure 11-4. Although the function $G(u)$ has real meaning only over the range $\pm 1/\lambda$ in u, we may still represent the function outside that range in a formal mathematical sense. This is shown by the dashed extension of the sinc function in Figure 11-4 for $|u| > 1/\lambda$.

The aperture function for the discrete line array may be thought of as a sampled version of the appropriate continuous aperture function, with the samples consisting of spatial impulse functions. For hydrophones spaced at equal increments d, we may use the comb sample function defined in Section 6.7. Thus the discrete form of the rectangular aperture function may be written as

$$g(x) = d \text{ comb}_d \left[\frac{1}{L} \text{ rect} \left(\frac{x}{L} \right) \right] = \frac{d}{L} \sum_{n=-\infty}^{+\infty} \delta(x - nd) \text{ rect} \left(\frac{x}{L} \right) \qquad (11\text{-}24)$$

For convenience, let $L = Nd$ with $N = 2M + 1$. This results in a total of N hydrophones within the interval L, with one located at the origin and M equally spaced elements on each side of the origin.

Sampling in one domain causes the transform to become repetitive in the opposite domain. In accordance with rule 13 from Table 6-1, we write

$$g(x) = d \text{ comb}_d \left[\frac{1}{L} \text{ rect} \left(\frac{x}{L} \right) \right] \longleftrightarrow [\text{rep}_{1/d} \text{ sinc} (Lu)] = G(u) \qquad (11\text{-}25)$$

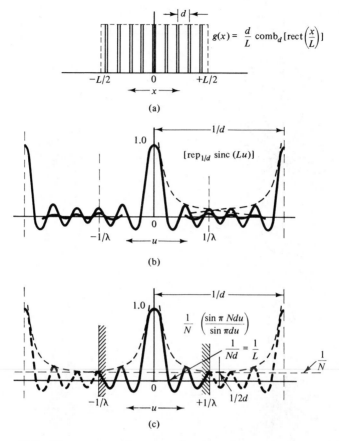

Figure 11-7   Discrete rectangular aperture function and resulting aperture function.

The sinc function pattern for the continuous aperture function is caused to repeat at intervals $1/d$ in the u-domain. The discrete aperture function and the resulting pattern function are shown in Figure 11-7. The discrete aperture function is normalized to have unit area, with the result that the peak of the pattern function is unity.

Figure 11-7(b) is a plot of several of the individual sinc functions contained in the repetitive pattern. Because of the infinite extent of each sinc function, the interval $(-1/\lambda, +1/\lambda)$ contains contributions from the "tails" of an infinite number of sinc functions centered outside this interval. This modifies the resultant real pattern in this region in a manner entirely analogous to aliasing effects encountered in the sampling of time-domain waveforms.

As demonstrated in Section 6.7, the transform of the discrete aperture function of (11-24) may also be expressed in closed form. Thus

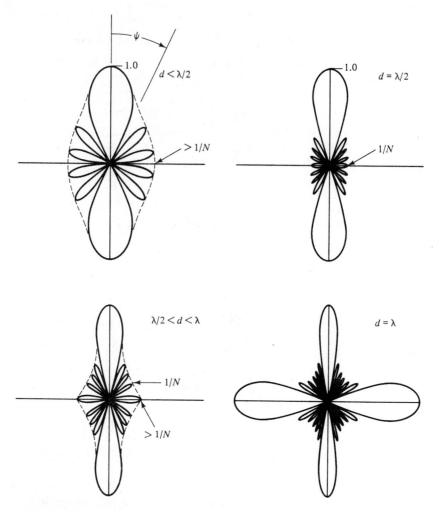

Figure 11-8 Discrete line array pattern functions for different element spacings relative to wave length.

$$[\text{rep}_{1/d} \text{sinc}\, (Lu)] = \frac{1}{N}\left[\frac{\sin{(\pi N du)}}{\sin{\pi du}}\right] \qquad (11\text{-}26)$$

The closed-form expression is shown graphically in Figure 11-7(c), and gives the total resulting pattern function in the interval $(-1/\lambda, +1/\lambda)$, including the aliasing effects caused by the discrete nature of the aperture function.

Notice in Figure 11-7(c) that the width of the main lobe, as measured by the position of the first zero crossing, is identical to that for the continuous aperture. That is, $G(u) = 0$ for $u = 1/Nd = 1/L$. Hence

lated shape if the predicted performance is to be achieved. The lower the value of the desired side lobes, the more accurate this match must be.

11.4 BEAM STEERING

The beamforming operation considered thus far results in the formation of a pattern function with a main response axis (MRA) perpendicular to the line defining the array. Steering the beam in other directions can be accomplished by physically changing the orientation of the acoustic aperture or, in the case of the array of discrete elements, electronically changing the direction of the MRA. Electronic beam steering of a discrete array is particularly useful because it permits the simultaneous formation of a number of receiving beams in different directions.

Using rule 6 for translation from Table 6-1, the required aperture function modification to obtain a shift in beam direction is easily seen. For instance, consider the aperture function and corresponding single-frequency pattern function,

$$g(x) \longleftrightarrow G(u)$$

Then $\hspace{10cm}$ (11-27)

$$g(x) \exp(j2\pi x u_0) \longleftrightarrow G(u - u_0)$$

Multiplication of the aperture function by the complex exponential causes a translation, or steering, of the pattern function by the amount u_0. For a discrete linear array of equally spaced hydrophones, (11-27) becomes

$$\sum_n g(nd)\delta(x - nd) \exp(j2\pi n d u_0) \longleftrightarrow G(u - u_0) \qquad (11\text{-}28)$$

Recalling that $u_0 = (\sin \psi_0)f/c$, the operation indicated by (11-28) is the addition to the output of each hydrophone of a phase shift that is a linear function of frequency and distance along the array. The phase shift added to the nth hydrophone is

$$\phi_n = \frac{2\pi f n d \sin \psi_0}{c} \qquad (11\text{-}29)$$

The relationship between the modified aperture function and the resulting pattern function at a particular frequency is shown in Figure 11-11.

For a single-frequency signal, the effect of a phase shift is equivalent to a time delay τ_n defined by

$$\phi_n = 2\pi f \tau_n$$

where

$$\tau_n = \frac{nd \sin \psi_0}{c} \tag{11-30}$$

By inserting the appropriate time delay in each hydrophone channel, the proper phase shift is assured regardless of signal frequency. The beam-steering time

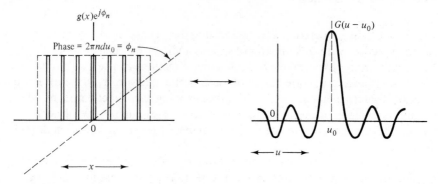

Figure 11-11 Effect on the pattern function of adding a linear phase term to the aperture function.

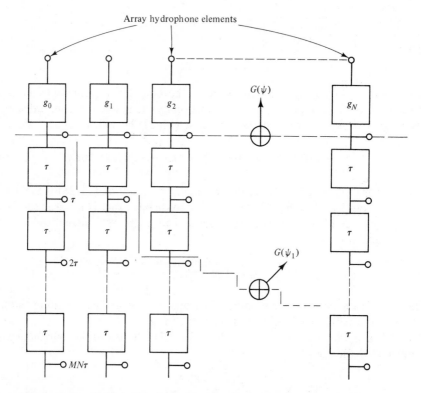

Figure 11-12 Multiple beamforming using tapped delay lines.

delay is just that required to cancel the geometric time delay at each array element for a plane wave signal arriving from the direction ψ_0. Thus the signal at all array elements will add in-phase when the signal direction is ψ_0.

By providing each hydrophone channel with a set of tapped delay elements, beams in several directions can be formed simultaneously. A line array using tapped delay elements to form multiple beams is shown schematically in Figure 11-12. The broadside beam $G(\psi)$ is formed by combining the element signals without any delays. Beams in other directions are formed by selecting delay line taps such that a linear progression of delays occurs from one end of the array to the other.

Assume that the slope of the delay versus hydrophone position is changed in equal-size steps to form the set of beams. This results in beams equally spaced in the u-domain, but not in the ψ-domain. That is, the delay slope is a linear function of $\sin \psi_0$. However, for the linear array this results in beams crossing over at the same level on the pattern, thus giving uniform coverage over the total

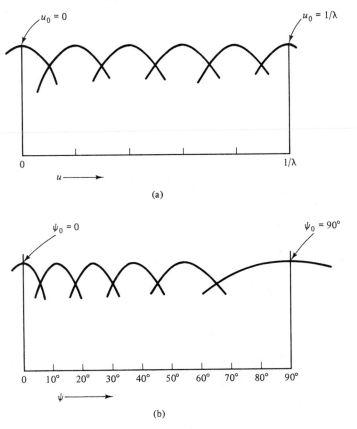

Figure 11-13 Multibeam pattern functions for a linear array with beams equally spaced in the u-domain.

angular range of interest. The multiple beams thus formed are sketched in Figure 11-13 both as a function of u and as a function of ψ. Figure 11-13(b) illustrates the fact that the beamwidth of the steered beams for a line array increases as the steering angle ψ_0 approaches 90°.

As shown in Figure 11-5, the three-dimensional pattern for a line array steered to broadside ($\psi_0 = 0$) is a disk-like figure in a plane normal to the array axis. When steered off broadside, the pattern resembles a conical shell, as shown in Figure 11-14. The side lobe structure is ignored in Figure 11-14 for clarity. The three-dimensional pattern for the line array is symmetrical with respect to the line-array axis for all steering angles. The pattern function for the discrete line array with a uniform aperture function steered to the angle ψ_0 is obtained with the help of (11-26) and (11-27). Thus if

$$G(u) = \frac{1}{N}\left[\frac{\sin(\pi N d u)}{\sin(\pi d u)}\right]$$

then

$$
\begin{aligned}
G(u - u_0) &= \frac{1}{N}\left\{\frac{\sin[\pi N d(u - u_0)]}{\sin[\pi d(u - u_0)]}\right\} \\
&= \frac{1}{N}\left\{\frac{\sin[\pi N d(\sin\psi - \sin\psi_0)/\lambda]}{\sin[\pi d(\sin\psi - \sin\psi_0)/\lambda]}\right\}
\end{aligned}
\qquad (11\text{-}31)
$$

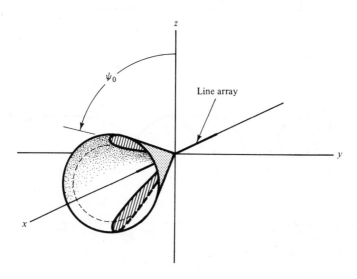

Figure 11-14 Three-dimensional line-array pattern for a beam steered off-broadside.

11.5 GENERALIZED ARRAY GAIN FOR THE DISCRETE LINE ARRAY

In Section 11.1.2 the directivity index for the line hydrophone was derived. By definition this derivation assumed an isotropic noise field with the line hydrophone providing a broadside-steered beam. We now extend this derivation to a discrete array of uniformly weighted hydrophones with arbitrary steering angles, considering both isotropic and nonisotropic noise fields. In the case of nonisotropic noise, the term *array gain* is used rather than directivity index. Recall that array gain is defined here as the ratio of noise power out of an omnidirectional sensor to the noise power out of the directional sensor, assuming the peak output of each is normalized to unity. This definition is appropriate when considering the performance of the array with respect to a single plane wave signal aligned with the maximum response direction of the array.

Consider first an isotropic noise field and a discrete array of equally spaced, uniformly weighted elements aligned with the z-axis. For a pattern function steered to the elevation angle ϕ_0, the discrete aperture function and corresponding pattern function at the wavelength λ are

$$g(z, \lambda) = d \ \text{comb}_d \left[\frac{1}{L} \text{rect} \left(\frac{z}{L} \right) \exp \left(j2\pi z u_0 \right) \right]$$

$$\longleftrightarrow \{\text{rep}_{1/d} \text{sinc} \left[L(u - u_0) \right] \} = G(u - u_0) \qquad (11\text{-}32)$$

where $u = (\sin \phi)/\lambda$. The aperture correlation function and squared pattern function are obtained from (11-32) as

$$R_g(z) = d \ \text{comb}_d \left[\frac{1}{L} \left(1 - \frac{|z|}{L} \right) \text{rect} \left(\frac{z}{2L} \right) \exp \left(j2\pi z u_0 \right) \right]$$

$$\longleftrightarrow \{\text{rep}_{1/d} \text{sinc}^2 \left[L(u - u_0) \right] \} \qquad (11\text{-}33)$$

The isotropic noise field at wavelength λ is described by the cross-spectral density and the angular intensity distribution. Thus

$$Q_n(z, \lambda) = 4\pi K(\lambda) \text{sinc} \left(\frac{2z}{\lambda} \right) \longleftrightarrow K(\lambda) = |N(\phi, \gamma, \lambda)|^2 \qquad (11\text{-}34)$$

where ϕ and γ are the elevation and azimuth angles, respectively.

The noise power delivered by an omnidirectional sensor or by the array can be determined using either the angle-space or linear-space representation of the functions. For the omnidirectional sensor it is convenient to perform the integration in angle space as follows:

$$P_o(\lambda) = K(\lambda) \int_{-\pi}^{+\pi} \int_{-\pi/2}^{+\pi/2} \cos \phi \, d\phi \, d\gamma$$

$$= K(\lambda) 2\pi\lambda \int_{-1/\lambda}^{+1/\lambda} du = 4\pi K(\lambda) \qquad (11\text{-}35)$$

For the vertical-line array in isotropic noise, the solution is obtained most readily in the opposite domain. Hence

$$P_a(\lambda) = \int_{-\infty}^{+\infty} R_g(z, \lambda) Q_n(z, \lambda) \, dz$$

$$= \frac{4\pi K(\lambda)}{L} \int_{-L}^{+L} d \operatorname{comb}_d \left[\left(1 - \frac{|z|}{L} \right) \exp(j2\pi z u_0) \right] \operatorname{sinc} \left(\frac{2z}{\lambda} \right) dz$$

$$(11\text{-}36)$$

Recall that the comb_d function is an impulse train with separation d. Also recognize that the integral of the imaginary part of (11-36) is zero and that the number of array elements is $N = L/d$. Equation (11-36) may therefore be written as

$$P_a(\lambda) = \frac{4\pi K(\lambda)}{N} \int_{-L}^{+L} \sum_{n} \delta(z - nd) \left(1 - \frac{|z|}{L} \right) \cos(2\pi z u_0) \operatorname{sinc} \left(\frac{2z}{\lambda} \right) dz$$

$$= \frac{4\pi K(\lambda)}{N^2} \sum_{n=-(N-1)}^{+(N-1)} (N - n) \cos(2\pi n d u_0) \operatorname{sinc} \left(\frac{2nd}{\lambda} \right) \qquad (11\text{-}37)$$

or

$$P_a(\lambda) = \frac{4\pi K(\lambda)}{N^2} \left[N + 2 \sum_{n=1}^{+(N-1)} (N - n) \cos(2\pi n d u_0) \operatorname{sinc} \left(\frac{2nd}{\lambda} \right) \right]$$

$$(11\text{-}38)$$

Combining (11-35) and (11-38), the directivity index as a numerical ratio for the uniform line array in isotropic noise is

$$\text{d.i.} = \frac{P_o(\lambda)}{P_a(\lambda)} = \frac{N^2}{N + 2 \sum_{n=1}^{N-1} (N - n) \cos(2\pi n d u_0) \operatorname{sinc}(2nd/\lambda)}$$

$$(11\text{-}39)$$

For the isotropic noise field, this result does not depend on the array orientation.

As an example let $u_0 = 0$ ($\phi_0 = 0$), and let the element spacing d equal $\lambda/2$. Because $\operatorname{sinc}(n)$ equals zero for all $n \neq 0$, the summation in the denominator is zero, with the result that d.i. $= N$. This result is also obtained for $d = m\lambda/2$, where m is any integer. As discussed in Section 11.2, if d exceeds $\lambda/2$, the pattern side-lobe characteristics are degraded, eventually resulting in full-amplitude ambiguous lobes.

Now let $u_0 = 1/\lambda$ ($\phi_0 = \pi/2$). This is often referred to as *end-fire* beam steering because the pattern function maximum response is aligned with the array axis. The summand in the denominator for this case contains the term $\cos(2\pi n d/\lambda) \operatorname{sinc}(2nd/\lambda)$, which is easily seen to be zero for d equal to any integral multiple of $\lambda/4$.

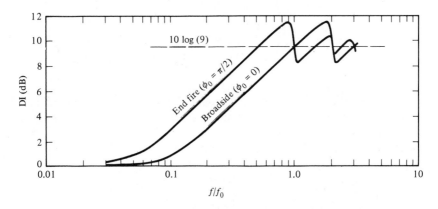

Figure 11-15   DI for discrete line array, broadside and end fire with $N = 9$.

Figure 11-15 presents the directivity index for a discrete line array of nine equally spaced elements as a function of frequency normalized to the frequency for which $d = \lambda/2$. The DI is shown for both the broadside-steered beam ($\phi_0 = 0$) and the end-fire beam ($\phi_0 = \pi/2$). As the frequency is decreased below the normalized value of unity, the DI for the broadside beam decreases approximately 3 dB per octave until the array length approaches one wavelength. As frequency decreases further, the DI asymptotically approaches zero. Notice that in the high-frequency region the DI oscillates about the value 10 log N. Values of DI in excess of 10 log N are possible because negative noise correlation between hydrophones can cause the total noise power to be *less* than the sum of the powers of the individual elements.

In the region below a normalized frequency of 0.5, the DI for the end-fire beam is approximately 3 dB higher than that for the broadside beam. This advantage for the end-fire beam is obtained at the expense of a more restricted useful frequency range. That is, for the end-fire beam the side-lobe characteristics start degrading for $d > \lambda/4$, with a full-amplitude ambiguous beam 180° from the steered direction when $d = \lambda/2$.

As a second example, consider a noise field resulting from surface-generated noise, including the component reflected from the ocean bottom. Using the model developed in Section 10.2.4, the noise angular density in the vertical plane is

$$|N(\phi, \gamma, \lambda)|^2 = \begin{cases} K(\lambda)\sin\phi & \text{for } 0 \le \phi \le \dfrac{\pi}{2} \\[2mm] K(\lambda)L_b\sin|\phi| & \text{for } \dfrac{-\pi}{2} \le \phi < 0 \end{cases} \qquad (11\text{-}40)$$

In terms of the variable $u = (\sin\phi)/\lambda$ this can be written

$$|N(u)|^2 = \begin{cases} K(\lambda)\lambda u & \text{for } 0 \le u \le \dfrac{1}{\lambda} \\[2ex] K(\lambda)L_b\lambda|u| & \text{for } \dfrac{-1}{\lambda} \le u < 0 \end{cases} \qquad (11\text{-}41)$$

We now wish to determine the array gain for a vertical line array of equally spaced and uniformly weighted hydrophones with arbitrary beam steering angle in this nonisotropic noise field.

The noise output from an omnidirectional sensor in this noise field is

$$P_o(\lambda) = K(\lambda)(1 + L_b) \int_{-\pi}^{+\pi} \int_0^{\pi/2} \sin\phi \cos\phi \, d\phi \, d\gamma$$

$$= K(\lambda)(1 + L_b)2\pi\lambda^2 \int_0^{1/\lambda} u \, du = K(\lambda)\pi(1 + L_b) \qquad (11\text{-}42)$$

To obtain the vertical array noise output it is convenient to perform the integration in the u-domain, representing the normalized power pattern function as

$$|G(u - u_0)|^2 = \frac{1}{N^2}\left|\sum_n \exp[j2\pi nd(u - u_0)]\right|^2$$

$$= \frac{1}{N^2}\sum_n \sum_m \exp[j2\pi(n - m)d(u - u_0)] \qquad (11\text{-}43)$$

where N is the number of hydrophones and u_0 is the steered direction in u-space.

In (11-43), let $r = (n - m)$ and recognize that the imaginary part of (11-43) is identically zero. We may then write

$$|G(u - u_0)|^2 = \frac{1}{N^2}\left\{N + 2\sum_{r=1}^{N-1}(N - r)\cos 2\pi rd(u - u_0)\right\} \qquad (11\text{-}44)$$

The array output noise is

$$P_a(\lambda) =$$

$$K(\lambda)\lambda^2\left[\int_{-\pi}^{+\pi}\int_0^{1/\lambda} u|G(u - u_0)|^2 \, du \, d\gamma - L_b\int_{-\pi}^{+\pi}\int_{-1/\lambda}^0 u|G(u - u_0)|^2 \, du \, d\gamma\right]$$

$$= 2\pi K(\lambda)\lambda^2\left[\int_0^{1/\lambda} u|G(u - u_0)|^2 \, du - L_b\int_{-1/\lambda}^0 u|G(u - u_0)|^2 \, du\right] \qquad (11\text{-}45)$$

Substitute (11-44) into (11-45) and integrate to obtain

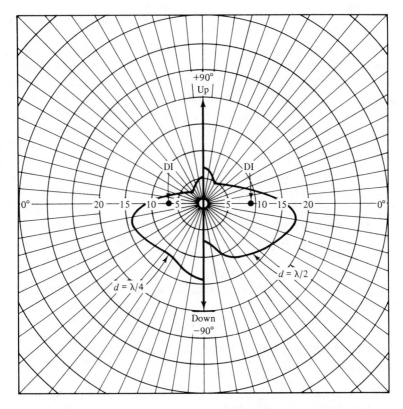

Figure 11-16 Nine-element vertical-line array gain versus elevation steering angle in a surface-generated noise field.

$$P_a(\lambda) = \frac{\pi K(\lambda)(1 + L_b)}{N^2}\left\{N + 4\sum_{r=1}^{N-1}(N - r)\cos(2\pi rdu_0)\left[\frac{\sin(2\pi rd/\lambda)}{2\pi rd/\lambda}\right.\right.$$

$$\left.\left. - \frac{1}{2}\frac{\sin^2(\pi rd/\lambda)}{(\pi rd/\lambda)^2}\right]\right\} + \frac{4\pi K(\lambda)(1 - L_b)}{N^2}\sum_{r=1}^{N-1}(N - r)\sin(2\pi rdu_0)$$

$$\left[\frac{\sin(2\pi rd/\lambda)}{(2\pi rd/\lambda)^2} - \frac{\cos(2\pi rd/\lambda)}{2\pi rd/\lambda}\right] \qquad (11\text{-}46)$$

From (11-42) and (11-46) the array gain is

$$\text{a.g.} = N^2\left/\left\{N + 4\sum_{r=1}^{N-1}(N - r)\cos(2\pi rdu_0)\left[\frac{\sin(2\pi rd/\lambda)}{2\pi rd/\lambda} - \frac{1}{2}\frac{\sin^2(\pi rd/\lambda)}{(\pi rd/\lambda)^2}\right]\right.\right.$$

$$\left.\left. + \frac{4(1 - L_b)}{(1 + L_b)}\sum_{r=1}^{N-1}(N - r)\sin(2\pi rdu_0)\left[\frac{\sin(2\pi rd/\lambda)}{(2\pi rd/\lambda)^2} - \frac{\cos(2\pi rd/\lambda)}{2\pi rd/\lambda}\right]\right\}\right.$$

$$(11\text{-}47)$$

345

Although (11-47) appears formidable, it is easily evaluated with the help of a computer. In Figure 11-16 the array gain in dB is plotted as a function of elevation steering angle ϕ_0 for a vertical array of nine elements in this surface-generated noise model. The right-hand half of the Figure assumes a frequency such that the element spacing is $d = \lambda/2$, while the left half assumes $d = \lambda/4$. The directivity index, DI, that would be obtained for the broadside beam in an isotropic noise field is noted in each case. The factor L_b is assumed to be 0.1 for Figure 11-16.

Note that the array gain exceeds the directivity index over a wide range of steering angles in the lower hemisphere. As expected, the array gain is typically less than the directivity index for beams steered in the direction of the surface. For $d = \lambda/2$ the array gain degrades significantly for the end-fire beam at $\phi_0 = -90°$. This is because of the ambiguous beam produced at $+90°$ at this frequency and steering direction. For $d = \lambda/4$ and $\phi_0 = -90°$ no ambiguous beam is produced in the vertical direction and the array gain remains high. This example illustrates the importance of considering the relationship of the array properties to the noise field characteristics when configuring the acoustic system.

11.6 USE OF ARRAY GAIN IN SYSTEM PERFORMANCE ANALYSIS

In preceding chapters the acoustic intensity of a source, the transmission properties of the medium, and the ambient noise properties have been discussed. With this information the signal-to-noise ratio (SNR) at the receiver location can be determined as sensed by an omnidirectional device. We shall now determine the SNR at the output of a directional sensor using the concept of array gain.

In the discussion that follows, intensity spectra for the signal at the source, $\Psi_s(f)$, and for noise, $\Psi_n(f)$, are assumed normalized by the appropriate reference intensity. Thus the source spectral level and noise spectral level at frequency f are given by

$$SL_s(f) = 10 \log \Psi_s(f)$$

and

$$NL_s(f) = 10 \log \Psi_n(f)$$

The transmission loss factor in general is a function of both range and frequency. At range r the signal spectral intensity at frequency f is related to the source intensity and the transmission loss factor by

$$\Psi_s(r,f) = \Psi_s(f)T(r,f)$$

Therefore,

$$TL(r,f) = -10 \log T(r,f) = 10 \log \left[\frac{\Psi_s(f)}{\Psi_s(r,f)}\right]$$

The signal spectrum at the receiver location may differ from the source signal spectrum in both amplitude and shape.

The received signal intensity over a given frequency band is given by the integral of the received signal intensity spectrum over that band. For the frequency range f_1, f_2 we have

$$I_s(r) = \int_{f_1}^{f_2} \Psi_s(r,f) \, df = \int_{f_1}^{f_2} \Psi_s(f) T(r,f) \, df$$

from which the total received signal level is

$$SL(r) = 10 \log I_s(r) = 10 \log \left[\int_{f_1}^{f_2} \Psi_s(f) T(r,f) \, df \right] \qquad (11\text{-}48)$$

If the transmission loss factor is essentially constant over the band of interest, it may be removed from the integral, giving the particularly simple result

$$SL(r) = 10 \log \left[\int_{f_1}^{f_2} \Psi_s(f) \, df \right] + 10 \log T(r)$$

$$= SL - TL(r) \qquad (11\text{-}49)$$

Where SL is the source level evaluated over the band of interest and $TL(r)$ is the transmission loss appropriate for this band. Assuming a plane wave signal aligned with the peak of the pattern function, the received signal levels indicated by (11-48) or (11-49) are also appropriate at the output of a normalized directional sensor.

The intensity spectrum of ambient noise is modified by the array gain of the directional sensor relative to the omnidirectional sensor. The noise intensity at the output of the directional sensor in the frequency range f_1, f_2 is, therefore,

$$I_n = \int_{f_1}^{f_2} \left[\frac{\Psi_n(f)}{A(f)} \right] df \qquad (11\text{-}50)$$

where $A(f)$ is the numerical array gain as a function of frequency. The output noise level over the band is $NL_o = 10 \log I_n$. If the frequency range f_1, f_2 is very narrow such that $\Psi_n(f)$ and $A(f)$ are nearly constant over this range, the total noise intensity level may be separated into terms related to the ambient spectral intensity, the bandwidth, and the array gain. Thus, for constant noise level and array gain,

$$NL_o = NL_s - AG + 10 \log \beta \qquad (11\text{-}51)$$

where NL_s = ambient noise spectral level in the band (constant)
  AG = array gain in the band (constant)
  $\beta = f_2 - f_1$ = bandwidth

In the general case, the signal-to-noise intensity ratio of the directional sensor output may be formulated using decibel notation as

$$\text{SNR}(r) = 10 \log\left\{ \int_{f_1}^{f_2} \Psi_s(f) T(r,f)\, df \right\}$$

$$- 10 \log\left\{ \int_{f_1}^{f_2} \left[\frac{\Psi_n(f)}{A(f)}\right] df \right\} = \text{SL}(r) - \text{NL}_o \qquad (11\text{-}52)$$

This expression may be further simplified only to the extent that simplifying assumptions such as transmission loss independent of frequency or very narrow bandwidth are valid. With the narrowband assumption we obtain the particularly simple result

$$\text{SNR}(r)_{\text{NB}} = \text{SL}_s - \text{TL}(r) - \text{NL}_s + \text{AG} \qquad (11\text{-}53)$$

If the bandwidth is not narrow, but the transmission loss is constant over the band, we get

$$\text{SNR}(r) = \text{SL} - \text{TL}(r) - \text{NL}_o \qquad (11\text{-}54)$$

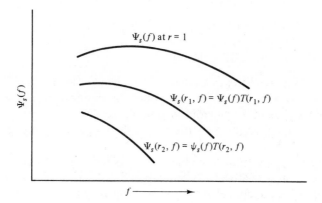

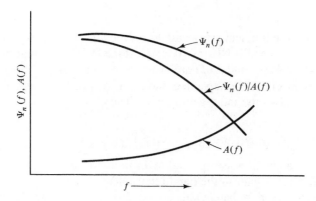

Figure 11-17 Spectral descriptions for source, transmission characteristics, and noise.

where NL_o = total noise level over the band at sensor output

   SL = source level over the band

   $TL(r)$ = transmission loss at range r appropriate for the band f_1, f_2.

In the more general case, the signal and noise terms for use in the integrals in (11-52) are sketched in Figure 11-17. Note that only in the narrowband case is it permissible to form the SNR directly at the spectral level. Otherwise, the signal and noise terms must be integrated separately over the frequency band prior to forming the ratio.

If the approximations involved in (11-53) or (11-54) are permissible, the functional dependence of SNR on range is separable from the integral involving signal spectra. Thus the integrals need only be evaluated once. If the transmission-loss dependence on frequency cannot be neglected, it is necessary to evaluate the signal intensity integral at each range of interest to obtain the SNR dependence on range.

11.7 TARGET ANGLE ESTIMATION

In addition to improving the SNR, the spatial filter plays a role in the estimation of the direction of arrival of a plane wave signal. Let Figure 11-18 represent the magnitude-squared response of a scanning beam in a noise field that contains a plane wave target signal at ψ_0. Based on the available data we may form an estimate of ψ_0 that will inevitably contain some uncertainty because of the presence of noise.

One approach to target angle estimation is to choose the peak output in Figure 11-18 as the estimate. Equivalently, we could differentiate the response in Figure 11-18 with respect to ψ, and select the location in the vicinity of ψ_0 where the derivative passes through zero as the estimate of target direction. The derivative of the spatial response is sketched in Figure 11-19.

Assuming that the pattern function is an even function about its main response axis (MRA), the derivative will be an odd function. Thus, in the vicinity of the MRA we have

$$\frac{d|G(\psi)|^2}{d\psi} \simeq K\psi \qquad (11\text{-}55)$$

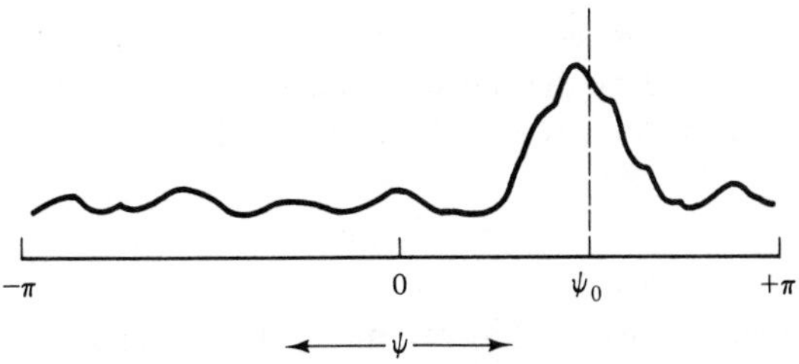

Figure 11-18   Scanning-beam output with a plane wave target at ψ_0.

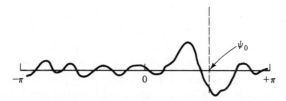

Figure 11-19   Derivative of the spatial response in Figure 11-18 with a target at ψ_0.

as the desired angular response to a plane wave target. A mechanization, referred to as a *split-aperture correlator,* will now be described that provides a response approximating the result in (11-55).

In Figure 11-20(a), a line array of length $L/2$ is shown centered at the origin having a pattern function $G(\psi)$. In part (b) of the figure the array is shifted to the left by $L/4$. The *shape* of the pattern function is not affected by this shift, but the *phase* of a signal received by the shifted aperture is changed relative to the signal at the centered aperture by an amount proportional to the distance moved. Similarly, in (c) the array is shifted to the right, resulting in a phase with the opposite sense. The left- and right-shifted apertures have pattern functions expressible in terms of the pattern of the centered aperture as follows:

$$G_L(\psi) = G(\psi) \exp\left(-j\frac{\pi L \sin \psi}{2\lambda}\right)$$
$$G_R(\psi) = G(\psi) \exp\left(+j\frac{\pi L \sin \psi}{2\lambda}\right) \tag{11-56}$$

In Figure 11-20(d), both shifted apertures are assumed present, providing the separate output signals e_L and e_R. For a plane wave single-frequency signal with amplitude a, the half-array signals (usually called half-beam signals) in complex form are

$$e_L(t, \psi) = aG(\psi) \exp\left[j\left(\omega t - \frac{\pi L \sin \psi}{2\lambda}\right)\right]$$
$$e_R(t, \psi) = aG(\psi) \exp\left[j\left(\omega t + \frac{\pi L \sin \psi}{2\lambda}\right)\right] \tag{11-57}$$

The left and right array signals in (11-57) are identical except for an electrical phase difference ϕ caused by the physical separation of the half-arrays and the spatial angle ψ between the MRA and the target. Thus

$$\phi = \frac{\pi L}{\lambda} \sin \psi \tag{11-58}$$

For ψ much less than unity, (11-58) becomes approximately

$$\phi \approx \frac{\pi L}{\lambda} \psi \tag{11-59}$$

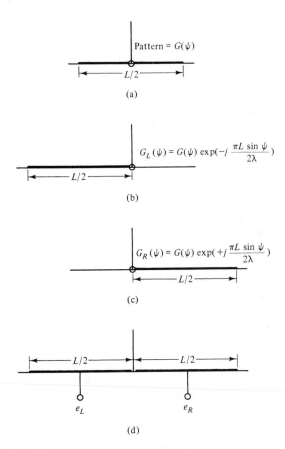

Figure 11-20   Effect of translation on the pattern function.

Hence the electrical phase between the half-array output signals is a measure of the target direction relative to the MRA. Notice that the constant of proportionality relating electrical and spatial angles involves the factor L/λ, suggesting that target direction can be determined with greater precision as L/λ is increased.

To obtain a measure of the phase difference ϕ, multiply the output of the left-half array by the conjugate of the right-half output. This gives

$$e_L(t, \psi)e_R^*(t, \psi) = a^2|G(\psi)|^2 \exp(-j\phi)$$

$$= a^2|G(\psi)|^2(\cos \phi - j \sin \phi) \qquad (11\text{-}60)$$

The real and imaginary parts of (11-60) are shown in Figure 11-21.

The ratio of the real and imaginary parts of (11-60) is explicitly the tangent of the electrical phase angle. Thus

$$\frac{\text{Im}[e_L e_R^*]}{\text{Re}[e_L e_R^*]} = \tan \phi = \tan\left(\frac{\pi L \sin \psi}{\lambda}\right) \qquad (11\text{-}61)$$

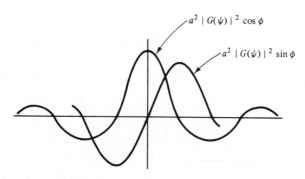

Figure 11-21 Real and imaginary parts of $e_L e_R^*$.

For ψ small such that $(L/\lambda) \sin \psi \ll 1$, we have approximately

$$\tan \phi \simeq \frac{\pi L}{\lambda} \psi \tag{11-62}$$

which is of the same form as (11-55).

In Figure 11-22 a mechanization is shown for achieving the result in (11-62) assuming a real, rather than complex, sinusoidal signal. The signals at the half-array outputs differ only in electrical phase and may be written as

$$e_L = aG(\psi) \cos\left(\omega t - \frac{\phi}{2}\right)$$
$$e_R = aG(\psi) \cos\left(\omega t + \frac{\phi}{2}\right) \tag{11-63}$$

The output of the left-half array is shifted in phase by $\pi/2$ (equivalent to forming the conjugate of the complex signal) and multiplied by e_R to give

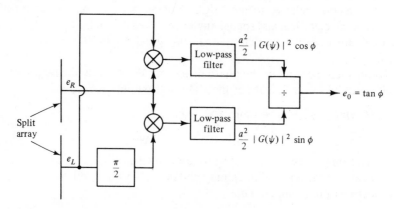

Figure 11-22 Narrowband cross-correlator for angle measurement.

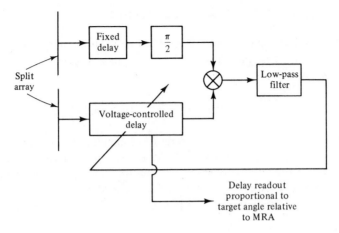

Figure 11-23   Closed-loop angle measurement using cross-correlator.

$$e_R e_L \underline{/\pi/2} = a^2 G^2(\psi) \cos\left(\omega t - \frac{\phi}{2} + \frac{\pi}{2}\right) \cos\left(\omega t + \frac{\phi}{2}\right)$$

$$= \frac{a^2}{2} G^2(\psi)(\sin \phi + \sin 2\omega t) \qquad (11\text{-}64)$$

A low-pass filter at the multiplier output removes the double-frequency component, providing an input to the divider proportional to $\sin \phi$.

The signals e_R and e_L are also multiplied directly, giving

$$e_R e_L = \frac{a^2}{2} G^2(\psi)(\cos \phi + \cos 2\omega t) \qquad (11\text{-}65)$$

A low-pass filter is again used to remove the double-frequency term, providing an input to the divider proportional to $\cos \phi$. The divider output is then equal to $\tan \phi$. The sequence of operations described to obtain $\sin \phi$ and $\cos \phi$ results in the formation of an estimate of the cross-correlation function of e_L and e_R—hence the name split-aperture correlator.

The presence of noise combined with the desired signal prevents a perfect measurement of target direction. The measurement error resulting from the presence of noise is discussed in Chapter 13.

As an alternative to the explicit measurement of angle described by Figure 11-22, a closed-loop solution may be used as shown in Figure 11-23. In this arrangement a variable delay is provided in one signal channel to cancel the effect of the delay caused by target angle relative to the MRA. The output of the low-pass filter, proportional to $\sin \phi$, is used to control the variable delay, driving $\sin \phi$ to zero. The amount of delay required to null the output is calibrated in terms of target angle relative to the MRA.

11.8 MULTIDIMENSIONAL SPATIAL FILTERS

In Section 11.1.1 the pattern function of a line hydrophone on the x-axis is defined by the transform relationship of (11-9). This is a special case of the more general relationship

$$G(\mathbf{u}) = \int g(\mathbf{r}) \exp\left[j2\pi\mathbf{r}\cdot(\mathbf{u} - \mathbf{u}_0)\right] dv \qquad (11\text{-}66)$$

where $\mathbf{u}$ and $\mathbf{u}_0$ are vectors with magnitude $1/\lambda$ indicating the direction of arrival of a plane wave and the beam steering direction, respectively, and $\mathbf{r}$ is the position vector of the volume element dv. The function $g(\mathbf{r})$ is the response per unit volume of the sensor to a unit signal incident at position $\mathbf{r}$.

For a discrete array of N hydrophones with arbitrary distribution, (11-66) becomes

$$G(\mathbf{u}) = \sum_n g(\mathbf{r}_n) \exp\left[j2\pi\mathbf{r}_n\cdot(\mathbf{u} - \mathbf{u}_0)\right] \qquad (11\text{-}67)$$

With a coordinate system as in Figure 10-6 the Cartesian components of $\mathbf{r}_n$ are x_n, y_n, and z_n. Using the polar angles ψ and θ, the components of $\mathbf{u} - \mathbf{u}_0$ are

$$u_x - u_{x0} = \frac{\sin\psi - \sin\psi_0}{\lambda}$$

$$u_y - u_{y0} = \frac{\cos\psi\sin\theta - \cos\psi_0\sin\theta_0}{\lambda} \qquad (11\text{-}68)$$

$$u_z - u_{z0} = \frac{\cos\psi\cos\theta - \cos\psi_0\cos\theta_0}{\lambda}$$

Performing the indicated dot-product operation in (11-67), we obtain

$$G(\mathbf{u}) = \sum_n g(x_n, y_n, z_n) \exp\left\{j2\pi[x_n(u_x - u_{x0}) + y_n(u_y - u_{y0}) + z_n(u_z - u_{z0})]\right\}$$

$$(11\text{-}69)$$

Equation (11-69) reduces to the line array case if y_n and z_n are zero for all n. Thus

$$G(u_x) = \sum_n g(x_n) \exp\left[\frac{j2\pi x_n(\sin\psi - \sin\psi_0)}{\lambda}\right] \qquad (11\text{-}70)$$

Equation (11-67) or (11-69) can be used to find the pattern function for an arbitrary volumetric distribution of hydrophones. Of course, closed-form solutions to these equations are not to be expected except for particularly simple geometric configurations for the array. However, given the hydrophone locations and weighting factors, (11-69) is easily solved by a computer.

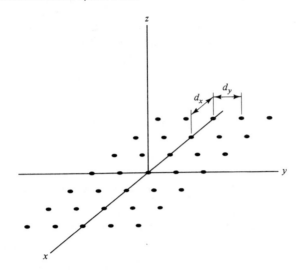

Figure 11-24 Planar array geometry.

One example of practical interest that may yield a simple mathematical solution is the planar array. Consider an array geometry as shown in Figure 11-24. Let the elements have equal spacing in the x direction equal to d_x and in the y-direction equal to d_y. Thus the coordinates of the mnth hydrophone are

$$x_m = md_x$$

$$y_n = nd_y$$

$$z_{m,n} = 0$$

Equation (11-69) may now be expressed as a double summation. With no loss in generality, we let $\mathbf{u}_0 = 0$ and write

$$G(\mathbf{u}) = \sum_m \sum_n g(x_m, y_n) \exp[j2\pi(md_x u_x + nd_y u_y)] \qquad (11\text{-}71)$$

A further simplification is possible if the planar aperture function is separable into the product of two line aperture functions. That is, assume that

$$g(x_m, y_n) = g_x(x_m)g_y(y_n) \qquad (11\text{-}72)$$

Equation (11-72) simply states that the shape of the amplitude weighting in the x-direction does not depend on y_n, and vice versa. The most obvious weighting function that satisfies (11-72) is the uniform or constant-amplitude weighting. Substitution of (11-72) in (11-71) permits us to write

$$G(\mathbf{u}) = \sum_m g_x(x_m) \exp(j2\pi md_x u_x) \sum_n g_y(y_n) \exp(2\pi nd_y u_y) \qquad (11\text{-}73)$$

Recognizing the individual summations as the pattern functions for the line
arrays along the x and y axes, (11-73) may be written as

$$G(\mathbf{u}) = G_x(u_x)G_y(u_y) \tag{11-74}$$

where

$$G_x(u_x) = \sum_m g_x(x_m) \exp(j2\pi m d_x u_x)$$

and

$$G_y(u_y) = \sum_n g_y(y_n) \exp(j2\pi n d_y u_y)$$

As an example, assume a square array with $d_x = d_y$ and uniform weighting
of all elements. With $u_x = \sin \psi/\lambda$ and $u_y = \cos \psi \sin \theta/\lambda$ the pattern function
is

$$G(\psi, \theta, \lambda) = \left[\frac{\sin(\pi Nd \sin \psi/\lambda)}{\sin(\pi d \sin \psi/\lambda)}\right] \left[\frac{\sin(\pi Nd \cos \psi \sin \theta/\lambda)}{\sin(\pi d \cos \psi \sin \theta/\lambda)}\right] \tag{11-75}$$

If $\theta = 0$ in (11-75) the pattern function in the z-x plane is obtained and is seen
to be the pattern of a simple line array. By letting $\psi = 0$, the identical result is
obtained in the z-y plane. Note, however, that the three-dimensional pattern
function is *not* obtained by rotating the line-array pattern about the z-axis. It is
left to the reader to show that the pattern measured in a plane containing the
z-axis and the diagonal of the square array is given by

$$G(\beta, \lambda) = \left[\frac{\sin(\pi Nd \sin \beta/\lambda\sqrt{2})}{\sin(\pi d \sin \beta/\lambda\sqrt{2})}\right]^2 \tag{11-76}$$

where β is measured from the z-axis in this plane. Figure 11-25 shows the
relationship between the pattern measured in a plane aligned with a principal axis
of the array and a pattern measured in a plane containing the array diagonal. As
measured by the location of the first null in the pattern, the beamwidth in the
diagonal plane is $\sqrt{2}\lambda/Nd$, as compared with λ/Nd, in the principal plane.
Note that in the diagonal-plane pattern there is no polarity reversal in the
side-lobe region. Also, the amplitude of the side lobes are quite small compared
with the principal-plane side lobes.

The directivity of a planar aperture can be expressed formally as

$$\text{d.i.}(\lambda) = 4\pi K \left[\int_0^{2\pi}\int_0^\infty R_g(\rho, \phi)Q_n(\rho, \lambda)\rho \, d\rho \, d\phi\right]^{-1} \tag{11-77}$$

where ρ and ϕ are the polar coordinates of a point in the x-y plane, R_g is the
aperture correlation function, and Q_n is the single-frequency correlation function
for isotropic noise.

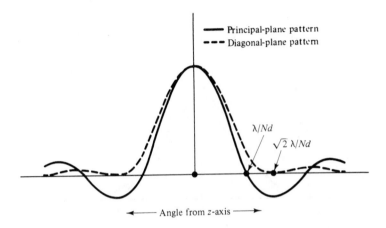

Figure 11-25   Principal-plane and diagonal-plane patterns for a square array.

Because R_g is a function of both ρ and ϕ for the square aperture, (11-77) does not have a simple exact solution. However, the effect of the lack of circular symmetry is slight, leading to a simple approximate result when the square dimensions are large compared to a wavelength and the aperture function is a constant over the plane surface. The aperture correlation function for this case is

$$R_g(\rho, \phi) = \frac{1}{L^2}\left[1 - \frac{\rho}{L}(|\cos \phi| + |\sin \phi|) + \frac{\rho^2}{L^2}|\cos \phi \sin \phi|\right] \qquad (11\text{-}78)$$

for ρ limited to the range where $R_g \geq 0$. With $\phi = 0$ or $90°$ we obtain a vertical cut through the function along one of the principal axes of the aperture. This is the familiar triangular correlation function associated with a rectangular aperture function. With $\phi = 45°$ we obtain the diagonal cut which shows the greatest departure from the triangular shape. These two cross sections are compared in Figure 11-26. The similarity of the two cross sections suggests that little error is introduced by assuming a correlation function independent of ϕ. Thus let

$$R_g(\rho) \simeq \frac{1}{L^2}\left(1 - \frac{\rho}{L}\right) \qquad \text{for } \rho \leq L \qquad (11\text{-}79)$$

Remembering that

$$Q_n(\rho, \lambda) = 4\pi K \frac{\sin(2\pi\rho/\lambda)}{2\pi\rho/\lambda}$$

and performing the integration over ϕ, (11-77) becomes

$$\text{d.i.}(\lambda) \simeq \left[\frac{\lambda}{L^2}\int_0^L \left(1 - \frac{\rho}{L}\right)\sin\left(\frac{2\pi\rho}{\lambda}\right) d\rho\right]^{-1} \qquad (11\text{-}80)$$

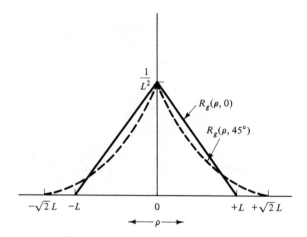

Figure 11-26 Principal-plane and diagonal-plane cross sections of the aperture correlation function for a uniformly weighted square planar aperture.

which if $L \gg \lambda$ is approximately

$$\text{d.i.}(\lambda) \simeq \frac{2\pi L^2}{\lambda^2} = \frac{\pi}{2}\left(\frac{2L}{\lambda}\right)^2 \tag{11-81}$$

This will be recognized as the square of the directivity of a line aperture of length L, multiplied by the factor $\pi/2$.

For the uniformly weighted rectangular aperture with dimensions L_x and L_y, both large compared with the wavelength, the directivity is approximately

$$\text{d.i.}(\lambda) \simeq \frac{\pi}{2}\left(\frac{4L_x L_y}{\lambda^2}\right) \tag{11-82}$$

The term in parentheses is the product of directivities of line apertures with lengths L_x and L_y.

With a plane aperture it is often desirable to baffle, or shield, one side of the plane so that the device is sensitive to signals on one side only. This reduces the noise power by a factor of 2, thereby doubling the directivity. Thus

$$\text{d.i.}(\lambda) = 4\pi \frac{L_x L_y}{\lambda^2}$$

$$= 4\pi \frac{A}{\lambda^2} \tag{11-83}$$

where A is the area of the rectangular aperture.

The second form of (11-83) is applicable to a wide class of nonrectangular planar arrays with uniform weighting, assuming that the minimum effective dimension is large compared to a wavelength. As an example, the directivity of a baffled circular piston with diameter $D \gg \lambda$ is approximately

$$\text{d.i.}(\lambda) = \left(\frac{\pi D}{\lambda}\right)^2 = \frac{4\pi}{\lambda^2}\left(\frac{\pi D^2}{4}\right) = \frac{4\pi A}{\lambda^2} \tag{11-84}$$

Equation (11-83) may also be applied to plane apertures with nonuniform weighting by defining an *effective* aperture area as

$$A_{\text{eff}} = \frac{\left|\iint g(x,y)\,dx\,dy\right|^2}{\iint |g(x,y)|^2\,dx\,dy} \tag{11-85}$$

Then

$$\text{d.i.}(\lambda) \simeq \frac{4\pi A_{\text{eff}}}{\lambda^2} \tag{11-86}$$

The effective area defined by (11-85) can easily be shown to be equal to or less than the actual area of the aperture by an application of the Schwarz inequality. For (11-86) to give accurate results, A_{eff} should be large compared with λ^2 and the least dimension associated with the planar region where $g(x,y)$ is appreciable should be large compared with λ.

Other examples of multidimensional spatial filters include cylindrical shell, spherical shell, spherical volumetric arrays, and conformal arrays matching the contour of a ship or submarine hull. For detailed analyses of the properties of such arrays, see, for example, Queen [4] and Anderson and Munson [5].

PROBLEMS

11.1. Two hydrophones with equal sensitivity and gain are located on the x-axis separated by a distance d. Determine the response of the sum of the hydrophone signals to a plane wave signal with wavelength λ arriving at an angle ψ relative to a plane perpendicular to the x-axis. Plot the polar pattern function assuming that $d = \lambda/4$, $\lambda/2$, and λ.

11.2. Repeat Problem 11.1 for three hydrophones along the x-axis with equal separation d.

11.3. In an isotropic noise field, the noise signals sensed by hydrophones separated by exactly $\lambda/2$ are statistically independent. Use this fact to calculate the directivity factor for the hydrophone arrays in Problems 11.1 and 11.2, assuming half-wavelength spacing.

11.4. Assume a line array of equally spaced hydrophones with spacing equal to a half-wavelength. It can be shown that amplitude shading of the hydrophone elements in accordance with the coefficients given by the binomial theorem results in zero side lobes. Demonstrate this by plotting the polar pattern function for an array of four elements with weights 1, 3, 3, and 1. Note the beamwidth of the resulting pattern and compare with the beamwidth of a uniformly weighted four-element array.

11.5. A continuous-line aperture has an aperture function given by

$$g(x) = \frac{1}{L}\left[1 + \cos\left(\frac{2\pi x}{L}\right)\right] \text{rect}\left(\frac{x}{L}\right)$$

Derive the expression for the pattern function $G(u)$ and sketch the pattern as a function of the variable u. Compare the beamwidth and peak side lobe with those for a uniformly weighted array.

11.6. Assuming isotropic noise and half-wavelength spacing, calculate the directivity factor for the binominal array in Problem 11.4.

11.7. Equation (11-47) gives the array gain for a vertical-line array in a surface-generated noise model, including the effect of a component reflected from the bottom. If the beam pattern is steered broadside to the array ($u_0 = 0$) notice that the array gain is independent of the bottom reflection factor L_b. Is this a reasonable result? Explain.

11.8. An angle tracking system is mechanized using a split-aperture correlator as shown in Figure 11-23. Assume a total array length L and a signal wavelength λ. What fixed delay and what range of variable delay are required to accommodate a target bearing in the range $-\lambda/2L \le \psi \le \lambda/2L$?

SUGGESTED READING

1. Horton, J. W., *Fundamentals of Sonar*, 2nd ed., United States Naval Institute, 1959.

2. Taylor, T. T., "Design of Line-Source Antennas for Narrow Beamwidth and Low Side Lobes," *IRE Trans.*, Vol. AP-3, p. 316 (1955).

3. Dolph, C. L., "A Current Distribution of Broadside Arrays Which Optimizes the Relationship between Beam Width and Side-Lobe Level," *Proc. IRE*, Vol. 34, p. 335 (June 1946).

4. Queen, W. C., "The Directivity of Sonar Receiving Arrays," *J. Acoust. Soc. Am.*, Vol. 47 No. 3, p. 711 (1970).

5. Anderson, V. C., and J. C. Munson, "Directivity of Spherical Receiving Arrays," *J. Acoust. Soc. Am.*, Vol. 35, No. 8, p. 1162 (1963).

6. Albers, V. M., *Underwater Acoustics Handbook—II*. University Park, Pa.: The Pennsylvania State University Press, 1965, Chap. 17.

7. Jordan, E. C., *Electromagnetic Waves and Radiating Systems*. New York: Prentice-Hall, Inc., 1950, Chap. 12.

12 Acoustic Characteristics of Targets

12.0 INTRODUCTION

Underwater acoustic systems detect the presence of objects in the water either by directly sensing acoustic energy radiated by the object, or by transmitting an acoustic signal and detecting the reflection, or *echo,* from the object. A system that relies on the target-generated acoustic signal is called a *passive* system, while a system relying on echo detection is an *active* system. It is of course possible for a system to simultaneously incorporate the capabilities of passive and active systems.

In this chapter the target characteristics of importance for passive and active detection systems are discussed. For the passive system, the target *acoustic signature* is characterized by the radiated acoustic spectrum level at a reference distance of 1 m from the effective acoustic center of the target. For the active system we define a *target strength* parameter to characterize the intensity of the signal reflected from the target as a result of an incident active signal.

Detailed measurements of both passive and active characteristics of ships and submarines were made during World War II and are now available in the unclassified literature [1–3]. Although this information relates to vessels now considered obsolete, the general characteristics are not unlike those of modern vessels in both form and magnitude. Detailed characteristics of present-day systems with military significance are understandably classified. We shall therefore concentrate on the general nature of target acoustic characteristics rather than attempting to catalog the characteristics of specific ship or submarine types.

12.1 PASSIVE ACOUSTIC SIGNATURES OF SHIPS AND SUBMARINES

Ships and submarines range in size from 100 to 1000 ft or more in length, and are propelled through the water with propulsion systems delivering from a few thousand to 100,000 or more shaft horsepower. Inevitably, some portion of the tremendous energy involved in this process is radiated from the vessel as acoustic energy in the water.

The principal sources of radiated acoustic noise are as follows [4,5]:

1. *Propulsion system*   This includes the engine, reduction gears (if any), drive shaft, bearings, and so on.
2. *Propeller*   Although a part of the propulsion system, the propeller is discussed separately because of the significant difference in the way it contributes to the generation of acoustic signals.
3. *Auxiliary machinery*   This includes nonpropulsion-related mechanical and electrical systems, such as air conditioning, electrical generators, and pumps.
4. *Hydrodynamic effects*   These include radiated flow noise and flow-induced excitation of plates or other structural features.

The radiated acoustic spectrum contains a broad, continuous spectral component as well as narrowband sinusoidal components often referred to as *discrete lines* or *tonals*. Depending on their origin, the various spectral components may or may not be functions of speed, depth, or other factors related to the operation of the vessel.

The acoustic signature is normally measured at a relatively close range, such as 100 or 200 m, and converted to the reference range of 1 m using the spherical spreading law. Because of the complexity of acoustic propagation in the ocean, the signature measured at even moderate ranges may be appreciably different from that obtained at close range. Certainly at long range, absorption effects attenuate the high-frequency components of the spectrum. In shallow channels the relative channel response to line components at different frequencies may alter considerably their measured relative amplitudes as compared with the short-range signature. The frequency dependence of the acoustic path should be included, if possible, when it is necessary to characterize an acoustic signature at a considerable range from the source.

12.1.1 Propulsion System Noise

The propulsion system contains large rotating shafts, bearings, gears, and depending on the particular design, reciprocating engines, turbines, or electric drive motors. Slight dynamic unbalances in these devices result in oscillating forces that are transmitted through the foundations or supports to the hull, which then radiates acoustic energy into the ocean. Signals generated in this manner are narrowband tonals at the system rotational frequencies and their harmonics. The forces involved typically increase as the square of the rotational velocity, so that tonals associated with the propulsion system often vary in intensity with the vessel speed.

Frictional forces are also present in the propulsion system. Linear friction forces result in a broadband component of the signature. This component is also speed dependent.

12.1.2 Propeller Noise

Propeller noise results from cavitation produced by the rotating blades. Rotation results in regions of low pressure on and around the blades. When the pressure drops below some critical value, the water ruptures and bubbles are formed. The bubbles initially grow in size and then rapidly collapse, generating a broadband-radiated acoustic noise signal. Because the onset of cavitation is related to depth as well as propeller speed, cavitation noise decreases with depth and increases with speed [4]. The general shape of the cavitation spectrum is shown in Figure 12-1. The frequency of the peak of the cavitation spectrum is a function of propeller size, as well as depth and speed, generally decreasing with increasing propeller diameter.

Cavitation noise is a major component of the signature of surface ships and for submarines operating near the surface. By traveling at sufficient depth a submarine may avoid propeller cavitation altogether and thereby reduce radiated noise significantly. At the onset of cavitation, the radiated noise of a submarine may increase 20 dB or more at the high-frequency end of the spectrum.

When cavitation is present, the propeller amplitude-modulates the radiated noise level. This modulation is at the propeller blade rate (shaft frequency times the number of blades) and provides a valuable clue for target classification.

In addition to propeller cavitation noise, tonal components may result from

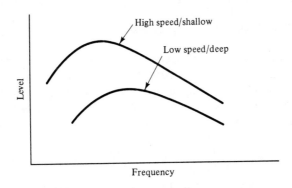

Figure 12-1 Propeller cavitation spectral shape.

the vibrational excitation of the propeller blades by the flow of water around them. This is referred to as propeller *singing*.

12.1.3 Auxiliary Machinery Noise

Auxiliary machines do the work required on a vessel that is not directly related to propulsion. Electrical generators, pumps, blowers, and similar devices operate more or less continuously, and generally rotate at constant speed.

The acoustic noise generated by auxiliary machinery is primarily tonal components caused by dynamic unbalance in rotating components. Because these machines are not involved in propulsion, the tonal components generated are relatively stable in amplitude and frequency.

12.1.4 Hydrodynamic Noise

The flow of water over the hull and appendages of a vessel results in radiated acoustic noise caused by several different hydrodynamic effects. The cavitation process has already been mentioned in connection with propeller noise. Cavitation can also take place on the hull or appendages wherever the pressure drops below the cavitation threshold. The onset of hull cavitation generally occurs at a much higher speed than propeller cavitation, and therefore does not add significantly to the noise signature.

Water flow past struts may induce structural vibrations through a process of unbalanced vortex shedding off the trailing edge of the strut. The vibrations in turn result in radiated noise. Water flow over the hull surface becomes turbulent above some critical speed. The turbulent and uneven flow may induce vibration in the hull structure, thus causing acoustic noise to be radiated.

Except for propeller cavitation, hydrodynamic noise is generally of secondary importance to machinery noise in the overall acoustic signature. However, if the vessel carries hull-mounted sonar receiving equipment, hydrodynamic noise may be a very important source of noise, limiting the detection performance of the on-board system.

12.1.5 Total Acoustic Signature

The total range of expected spectral intensity levels is illustrated in Figure 12-2. A decreasing spectrum at a rate of about 20 dB/decade is indicated above 1 kHz, with a flattening off somewhere below 100 Hz depending on vessel size, speed, and depth. The lower boundary in Figure 12-2 corresponds to a submerged noncavitating submarine on electric drive, while the upper boundary is representative of a battleship at 20 knots. At any one frequency the spread between these extremes is 50 to 70 dB. For comparison, a typical ambient noise spectrum is also shown in Figure 12-2.

It is interesting to determine the total acoustic power represented by the extremes in Figure 12-2. For the large surface vessel we find a total radiated

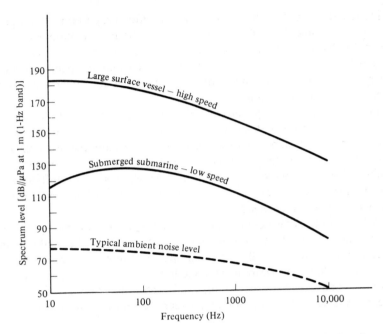

Figure 12-2   Range of expected spectrum levels for surface vessels and submarines.

power on the order of 1 kW, whereas for the submarine the total is a few milliwatts. In either case, the radiated acoustic power is a very small fraction of the total mechanical power used to propel the vessel. In the case of the submarine, the very small radiated acoustic power provides a real challenge to the designers of acoustic systems intended for submarine detection.

In Figure 12-3 the more detailed structure of the signature of a submarine is shown at two different speeds. The total signature includes line components related to the propulsion system, the propeller, and auxiliary machinery as well as a broadband spectrum.

The auxiliary line components are typically quite stable and not related to

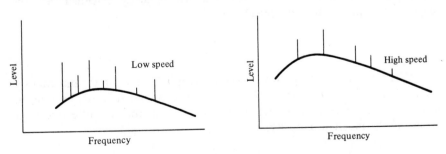

Figure 12-3   Broadband and narrowband components of a submarine signature at low speed and high speed.

vessel speed. The bandwidths of such lines are often proportional to frequency, with values that range from about 0.3 to 0.03% of frequency. Propulsion system and propeller lines may vary in both amplitude and frequency with vessel speed. The bandwidths of these lines are generally greater than the bandwidths of the auxiliary lines, and it is not uncommon to have tonal components that vary in frequency in a periodic manner.

In the absence of propeller cavitation the line structure of the submarine signature is quite prominent, including both auxiliary and propulsion lines. As the submarine speed increases to where cavitation occurs, the broadband noise component increases to the point where some of the tonal components become obscured. The lines associated with the propulsion system shift upward in frequency and may increase in amplitude while the auxiliary machinery line components remain unchanged.

As mentioned at the beginning of this section, acoustic signature character-istics may be altered by the propagation characteristics of the medium. In the case of narrowband tonals the time-varying and multipath properties of the medium may alter the apparent bandwidth of the received signal as well as modify the relative amplitude. To this modification must be added any shift in frequency caused by relative motion between the source and the receiver.

12.2 TARGET STRENGTH FOR ACTIVE SYSTEMS

In active sonar systems, a finite-duration pulse is transmitted into the medium by the projector. The acoustic target represents a discontinuity in the medium, resulting in a reflection of a portion of the energy of the incident transmitted waveform. This reflected *echo* is the signal of interest at the receiver location. In a *monostatic* sonar system the receiver and projector are at the same location. The receiver and projector are not at the same location in a *bistatic* system.

The target characteristic of interest with active systems is the ratio of the intensity of the reflected signal to the intensity of the incident signal. This is illustrated in Figure 12-4 for both the monostatic and bistatic cases. The source intensity is I_0, measured 1 m from the source. The intensity incident at the target range is (assuming spherical spreading)

$$I_i = \frac{I_0}{r_1^2}$$

The target reflects a portion of the intercepted energy in the directions of the receivers. The *target strength*, t_s, is defined as the ratio of the reflected intensity I_r in the receiver direction, measured 1 m from the effective target center, to the incident intensity.

$$t_s = \frac{I_r}{I_i} \tag{12-1}$$

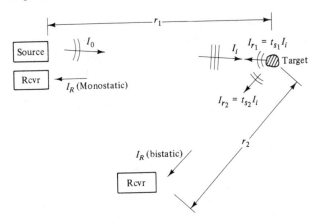

Figure 12-4 Active sonar geometry.

Because of the target geometry and acoustic properties, the target strength for the bistatic system is not in general equal to the target strength for the monostatic system. The monostatic target strength is sometimes called the *back-scattering* target strength.

The intensity of the echo at the receiver location is now easily obtained. For the monostatic system we have

$$I_R \text{ (monostatic)} = \frac{I_r}{r_1^2} = \frac{t_{s_1} I_i}{r_1^2}$$

$$= t_{s_1} \frac{I_0}{r_1^4} \tag{12-2}$$

and for the bistatic system

$$I_R \text{ (bistatic)} = \frac{t_{s_2} I_0}{r_1^2 r_2^2} \tag{12-3}$$

In decibel notation, (12-2) and (12-3) become

$$\text{EL (monostatic)} = \text{SL} + \text{TS}_1 - 2\text{TL}$$

$$\text{EL (bistatic)} = \text{SL} + \text{TS}_2 - \text{TL}_1 - \text{TL}_2 \tag{12-4}$$

where

$$\text{EL} = \text{received echo level}$$

$$\text{TS} = 10 \log \left(\frac{I_r}{I_i} \right) \tag{12-5}$$

The target strength of an acoustic target is a function of geometry, size, acoustic impedance, and frequency of the incident signal. The target strengths of simple geometric shapes, assuming large impedance mismatch, have been derived and are tabulated in a number of places [3,4,6,7]. The target strength for more complicated structures such as ships and submarines is best determined experimentally [3]. To obtain a feel for the magnitude of target strength, consider a perfectly rigid sphere with radius $a \gg \lambda$ in a plane wave field of intensity I_i. The acoustic power intercepted from the incident field is

$$P_i = \pi a^2 I_i$$

For a large perfectly rigid sphere, the intercepted power will be reradiated uniformly in all directions. The intensity of the reradiated field at a range of 1 m is

$$I_r = \frac{P_i}{4\pi} = \frac{\pi a^2 I_i}{4\pi} = \frac{a^2 I_i}{4} \tag{12-6}$$

from which

$$TS = 10 \log \left(\frac{I_r}{I_i}\right) = 10 \log \left(\frac{a^2}{4}\right) \tag{12-7}$$

Thus a target strength of 0 dB corresponds to the reflection from a rigid sphere with a radius of 2 m.

The general character of the target strength of a submarine as a function of azimuth aspect is shown in Figure 12-5. The target strength is a maximum at broadside because of the large cross section and simple geometric shape. Conversely, the strength tends to be a minimum for bow and stern aspects. The effect of the sail and other appendages, as well as the effect of ballast tanks and other internal features, cause additional variations in the detail of the target strength aspect dependence.

Tabulated values for target strength for various targets assume that the transmitted pulse length is large enough to ensonify the entire target simulta-

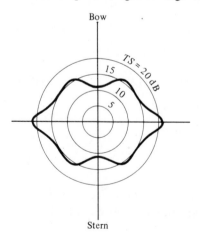

Figure 12-5 Typical submarine target strength.

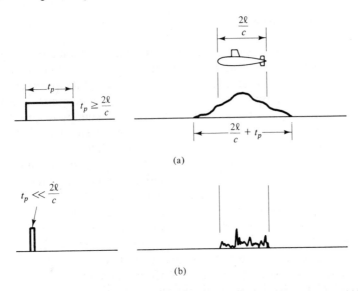

(a)

(b)

Figure 12-6   Active return from a bow-aspect submarine assuming (a) a pulse width matched to the target length; (b) a pulse width small compared with target length.

neously. The increment of range ensonified by a pulse at any one time with a monostatic sonar system is

$$\Delta r = \frac{ct_p}{2}$$

where t_p is the pulse width and c is the sound speed. Thus for a target dimension of 100 m in the direction of acoustic propagation, the pulse width must be at least 133 msec to ensonify the entire target range extent at one time. The total duration of the reflected signal will be $t_p + 2\ell/c$, where ℓ is the range dimension of the target. In Figure 12-6 the transmitted and received waveforms are sketched for a wide pulse and a narrow pulse assuming that the target is a submarine at bow aspect. With the wide pulse the reflected signal increases steadily to a peak and then decays as the pulse propagates past the target. With the narrow pulse, a series of short-duration reflections is received, representing contributions from the discrete important reflecting regions, or highlights, on the submarine. In the language of network theory, the reflected signal is the convolution of the transmitted waveform with the impulse response of the target. The target impulse response is of course a function of aspect angle.

Other reflecting objects of military or commercial interest include mines, torpedoes, and fish. Mines and torpedoes have simple geometric shapes, resulting in target strengths that may be calculated from the equations tabulated in Urick [4, Table 9.1]. Values ranging from about −15 dB to +15 dB are obtained depending on size and aspect. The target strength of fish is a function of their size and physiology. For fish of commercial interest the target strength is generally in the range −15 to −40 dB [8,9].

12.3 REVERBERATION

With an active system *reverberation* results from the scattering of energy from the propagating pulse as a result of inhomogeneities in the ocean and its boundaries. In most cases this results in an *undesired* signal at the receiver that may obscure the *desired* target echo. Thus an active system must contend with reverberation as well as with ambient noise when attempting to detect a target signal. A similar problem exists with radar systems when attempting to detect targets in the presence of rain "clutter" or ground "clutter."

There are several significant causes of scattering that result in reverberation. In the body of the ocean, fish, plankton, and other biological sources are probably the main contributors. Reverberation originating from these sources is called *volume reverberation*. Inhomogeneities and roughness at the top and bottom surfaces of the ocean give rise to *surface reverberation*. For simplicity the sources of volume reverberation are assumed uniformly distributed throughout the volume region of interest, and the sources of surface reverberation are assumed uniformly distributed over the surface region of interest. In practice, this may not be so. For instance, biological sources often concentrate in layers within the body of the ocean, resulting in significant variations in the volume-scattering properties.

To describe quantitatively the effect of reverberation, we define a *scattering strength* per unit volume, or per unit surface area, in a manner similar to that used to define target strength with respect to a point target. For instance, volume scattering strength is defined as the ratio of the intensity of scattered energy at 1 m from a unit volume, to the incident energy at the volume element location. Thus

$$s_v = \frac{I_r/\text{unit volume}}{I_i}$$

or in decibel notation,

$$S_v = 10 \log s_v = 10 \log \left(\frac{I_r/\text{unit vol}}{I_i}\right) \tag{12-8}$$

Now assume an active source radiating a pulse of duration t_p through a directional power pattern of normalized shape $b(\psi, \theta)$. If the intensity 1 m from the source is I_0, the intensity at range r and polar angles (ψ, θ) is

$$I_i = \frac{I_0\, b(\psi, \theta)}{r^2} \tag{12-9}$$

By definition, the intensity scattered by an elemental volume dv at this range and bearing is

$$dI_r = s_v I_i\, dv = \frac{s_v I_0 b(\psi, \theta)\, dv}{r^2} \tag{12-10}$$

With a monostatic system we are interested in the scattered energy that returns to the vicinity of the source. If the receiving pattern is identical to the transmit pattern, the intensity at the receiver resulting from the element dv is

$$dI_R = \frac{s_v I_0 b^2(\psi,\ \theta)\ dv}{r^4} \tag{12-11}$$

If the receiving pattern is not identical to the transmit pattern, the factor $b^2(\psi,\ \theta)$ is replaced by the product of the two patterns. For simplicity we assume identical patterns for the remainder of this discussion.

The total volume reverberation from range r is obtained by integrating over all possible volume elements at that range. Thus

$$I_R = \frac{s_v I_0}{r^4} \int b^2(\theta,\ \psi)\ dv \tag{12-12}$$

The volume element in (12-12) is defined with the help of Figure 12-7. The element consists of a cylinder of length $ct_p/2$ with an end-face area given by $r^2\ d\Omega$, where $d\Omega$ is the elemental solid angle. The volume of the element is

$$dv = \frac{ct_p}{2} r^2 d\Omega \tag{12-13}$$

Substitution of (12-13) in (12-12) gives

$$I_R = \frac{s_v I_0 ct_p}{2r^2} \int b^2(\psi,\ \theta)\ d\Omega \tag{12-14}$$

The integral in (12-14) may be thought of as an equivalent solid-angle measure of beamwidth for the two-way power pattern. Define this beamwidth as Ω_B and write (12-14) as

$$I_R = \frac{s_v I_0 ct_p \Omega_B}{2r^2} \tag{12-15}$$

Converting (12-15) to decibel notation, we obtain

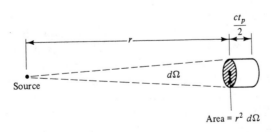

Figure 12-7   Volume element for defining volume scattering field.

$$RL_v = S_v + SL - TL + 10 \log\left(\frac{ct_p \Omega_B}{2}\right) \qquad (12\text{-}16)$$

where RL_v = volume reverberation level from range r
    S_v = volume scattering coefficient
    SL = source level corresponding to I_0
    TL = spherical spreading one-way transmission loss at range r

The magnitude of S_v is a function of depth, season, time of day, frequency, and the degree of concentration of biological organisms. Typically, the magnitude of S_v will be found in the range -60 dB to -90 dB.

Notice that the reverberation level is reduced by reducing the transmitted pulse duration or by reducing the effective beamwidth. Notice also that the reverberation level decreases as 20 log r rather than the 40 log r experienced with a point target. This is because the volume of the elemental volume element increases as the square of range, partially offsetting the two-way path loss. The value of Ω_B for some commonly encountered array types is tabulated in convenient form in Urick [4, Table 8.1].

The expression for surface reverberation is derived in the same manner as that for volume reverberation. For simplicity assume a beam pattern with its major axis parallel to the surface. The elemental surface area at range r is

$$da = \frac{ct_p}{2} r \, d\gamma \qquad (12\text{-}17)$$

where $d\gamma$ is now a plane angle measured in the azimuth plane. Proceeding as before, and defining γ_B as the effective two-way azimuth beamwidth, the expression for surface reverberation level is

$$RL_s = S_s + SL - \tfrac{3}{2} TL + 10 \log_{10}\left(\frac{ct_p \gamma_B}{2}\right) \qquad (12\text{-}18)$$

where S_s is the *surface scattering* coefficient.

Notice that surface reverberation falls off more rapidly with range than volume reverberation. This is because the elemental surface area increases as r rather than as r^2.

Values for surface scattering strength have been obtained experimentally for both the ocean surface and bottom [10–12]. The values obtained are functions of frequency, grazing angle, wind conditions, and in the case of the ocean bottom, the bottom material and roughness. The reported experimental values for S_s generally lie in the range -10 to -50 dB. The surface scattering strength generally increases with frequency, grazing angle, and surface roughness.

In system performance calculations we are interested in the ratio of the received intensity from a point target to the reverberation intensity received from the target range. For example, assume a target at range r in volume reverberation. Using (12-4) for a monostatic system and (12-16), we write

$$EL = TS + SL - 2TL$$

$$RL_v = S_v + SL - TL + 10 \log\left(\frac{ct_p \Omega_B}{2}\right)$$

$$EL - RL_v = TS - S_v - TL - 10 \log\left(\frac{ct_p \Omega_B}{2}\right) \qquad (12\text{-}19)$$

Notice that echo-to-reverberation ratio does not depend on source level. The ratio in (12-19) is inversely proportional to the square of range. Equation (12-19) also implies that the ratio is improved by decreasing the pulse width or the effective beamwidth. This conclusion is valid provided the target dimensions are small compared with the range extent and angular extent of the acoustic signal in the water at the target location.

Using (12-4) and (12-18) the echo-to-reverberation ratio for surface reverberation is

$$EL - RL_s = TS - S_s - \tfrac{1}{2} TL - 10 \log\left(\frac{ct_p \, \gamma_B}{2}\right) \qquad (12\text{-}20)$$

With surface reverberation, the echo-to-reverberation ratio decreases inversely with r rather than r^2. To demonstrate the potentially limiting nature of reverberation on the performance of an active system, assume that a target echo and surface reverberation are equal at a range of 20 nm. However, for reliable detection assume that we require the target echo to be 10 dB above the surface reverberation level. From (12-20), this requires that the range be decreased by a factor of 10 to achieve the 10-dB advantage. Thus the echo would be reliably detected at 2 nm.

12.3.1 Statistical and Spectral Properties of Reverberation

Reverberation occurs as a result of a random distribution of acoustic scattering elements throughout the region ensonified by the active sonar. To examine the properties of reverberation, assume first that both the sonar and the scattering elements are immobile. The projected waveform is a finite-duration pulse with a bandwidth typically small compared with the center frequency. We wish to describe the statistical properties of the envelope of the reverberation waveform and its spectral density. A limited range interval will be considered centered at a range much larger than the duration of the interval.

Let the transmitted pressure waveform be $s(t)$ and the signal received from the nth scattering element in the limited range interval be $a_n s(t - t_n)$. This received signal can also be expressed as the convolution of the transmitted signal with an impulse of strength a_n located at $t = t_n$. Thus

$$a_n s(t - t_n) = s(t) \otimes a_n \delta(t - t_n) \qquad (12\text{-}21)$$

The total received signal from the limited range interval may be expressed as

$$s_R(t) = s(t) \otimes \sum_n a_n \delta(t - t_n) = s(t) \otimes c(t) \qquad (12\text{-}22)$$

where the values of t_n are restricted to the desired range interval. The function $c(t)$ is the impulse response of the collection of scattering elements in the selected interval.

We now assume that the number of elements in the interval is large, that t_n is more or less uniformly distributed over the interval and that a_n is a finite-variance zero-mean random variable with a_n independent of a_m for all $n \neq m$. Let $s(t)$ be expressed in analytic form as

$$s(t) = \mu(t) \exp(j2\pi f_0 t)$$

where $\mu(t)$ is the envelope function and f_0 is the carrier frequency. Equation (12-22) now becomes

$$s_R(t) = \sum_n a_n \mu(t - t_n) \exp[j2\pi f_0 (t - t_n)] \qquad (12\text{-}23)$$

from which the envelope of the received signal is by definition

$$\mu_R(t) = s_R(t) \exp(-j2\pi f_0 t)$$

$$= \sum_n a_n \mu(t - t_n) \exp(-j2\pi f_0 t_n)$$

$$= \sum_n a_n \mu(t - t_n) \cos 2\pi f_0 t_n - j \sum_n a_n \mu(t - t_n) \sin 2\pi f_0 t_n$$

$$(12\text{-}24)$$

The summations in (12-24) have the same variance, and by the central limit theorem are assumed Gaussian with zero mean. Furthermore, because of the orthogonality of the cosine and sine functions, the real and imaginary parts of (12-24) are statistically independent. The received envelope function $\mu_R(t)$ is therefore identified as a complex Gaussian variable. The statistical properties of such a variable are derived in Section 9.7.4. In particular, the squared magnitude of $\mu_R(t)$ has a p.d.f. of the form of a two-degree-of-freedom chi-squared variable, and the magnitude of the envelope is a Rayleigh variable.

Because the apparent strength of the scattering elements and the number of elements that must be considered at any one time are functions of range, the mean value and variance associated with the envelope function are also functions of range.

To determine the spectral density of the received reverberation signal, we form the Fourier transform of (12-22). Let

$$s(t) \longleftrightarrow S(f)$$

$$c(t) \longleftrightarrow C(f) = \sum_n a_n \exp(-j2\pi f t_n)$$

$$s_R(t) = s(t) \otimes c(t) \longleftrightarrow S(f) \sum_n a_n \exp(-j2\pi f t_n) \qquad (12\text{-}25)$$

By definition, the spectral density is the expected value of the squared magnitude of the frequency-domain function in (12-25). Thus

$$\Psi_R(f) = E[|S(f)C(f)|^2]$$

$$= |S(f)|^2 E[|C(f)|^2]$$

$$= |S(f)|^2 E\left[\sum_m \sum_n a_m a_n \exp[-j2\pi f(t_m - t_n)]\right] \qquad (12\text{-}26)$$

Because of the independence of a_n and a_m, the expectation in (12-26) is zero unless $n = m$. Hence

$$\Psi_R = |S(f)|^2 E\left[\sum_n a_n^2\right] = |S(f)|^2 N\sigma_{a_n}^2 \qquad (12\text{-}27)$$

where N is the total number of elements in the volume (or surface) interval considered.

Equation (12-27) indicates that the spectral density of the reverberation signal received from a limited interval has the same form as the power spectrum of the transmitted waveform. The magnitude of the reverberation spectral density will vary with range for the reasons mentioned above.

Now assume the sonar is moving with constant velocity in a field of stationary scattering elements. This situation is depicted two-dimensionally in Figure 12-8. Because of the Doppler effect, the received signal from each scattering element is shifted in frequency by an amount proportional to its radial velocity relative to the sonar. For a simple sinusoidal transmission at a frequency f_0, the Doppler shift for the nth element is approximately

$$f_{d_n} \simeq \frac{2V \cos \psi_n}{\lambda_0} \qquad (12\text{-}28)$$

Considering the signals received from an annular ring centered on the moving sonar platform, the Doppler shift must range from $+2V/\lambda_0$ to $-2V/\lambda_0$.

For a transmitted waveform with a bandwidth narrow compared with its center frequency, the received spectrum from a single scattering element is approximately a replica of the transmitted spectrum shifted in frequency by the Doppler frequency $f_{d_n} = 2V_n/\lambda_0$, where V_n is the relative radial velocity and λ_0 is the wavelength corresponding to the center frequency of the transmitted waveform. For a transmitted signal and its spectrum given by

$$s(t) \longleftrightarrow S(f)$$

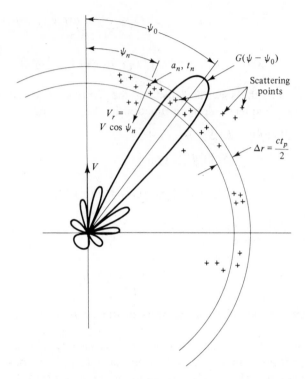

Figure 12-8   Moving sonar platform in stationary scattering field.

the received signal and spectrum, assuming an omnidirectional projector and receiver, from the nth scatterer are

$$a_n s(t - t_n) \exp(+j2\pi f_{d_n} t) \longleftrightarrow a_n S(f - f_{d_n}) \exp\left[-j2\pi(f - f_{d_n})t_n\right]$$

$$(12\text{-}29)$$

With a directional projector and receiver with identical patterns $G(\psi - \psi_0)$, the total received spectrum from the annular range element is

$$S_R(f) = \sum_n a_n |G(\psi_n - \psi_0)|^2 S(f - f_{d_n}) \exp\left[-j2\pi(f - f_{d_n})t_n\right]$$

$$(12\text{-}30)$$

Again assuming that the individual scattering elements are statistically independent, the average received power spectrum corresponding to (12-30) is

$$\Psi_R(f) = \sum_n \sigma_{a_n}^2 |G(\psi_n - \psi_0)|^4 |S(f - f_{d_n})|^2 \qquad (12\text{-}31)$$

The total power spectrum is thus the sum of the spectra from the individual elements, each weighted by the square of the two-way pattern function. In

Figure 12-9 the received power spectrum is sketched for several assumed values of ψ_0. In Figure 12-9(a), the beam direction is aligned with the velocity vector so that the received power is concentrated about the frequency $+2V/\lambda$. The width of the main spectral lobe is determined by the bandwidth of the transmitted waveform and the angular width of spatial pattern function. The change in Doppler shift over the width of the pattern function will cause some increase in the received signal bandwidth. As an example, assume a transmitted pulse width of 0.1 sec, a pattern beamwidth of 0.1 rad, and $2V/\lambda = 100$ Hz. With $\psi_0 = 0$, the Doppler spread within the pattern beamwidth is

$$\Delta f_d = \frac{2V}{\lambda}\left[1 - \cos\left(\frac{\psi_B}{2}\right)\right] = 0.125 \text{ Hz}$$

which is quite small compared with the 10-Hz bandwidth of the original signal.

Between the frequencies $f_0 + 2V/\lambda$ and $f_0 - 2V/\lambda$, the received reverberation power is reduced by the side lobe characteristics of the pattern function. Because the Doppler shift cannot exceed $2V/\lambda$, the spectrum above the frequency $f_0 + 2V/\lambda$ is determined primarily by the shape of the transmitted spectrum centered at $f_0 + 2V/\lambda$.

In Figure 12.9(b), the beam pointing angle is 45° off of the velocity vector. The main spectral lobe is centered at $0.707(2V/\lambda)$, with some energy still spread

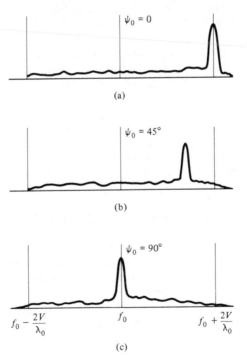

(a)

(b)

(c)

Figure 12-9 Reverberation spectra for several beam pointing angles on a moving platform.

over the range $f_0 \pm 2V/\lambda$ as a result of signals received through the spatial side lobes. Finally, in Figure 12-9(c) the beam is pointed at right angles to the velocity vector. This results in no net Doppler shift of the spectrum, although reverberation energy is spread symmetrically over the full Doppler range.

PROBLEMS

12.1. Using the spectral characteristics given in Figure 12-2, calculate the band source levels for the low-speed submerged submarine for the decade bands 10 to 100 Hz, 100 to 1000 Hz, and 1000 to 10,000 Hz.

12.2. Using the ambient noise curve given in Figure 12-2, calculate the noise band levels in each of the bands specified in Problem 12.1. At a range of 1000 m determine the signal-to-noise ratios for the signal and bands specified in Problem 12.1. Assume simple spherical spreading.

12.3. Assume a monostatic active system with an omnidirectional projector radiating a pulsed sinusoidal waveform with an acoustic power of 1 kW. With a target at a range of 2000 m, calculate the received echo level assuming spherical spreading and a target strength of +15 dB.

12.4. The received echo intensity in a bistatic system is inversely proportional to $r_1^2\, r_2^2$, where r_1 and r_2 are the ranges from the target to projector and target to receiver, respectively. Assume that the target moves such that $r_1 + r_2 = K$. Determine the relationship between r_1 and r_2 that minimizes the received echo assuming that target strength is constant.

12.5. With the monostatic system of Problem 12.3, calculate the volume reverberation level and the echo-to-reverberation ratio for a target at 2000 m assuming that $S_v = -80$ dB, $t_p = 0.1$ sec, $\Omega_B = 0.1$ rad, and TS = +10 dB.

12.6. A sonar platform is moving at a constant speed of 40 knots in a uniform field of stationary scattering elements. The projected pulse has a simple rectangular envelope with $t_p = 10$ msec and a carrier frequency $f_0 = 20$ kHz. Assume an effective beamwidth of 0.3 rad and a beam-steering angle of 30° relative to the platform velocity vector. Sketch the approximate received reverberation spectral density, including the effect of the Doppler spreading over the effective beamwidth.

SUGGESTED READING

1. Dow, M. T., J. W. Emling, and V. O. Knudsen, "Survey of Underwater Sound No. 4: Sounds from Surface Ships," NDRC Division 6, Section 6.1, NDRC-2124 (1945); available through U.S. Department of Commerce, Federal Clearinghouse for Scientific and Technical Information, Springfield, Va.

2. Knudsen, V. O., R. S. Alford, and J. W. Emling, "Survey of Underwater Sound No. 2: Sounds from Submarines," NDRC Division 6, Section 6-1, NDRC-1306 (1943).

3. *Physics of Sound in the Sea*, Part III: *Reflection of Sound from Submarines and Surface Vessels*, NDRC Division 6, Vol. 8, Summary Technical Reports (1946); reprinted in 1969 by the U.S. Government Printing Office, Washington, D.C.

4. Urick, R. J., *Principles of Underwater Sound for Engineers,* 2nd ed. New York: McGraw-Hill Book Company, 1975, Chaps. 8, 9, and 10.

5. Albers, V. M., *Underwater Acoustics Handbook—II.* University Park, Pa.: The Pennsylvania State University Press, 1965, Chap. 17.

6. *Principles of Sonar Installation,* U.S. Naval Underwater Systems Center, Technical Document 6059.

7. Kerr, D. E., Ed., *Propagation of Short Radio Waves,* MIT Radiat. Lab. Ser., Vol. 13. New York: McGraw-Hill Book Company, 1951.

8. Cushing, D. H., et al., "Measurements of the Target Strength of Fish," *J. Br. IRE,* Vol. 25, p. 299 (1963).

9. Volberg, H. W., "Acoustic Properties of Fish," 14th Pacific Tuna Conference, Sept. 1963.

10. Marsh, H. W., "Sound Reflection and Scattering from the Sea Surface," *J. Acoust. Soc. Am.,* Vol. 35, p. 240 (1963).

11. McKinney, C. M., and C. D. Anderson, "Measurements of Backscattering of Sound from the Ocean Bottom," *J. Acoust. Soc. Am.,* Vol. 36, p. 153 (1964).

12. Burstein, A. W., and J. J. Keane, "Backscattering of Explosive Sound from Ocean Bottoms," *J. Acoust. Soc. Am.,* Vol. 36, p. 1596 (1964).

13 Statistical Basis for Performance Analysis

13.0 INTRODUCTION

The primary objectives of underwater acoustic systems are to detect the presence of acoustic signals emitted by or reflected from underwater objects, and to provide estimates of certain parameters associated with these objects, such as size, range, bearing, and relative velocity. Because of the inevitable presence of noise and other interfering signals, the definition of system performance relative to these objectives requires the application of the methods of statistical hypothesis testing and parameter estimation. An exhaustive treatment of these subjects is beyond our present scope. Therefore, our discussion is limited to those aspects of these disciplines of direct significance to the understanding of mechanizations and performance analysis methods to follow in Chapter 14. The reader is referred to the Suggested Reading list for further study of these subjects.

13.1 HYPOTHESIS TESTING

To introduce the concept of hypothesis testing, assume that a received signal $x(t)$ consists of a random waveform $n(t)$ which may or may not include a signal with a known constant amplitude μ. These two possibilities are identified as

$$x_1(t) = n(t) + \mu$$
$$x_0(t) = n(t) \tag{13-1}$$

Based on a limited record of the received waveform, the detection process consists of choosing between the hypotheses that the signal is $x_1(t)$ or $x_0(t)$. The signal-plus-noise hypothesis is labeled H_1 and the noise-only hypothesis is H_0.

For simplicity we first assume a single sample of the received waveform. Based on some statistical model of the situation, it is reasonable to choose the hypothesis that is the most probable, or likely, based on the measured amplitude of the single sample. Thus we have the two conditional probabilities,

$$P(H_1 \mid x) = \text{probability of } H_1, \text{ given } x$$

$$P(H_0 \mid x) = \text{probability of } H_0, \text{ given } x \tag{13-2}$$

Because these probabilities require a knowledge of the received sample amplitude, they are called *a posteriori* probabilities. The detection rule consists of choosing the hypothesis corresponding with the maximum a posteriori probability. Thus H_1 is chosen if

$$\frac{P(H_1 \mid x)}{P(H_0 \mid x)} \geq 1 \tag{13-3}$$

Using the rules relating to conditional probabilities developed in Chapter 9, we write

$$P(H \mid x)P(x) = P(x \mid H)P(H)$$

from which the ratio in (13-3) can be written as

$$\frac{P(H_1 \mid x)}{P(H_0 \mid x)} = \frac{P(x \mid H_1)P(H_1)}{P(x \mid H_0)P(H_0)} \geq 1 \tag{13-4}$$

where $P(H_1)$ and $P(H_0)$ are the *a priori* probabilities associated with the presence or absence of signal.

Because $P(H_1) = [1 - P(H_0)]$, (13-4) has the equivalent form

$$\frac{P(x \mid H_1)}{P(x \mid H_0)} \geq \frac{P(H_0)}{1 - P(H_0)} \tag{13-5}$$

It is easily shown that the ratio on the left in (13-5) is also equal to the ratio of the corresponding conditional probability density functions. Thus

$$\frac{p(x \mid H_1)}{p(x \mid H_0)} \geq \frac{P(H_0)}{1 - P(H_0)} \tag{13-6}$$

This ratio of conditional p.d.f.'s is called a *likelihood ratio* and the functions $p(x \mid H_1)$ and $p(x \mid H_0)$ are *likelihood functions*. Hypotheses tests based on the likelihood ratio are *likelihood ratio tests*. The test in (13-6) is the *maximum a posteriori* likelihood ratio test. For convenience, $p(x \mid H_0)$ and $p(x \mid H_1)$ will be designated $p_0(x)$ and $p_1(x)$ in the following discussions.

13.1.1 Neyman–Pearson Criterion

In Figure 13-1 are plotted the likelihood functions for a hypothetical situation, together with a division of the x-axis into regions where H_1 and H_0 would be the accepted hypotheses based on the maximum a posteriori likelihood criterion.

Regardless of whether H_1 or H_0 is selected, there is a finite probability of error. In Figure 13-1 this is demonstrated by regions 1 and 2. The probability of choosing H_1 when H_0 is true is called a *false alarm* or an *error of the first kind,* and is equal to the area in region 1. Thus, if γ is the value of x separating the two regions, the *false-alarm probability* is

$$P_{\text{fa}} = \int_{\gamma}^{\infty} p_0(x) \, dx \tag{13-7}$$

If hypothesis H_0 is chosen when H_1 is true, we have a *false dismissal* or an *error of the second kind.* The probability of this type of error, P_{fd}, is given by the area in region 2. The *probability of detection, P_D,* is 1 minus the probability of a false dismissal. Thus

$$P_D = \int_{\gamma}^{\infty} p_1(x) \, dx \tag{13-8}$$

and

$$P_{\text{fd}} = \int_{-\infty}^{\gamma} p_1(x) \, dx = 1 - P_D \tag{13-9}$$

To establish the threshold γ for which (13-6) becomes an equality we require the prior probability $P(H_0)$ as well as the conditional density functions. In systems such as radar or sonar the a priori probabilities $P(H_0)$ and $P(H_1)$ are typically not available. That is, there is generally no statistical basis for assign-

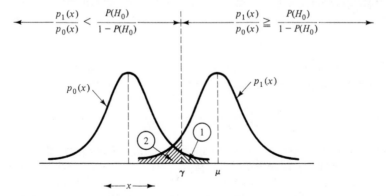

Figure 13-1 Plot of likelihood functions and division of space into regions representing the choices H_0 and H_1 using the maximum a posteriori likelihood criterion.

ing a number to the probability that a target will be present at some given location at a particular time in the future. Furthermore, given that H_1 is true, the conditional probability $p_1(x)$ depends on the unknown amplitude of the target signal component of x relative to the noise component. Thus, we are left with, at best, the conditional probability of $x \mid H_0$ as the only known term in (13-6).

The *Neyman–Pearson criterion* avoids the unknown signal amplitude and a priori difficulties by requiring only that the probability of false alarm be fixed at some acceptable level. Assuming that the noise is stationary, this is accomplished by applying a fixed threshold γ to x in accordance with (13-7). The detection probability is then the maximum attainable with the given false-alarm constraint. In general, if the prior probabilities for H_0, H_1 and the signal amplitude were known, the maximum a posteriori likelihood criterion would result in a higher probability of detection than the Neyman–Pearson criterion only at the expense of a higher false-alarm probability.

For a given signal amplitude, the Neyman–Pearson criterion is a likelihood-ratio test of the form

$$\frac{p_1(x)}{p_0(x)} \geq \lambda_0 \tag{13-10}$$

where λ_0 is a function of the threshold setting γ and the signal and noise amplitudes. As an example, let

$$p_1(x) = \frac{1}{\sigma_n \sqrt{2\pi}} \exp\left[-\frac{(x - \mu)^2}{2\sigma_n^2}\right]$$

$$p_0(x) = \frac{1}{\sigma_n \sqrt{2\pi}} \exp\left(\frac{-x^2}{2\sigma_n^2}\right) \tag{13-11}$$

Define the likelihood ratio as $\lambda(x)$ and write

$$\lambda(x) = \frac{p_1(x)}{p_0(x)} = \exp\left[\frac{(2\mu x - \mu^2)}{2\sigma_n^2}\right] \geq \lambda_0 \tag{13-12}$$

Equation (13-12) becomes an equality for $x = \gamma$. Thus

$$\lambda_0 = \exp\left(\frac{2\mu\gamma - \mu^2}{2\sigma_n^2}\right) \tag{13-13}$$

It is convenient to take the natural logarithm of (13-12), which can then be put in the form

$$x \geq \frac{\sigma_n^2}{\mu} \ln(\lambda_0) + \frac{\mu}{2} = \gamma \tag{13-14}$$

The received sample x in this case is called the *test statistic*, which is compared with the fixed Neyman–Pearson threshold γ.

13.1.2 Multiple Sample Testing

If many samples of the received signal are available, they may be used to improve the accuracy of the test. Assume a set of m samples $x_1, x_2, \ldots, x_m$ represented by the vector $\mathbf{x}$. The likelihood ratio is then the ratio of the conditional joint probabilities. Thus

$$\lambda(\mathbf{x}) = \frac{p_1(x_1, x_2, \ldots, x_m)}{p_0(x_1, x_2, \ldots, x_m)} = \frac{p_1(\mathbf{x})}{p_0(\mathbf{x})} \tag{13-15}$$

Let $\mathbf{x}$ be a sample function of a Gaussian process with x_i independent of x_j for $i \neq j$. The joint density functions are then the products of the individual density functions, given by

$$p_1(\mathbf{x}) = \left(\frac{1}{2\pi\sigma_x^2}\right)^{m/2} \prod_{i=1}^{m} \exp\left[-\frac{(x_i - \mu)^2}{2\sigma_x^2}\right]$$

$$= \left(\frac{1}{2\pi\sigma_x^2}\right)^{m/2} \exp\left[-\sum_{i=1}^{m} \frac{(x_i - \mu)^2}{2\sigma_x^2}\right]$$

$$p_0(\mathbf{x}) = \left(\frac{1}{2\pi\sigma_x^2}\right)^{m/2} \exp\left(-\sum_{i=1}^{m} \frac{x_i^2}{2\sigma_x^2}\right) \tag{13-16}$$

from which

$$\lambda(\mathbf{x}) = \exp\left[\sum_{i=1}^{m} \frac{(2\mu x_i - \mu^2)}{2\sigma_x^2}\right] \tag{13-17}$$

Using the log-likelihood ratio and a Neyman–Pearson threshold, we obtain

$$y = \frac{1}{m} \sum_{i=1}^{m} x_i \geq \gamma \tag{13-18}$$

or alternatively,

$$y' = \frac{1}{m\sigma_x^2} \sum_{i=1}^{m} \mu x_i \geq \gamma' \tag{13-19}$$

where y or y' is the test statistic, formed by calculating the sample mean of the available data set.

The false-alarm and detection probabilities may be calculated using the joint likelihood functions $p_0(\mathbf{x})$ and $p_1(\mathbf{x})$, or more easily by using the density function for y. The random variable y has a variance equal to σ_x^2/m and a mean value of zero or μ depending on whether H_0 or H_1 is true. Thus

$$P_{fa} = \int_{\gamma}^{\infty} p_0(y)\, dy$$

and   (13-20)

$$P_D = \int_{\gamma}^{\infty} p_1(y)\, dy$$

13.1.3 Correlation Receiver: Matched Filters

In the previous sections the signal portion of x was assumed to be either zero or a constant. We now let the signal be a time function of completely known form such that

$$x(t) \mid H_1 = s(t) + n(t)$$

The noise waveform is assumed to be bandlimited white noise with a spectral level equal to $N_0/2$. With time samples spaced such that the noise samples are independent, the likelihood function takes on the form

$$\lambda(\mathbf{x}) = \exp\left[\frac{\sum_{i=1}^{m}(2s_i x_i - s_i^2)}{2\sigma_n^2}\right] \tag{13-21}$$

where s_i and x_i are time samples of the signal and the received waveform at time t_i.

By letting the time sample interval approach zero while letting the noise bandwidth approach infinity so that the noise-sample independence is maintained, the summation in (13-21) passes to an integral, resulting in a likelihood function for continuous waveforms. It is easy to show that a test statistic in this case can be expressed as

$$y(T) = \int_0^T x(t)s(t)\, dt \geq \gamma \tag{13-22}$$

The threshold γ is chosen to provide the desired false-alarm probability.

A receiver designed to provide the test statistic in (13-22) is called a *correlation receiver* because the integral is the cross-correlation of the received waveform with the known signal waveform over the interval T. A simple mechanization diagram for this type of receiver is shown in Figure 13-2.

A signal-to-noise ratio can be associated with the test statistic in (13-22) by defining the desired output signal as the shift in the mean value of y caused by the presence of signal. The signal-to-noise ratio is then defined as

$$\text{SNR}(y) = \frac{\{E[y_1] - E[y_0]\}^2}{\text{Var}[y_0]} \tag{13-23}$$

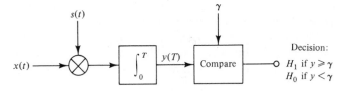

Figure 13-2   Correlation receiver.

The expected value of y_0 is zero because the signal and noise are statistically independent and noise is zero mean. With signal present we obtain

$$E[y_1] = \int_0^T [s^2(t) + \overline{s(t)n(t)}]\, dt = \int_0^T s^2(t)\, dt = E \qquad (13\text{-}24)$$

where E is the signal energy in time T. The variance of y_0 is equal to its mean-squared value, which is

$$\text{Var}[y_0] = \int_0^T \int_0^T s(t)s(t')E[n(t)n(t')]\, dt\, dt' \qquad (13\text{-}25)$$

With the assumption that $n(t)$ is white noise, the expectation in (13-25) is $(N_0/2)\delta(t - t')$, which reduces the double integral to a single integral. Hence

$$\text{Var}[y_0] = \frac{N_0}{2} \int_0^T s^2(t)\, dt = \frac{N_0 E}{2} \qquad (13\text{-}26)$$

Substitution of (13-24) and (13-26) in (13-23) gives

$$\text{SNR}(y) = \frac{2E}{N_0} \qquad (13\text{-}27)$$

indicating that the signal-to-noise ratio at the correlator output is the ratio of signal energy in time T to the white-noise spectral level, $N_0/2$.

It is easily shown that a properly chosen linear filter may be substituted for the multiplier and integrator in the correlation receiver. Assume a filter with impulse response $h(t)$. With the input signal $x(t)$ the output at time T is

$$y(T) = \int_0^T h(\tau)x(T - \tau)\, d\tau \qquad (13\text{-}28)$$

Now let

$$h(t) = s(T - t) \qquad \text{for } 0 \leqslant t \leqslant T \qquad (13\text{-}29)$$

and substitute in (13-28) to obtain

$$y(T) = \int_0^T s(T - \tau)x(T - \tau)\, d\tau = \int_0^T s(t)x(t)\, dt \qquad (13\text{-}30)$$

This result is obviously identical to that obtained with the correlation receiver. A filter chosen in accordance with (13-29) is called a *matched filter* and provides the optimum performance assuming a known signal in white noise.

A simple application of the Schwarz inequality shows that in the presence of white noise the matched filter provides the maximum possible signal-to-noise ratio as defined by (13-23). For an arbitrary filter with impulse response $h(t)$, the square of the expected value of y_1 at time T is limited by the inequality

$$E^2[y_1] = \left[\int_0^T h(\tau)s(T - \tau)\, d\tau \right]^2 \leqslant \int_0^T h^2(\tau)\, d\tau \int_0^T s^2(T - \tau)\, d\tau \qquad (13\text{-}31)$$

The variance of y_0 is

$$\text{Var}[y_0] = \frac{N_0}{2} \int_0^T h^2(\tau) \, d\tau \qquad (13\text{-}32)$$

The signal-to-noise ratio is therefore

$$\text{SNR}[y] = \frac{\left[\int_0^T h(\tau)s(T-\tau)\, d\tau\right]^2}{\frac{N_0}{2} \int_0^T h^2(\tau) \, d\tau} \leq \frac{2\int_0^T s^2(T-\tau)\, d\tau}{N_0} = \frac{2E}{N_0} \qquad (13\text{-}33)$$

Furthermore, in (13-33) the inequality becomes an equality if $h(t) = s(T - t)$.

The Fourier transform of the matched filter is related to the signal Fourier transform using the rules in Table 6-1. Thus

$$s(t) \longleftrightarrow S(f)$$

$$s(-t) \longleftrightarrow S(-f) = S^*(f)$$

$$h(t) = s(T - t) \longleftrightarrow S^*(f) \exp(j2\pi fT) \qquad (13\text{-}34)$$

13.1.4 Signals with Unknown Parameters

In preceding sections it was assumed that the signal portion of a received waveform, if present, was precisely known in all respects. Only the presence or absence of the signal was to be determined. In underwater acoustic systems this is seldom the case. In active systems the approximate shape of the desired signal is known, but the amplitude and time of arrival are unknown. In addition, the signal frequency may be modified by the Doppler effect. In passive systems, the signal to be detected is often a random function so that at best only the shape of its power spectral density is known.

Complete treatments of the detection of signals with unknown parameters can be found in Whalen [1] or Helstrom [2] as well as elsewhere in the literature. For our purposes we take a heuristic approach to extend the results for the known signal to define practical receiver structures for commonly encountered situations. In general, these structures involve approximations as compared with rigorously derived detection strategies.

Active system—single pulse.   Consider first an active system with a known signal shape but unknown amplitude and time location. That is, with a transmitted waveform $s(t)$, the received signal is $as(t - t_0)$, where a and t_0 are unknown. In this case a filter matched to the shape of the transmitted waveform may still be used, but the time location of the peak filter output is not known in advance. The filter output is in the form of the cross-correlation function of the transmitted and received waveforms. Thus, for $h(t) = s(T - t)$,

$$y(t) = \int_0^t h(\tau)x(t - \tau)\, d\tau$$

$$= a \int_0^t s(T - \tau)s(t - \tau + t_0)\, d\tau + \int_0^t s(T - \tau)n(t - \tau)\, d\tau$$

$$= a\mathfrak{R}_s(t - T - t_0) + \int_0^t s(T - \tau)n(t - \tau)\, d\tau \qquad (13\text{-}35)$$

where $\mathfrak{R}_s(t)$ is the autocorrelation function of the known transmitted waveform.

Because of the uncertainty with regard to t_0, $y(t)$ must be examined as a function of t to decide on the presence or absence of signal. However, in this case we are not interested in the fine structure in $\mathfrak{R}_s(t)$ caused by the carrier frequency, but only in the *envelope* characteristics. From a statistical detection standpoint there is little difference between the envelope and the squared magnitude of the envelope. Because the mathematics are significantly simplified by using the squared magnitude, we shall confine our attention to that case.

By using the envelope of the received waveform it is necessary only to observe the magnitude at time intervals approximately equal to the width of the signal autocorrelation function. This time interval is also equal to the reciprocal of the effective envelope bandwidth as defined in Section 8.4.1.

The test statistic for the active system is now defined as

$$z_k = |y(kt_s)|^2 = \left| \int_0^{kt_s} h(\tau)x(kt_s - \tau)\, d\tau \right|^2 \qquad (13\text{-}36)$$

where t_s is the sample interval, equal to or less than the reciprocal of the envelope bandwidth at the filter output. A threshold γ is established to achieve the desired false-alarm probability. The receiver structure for this case is shown in Figure 13-3.

Notice that with the Neyman–Pearson criterion the lack of knowledge concerning the signal amplitude does not affect the threshold setting. The use of the filter matched to the transmitted waveform provides the best possible signal-to-noise ratio as defined by (13-23). The SNR at the actual target location is easily shown to be

$$\text{SNR}(y) = \frac{2a^2E}{N_0} \qquad (13\text{-}37)$$

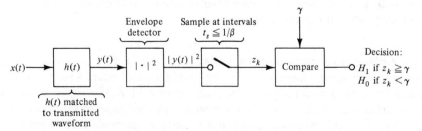

Figure 13-3 Receiver structure for detecting signal with unknown amplitude and location.

In practice the sampling and threshold-comparison operations shown in Figure 13-3 may actually be performed by a human operator observing the amplitude–time characteristics of the received waveform on some form of display surface.

To develop the statistical properties of the squared magnitude of the envelope function it is convenient to consider the analytic form of $y(t)$. Following the methods developed in Section 6.10, we write

$$y_a(t) = y(t) + j\hat{y}(t) = |y_a(t)| \exp [j\omega_0 t + \phi(t)] \qquad (13\text{-}38)$$

where $\hat{y}(t)$ is the Hilbert transform of $y(t)$. Alternatively, the real waveform is the real part of (13-38). Thus

$$y(t) = |y_a(t)| \cos [\omega_0 t + \phi(t)]$$
$$= [y^2(t) + \hat{y}^2(t)]^{1/2} \cos [\omega_0 t + \phi(t)] \qquad (13\text{-}39)$$

from which

$$y^2(t) = \tfrac{1}{2}[y^2(t) + \hat{y}^2(t)]\{1 + \cos [2\omega_0 t + 2\phi(t)]\} \qquad (13\text{-}40)$$

We shall assume that the squared-magnitude envelope detector includes a filter to remove the double-frequency component of (13-40). Using the $|\cdot|^2$ notation to signify that the high-frequency components have been removed, we obtain finally

$$z = |y(t)|^2 = \tfrac{1}{2}[y^2(t) + \hat{y}^2(t)] = \tfrac{1}{2}|y_a(t)|^2 \qquad (13\text{-}41)$$

With the H_0 hypothesis, $y_0(t)$ is assumed to be zero-mean Gaussian. By definition, a function and its Hilbert transform are orthogonal, which in this case also guarantees their statistical independence. The function z_0 is therefore the sum of the squares of independent ZMG variables with equal variances. The density function has the form of a two-degree-of-freedom chi-squared density function. In particular, for $z \geq 0$,

$$p_0(z) = \frac{1}{\sigma_{y_0}^2} \exp \left(\frac{-z}{\sigma_{y_0}^2} \right) \qquad (13\text{-}42)$$

where $\sigma_{y_0}^2 = \bar{z}_0$. Assuming that the input noise is white with spectral density $N_0/2$, we have

$$\bar{z}_0 = \sigma_{y_0}^2 = \frac{N_0}{2} \int_{-\infty}^{+\infty} h^2(\tau) \, d\tau \qquad (13\text{-}43)$$

With signal present the probability density function for z_1 is said to have a *noncentral* chi-squared distribution. A complete discussion of the noncentral chi-squared distribution is contained in Whalen [1], Chapters 4 and 8. Let the signal-to-noise power ratio at the filter output be S. The resulting noncentral chi-squared distribution is

$$p_1(z) = \frac{1}{\sigma_{y_0}^2} \exp \left(-S - \frac{z}{\sigma_{y_0}^2} \right) I_0 \left[2 \left(\frac{zS}{\sigma_{y_0}^2} \right)^{1/2} \right] \qquad (13\text{-}44)$$

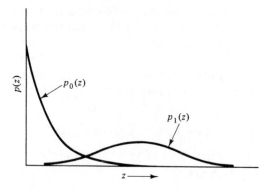

Figure 13-4 Probability density functions at output of a squared-magnitude envelope detector for noise only and signal plus noise.

where $I_0(\,\cdot\,)$ is the zero-order modified Bessel function. In the case of a matched filter, $S = 2E/N_0$. The density functions for signal-present and signal-absent are sketched in Figure 13-4.

Active system—multiple pulses.   An active system typically transmits a periodic train of pulses. Assume that a group of M received pulses from such a periodic train is available to the receiver. The detection performance can be improved, relative to the single-pulse case, by the addition of a *post-detection* processing section to the receiver structure. At the output of the envelope detector the received signals at each range are summed for a total of M transmission periods to form the test statistic. Let $|y_{ik}|^2$ be the squared-magnitude envelope for the ith transmission period at the kth range. The test statistic z_k at this range is defined as

$$z_k = \sum_{i=1}^{M} |y_{ik}|^2 \tag{13-45}$$

The resulting receiver structure is shown in Figure 13-5. With this mechanization, the filter impulse response $h(t)$ is matched to the single pulse waveform as before.

Assuming that the noise in a given range bin is statistically independent

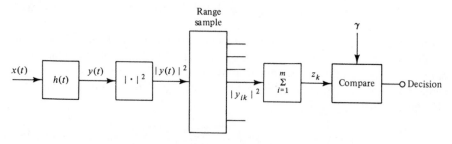

Figure 13-5 Receiver for multiple-pulse detection.

from one transmission period to the next, z_k is seen to be a $2M$-degree-of-freedom chi-squared variable. The noise-only density function can be put in the form

$$p_0(z) = \frac{1}{\sigma_{y0}^2 \Gamma(M)} \left(\frac{z}{\sigma_{y0}^2}\right)^{M-1} \exp\left(\frac{-z}{\sigma_{y0}^2}\right) \tag{13-46}$$

The moments of z_0 are

$$\bar{z}_0 = M\sigma_{y0}^2$$

$$E[z_0^2] = M(M+1)\sigma_{y0}^4$$

$$\text{Var}[z_0] = M\sigma_{y0}^4 \tag{13-47}$$

If $M = 1$, note that these results reduce to the single-pulse case. Also notice that the ratio of the mean to standard deviation is $\sqrt{M}$, indicating that the noise fluctuation relative to the mean value is reduced by the averaging process.

With signal present the density function for z_1 is given by

$$p_1(z) = \frac{1}{\sigma_{y0}^2} \left(\frac{z}{S\sigma_{y0}^2}\right)^{(M-1)/2} \exp\left(-S - \frac{z}{\sigma_{y0}^2}\right) I_{M-1}\left[2\left(\frac{zS}{\sigma_{y0}^2}\right)^{1/2}\right] \tag{13-48}$$

The mean value of z_1 is increased relative to $\bar{z}_0$ by a factor related to the signal-to-noise ratio at the envelope detector input. Thus

$$\bar{z}_1 = \bar{z}_0(1 + S) \tag{13-49}$$

For M large and S small, the variance of z_1 is approximately the same as for z_0. For this case it is useful to define a signal-to-noise ratio for z similar to that used for y in (13-23). Hence

$$\text{SNR}(z) = \frac{\{E[z_1] - E[z_0]\}^2}{\text{Var}[z_0]} \tag{13-50}$$

Substitution from (13-47) and (13-49) into (13-50) gives

$$SNR(z) = MS^2 \tag{13-51}$$

Typical plots of $p_0(z)$ and $p_1(z)$ for the multiple pulse case are shown in Figure 13-6. For M large both $p_0(z)$ and $p_1(z)$ approach the Gaussian shape in accordance with the central limit theorem.

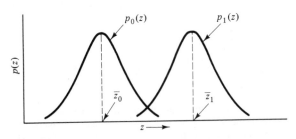

Figure 13-6   Typical density functions for the multiple-pulse case.

Active system with Doppler shift.   Relative motion between the sonar system and the target results in a Doppler shift of the received signal relative to the transmitted signal. The amount of frequency shift is typically unknown a priori, requiring a modification of the receiver mechanization to avoid a serious loss in detection performance.

Limiting our discussion to the single-pulse case, two possible approaches to the receiver mechanization with unknown Doppler shift are depicted in Figure 13-7. In Figure 13-7(a) parallel filters are used, each matched to the received signal at slightly different frequencies. The total frequency range covered by the filter bank equals the expected total range of Doppler frequency shift.

Each channel in mechanization (a) has a performance equal to that discussed for the single pulse matched filter system. The penalty paid for the unknown Doppler shift lies in the added receiver complexity and in the fact that the number of possibilities for false alarm is increased by the number of parallel channels required. This results in a requirement to raise the threshold on each channel to maintain the overall desired false alarm characteristics.

In mechanization (b) the single predetection filter is matched to the overall Doppler spread rather than to the transmitted signal shape. This results in a value of SNR(y) less than the optimum value because the noise bandwidth exceeds the bandwidth of an individual received signal.

The postdetection filter is matched in some sense to the signal pulse wave-

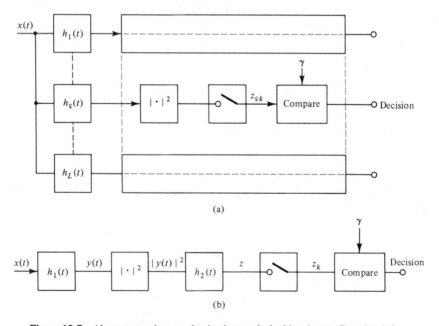

(a)

(b)

Figure 13-7   Alternate receiver mechanizations to deal with unknown Doppler shift: (a) receiver mechanization with parallel channels of matched filters; (b) receiver mechanization with wideband predetection filter $h_1(t)$ followed by narrowband postdetection filter.

form at the output of the envelope detector, thereby providing a noncoherent gain in signal-to-noise ratio relative to the detector output. Assume a rectangular signal pulse envelope and a predetection bandwidth β such that $\beta t_p \gg 1$, where t_p is the pulse width. At the filter output, the SNR is reduced by approximately βt_p relative to a matched filter. The gain provided by the postdetection filter partially compensates for this loss, resulting in a net effective loss in SNR(y) equal to $\sqrt{\beta t_p}$. In Chapter 14 we shall see that this result is further modified by the differences in the noise-only probability density functions for the mechanizations in Figure 13-7. For βt_p products less than 10, the relative performances of (a) and (b) as measured by required input SNR for fixed P_D and P_{fa} differ by less than 1.5 dB.

Passive broadband system.   The received target signal in a passive system is typically a random function with a wide bandwidth and a time duration limited only by the geometry and relative motion involved in the encounter. In this case only certain statistical properties of the signal are known. For instance, we assume that the shape of the signal power spectral density is known and that some idea of total signal duration, when present, is known. Also, the noise power spectral density is known and is not necessarily white.

A receiver structure as shown in Figure 13-8 is assumed. We wish to determine the form of the predetection and postdetection filters that will in some sense optimize the detection performance of the system. As a criterion for optimization we choose to maximize the value of SNR(z) based on the available information.

As an example, let the signal, when present, be turned on with a constant level for a time $T \gg 1/\beta_s$, where β_s is a measure of the signal bandwidth. We then define

$$\Psi_{x_1}(f) = \Psi_s(f) + \Psi_n(f)$$

$$\Psi_{x_0}(f) = \Psi_n(f)$$

where $\Psi_s(f)$ is the target-signal power spectral density measured at the receiver terminals and $\Psi_n(f)$ is the noise power spectral density. Both the target signal and the noise are assumed to be ZMG processes, independent of each other.

The signal-to-noise ratio at the output of the predetection filter is

$$\text{SNR}(y) = \frac{\int \Psi_s(f)|H_1(f)|^2 \, df}{\int \Psi_n(f)|H_1(f)|^2 \, df} = \frac{\int [\Psi_{y_1}(f) - \Psi_{y_0}(f)] \, df}{\int \Psi_{y_0}(f) \, df} \qquad (13\text{-}52)$$

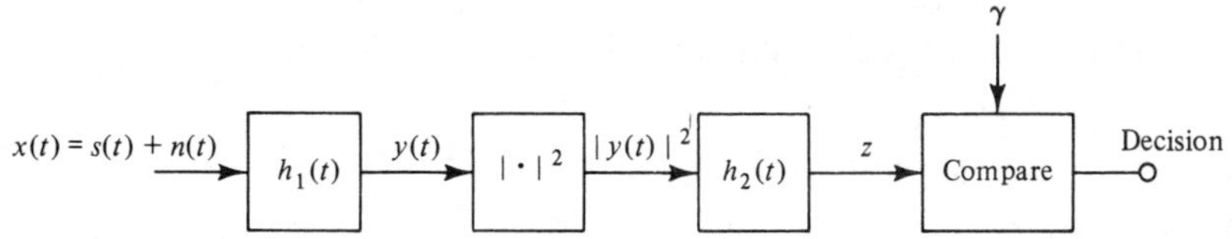

Figure 13-8 Passive broadband receiver mechanization.

where $H_1(f)$ is the transform of $h_1(t)$ and the limits of integration are assumed infinite in this and subsequent expressions unless otherwise stated.

At the output of the square-law envelope detector we have

$$E[|y_1^2|] = E[|y_s|^2] + E[|y_n|^2]$$
$$E[|y_0|^2] = E[|y_n|^2]$$

where y_s and y_n are the target-signal and noise components at the filter output.

To form the signal-to-noise ratio at the detector output we require the variance of $|y_0|^2$. By definition we write

$$\mathrm{Var}[|y_0|^2] = E[|y_0|^4] - E^2[|y_0|^2]$$

By the Gaussian assumption, the fourth moment can be expressed in terms of the second moment. Because the envelope detector is presumed to have removed the double-frequency components of $|y^2(t)|$, the fourth moment of the envelope is equal to twice the square of the second moment [rather than three times, as would be true for $y^2(t)$]. Thus

$$\mathrm{Var}[|y_0|^2] = E^2[|y_0|^2]$$

and

$$\mathrm{SNR}(|y|^2) = \frac{\{E[|y_1|^2] - E[|y_0|^2]\}^2}{\mathrm{Var}[|y_0|^2]} = \frac{E^2[|y_s|^2]}{E^2[|y_0|^2]}$$

$$= \mathrm{SNR}^2(y) \tag{13-53}$$

For convenience, the response of the postdetection filter is assumed normalized so that the expected value of z is identical to the expected value of $|y|^2$. This simply requires that the integral of $h_2(t)$ be unity.

The variance of z_0 is obtained as follows:

$$E[|z_0|^2] = \int \Psi_{|y_0|^2}(f)|H_2(f)|^2 \, df = \int R_{|y_0|^2}(\tau)R_{h_2}(\tau) \, d\tau \tag{13-54}$$

Using the properties of the Gaussian envelope function, we obtain

$$R_{|y_0|^2}(\tau) = R_{|y_0|}^2(0) + R_{|y_0|}^2(\tau)$$
$$= E^2[|y_0|^2] + R_{|y_0|}^2(\tau)$$
$$= E^2[z_0] + R_{|y_0|}^2(\tau) \tag{13-55}$$

where $|y_0|$ is understood to be the noise-only envelope function at the output of the predetection filter. Substitution of (13-55) in (13-54) gives

$$\mathrm{Var}[z_0] = \int R_{|y_0|}^2(\tau)R_{h_2}(\tau) \, d\tau \tag{13-56}$$

With the assumption of a signal at a constant power level for a time T and zero elsewhere, a reasonable choice for the impulse response $h_2(t)$ is a rect function of duration T. Thus let

$$h_2(t) = \frac{1}{T} \text{rect}\left(\frac{t}{T}\right)$$

from which

$$R_{h_2}(\tau) = \frac{1}{T}\left(1 - \frac{|\tau|}{T}\right) \text{rect}\left(\frac{\tau}{2T}\right) \qquad (13\text{-}57)$$

and

$$\text{Var}[z_0] = \frac{1}{T}\int_{-T}^{+T}\left(1 - \frac{|\tau|}{T}\right) R_{|y_0|}^2(\tau)\, d\tau \qquad (13\text{-}58)$$

The width of the noise-only autocorrelation function at the output of the predetection filter is typically small compared with T because of the assumption of a large signal time–bandwidth product. Hence

$$\text{Var}[z_0] \simeq \frac{1}{T}\int_{-\infty}^{+\infty} R_{|y_0|}^2(\tau)\, d\tau \qquad (13\text{-}59)$$

Using the effective envelope bandwidth definition from Section 8.4.1, (13-59) can be written as

$$\text{Var}[z_0] = \frac{R_{|y_0|}^2(0)}{T\beta_n} = \frac{E^2[|y_0|^2]}{T\beta_n} \qquad (13\text{-}60)$$

where β_n is the effective noise bandwidth at the output of the predetection filter. The expression for $\text{SNR}(z)$ now becomes

$$\text{SNR}(z) = \frac{\{E[z_1] - E[z_0]\}^2}{\text{Var}[z_0]}$$

$$= T\beta_n \frac{E^2[|y_s|^2]}{E^2[|y_0|^2]} = T\beta_n \text{SNR}^2(y) \qquad (13\text{-}61)$$

The postdetection filter provides a signal-to-noise improvement by a process equivalent to the averaging of $T\beta_n$ independent noise samples.

We now examine the possibility of selecting the form of the predetection filter to optimize the output signal-to-noise ratio. From (13-52) and using the frequency-domain expression for effective bandwidth, we write (13-61) as

$$\text{SNR}(z) = \frac{T}{2}\left\{\frac{\left[\int \Psi_n(f)|H_1(f)|^2\, df\right]^2}{\int \Psi_n^2(f)|H_1(f)|^4\, df}\left[\frac{\int \Psi_s(f)|H_1(f)|^2\, df}{\int \Psi_n(f)|H_1(f)|^2\, df}\right]^2\right\}$$

$$= \frac{T}{2}\frac{\left[\int \Psi_s(f)|H_1(f)|^2\, df\right]^2}{\int \Psi_n^2(f)|H_1(f)|^4\, df} \qquad (13\text{-}62)$$

In the integral in the numerator of (13-62) we now multiply and divide by the noise spectral density and apply the Schwarz inequality as follows:

$$\left\{ \int \left[\frac{\Psi_s(f)}{\Psi_n(f)} \right] [\Psi_n(f) |H_1(f)|^2] \, df \right\}^2 \leq \int \frac{\Psi_s^2(f)}{\Psi_n^2(f)} \, df \int \Psi_n^2(f) |H_1(f)|^4 \, df \quad (13\text{-}63)$$

Substitution of (13-63) into (13-62) provides the inequality

$$\text{SNR}(z) \leq \frac{T}{2} \int \frac{\Psi_s^2(f)}{\Psi_n^2(f)} \, df \qquad (13\text{-}64)$$

It is easily verified that (13-64) becomes an equality if we select $H_1(f)$ such that

$$|H_1(f)|^2 = \frac{\Psi_s(f)}{\Psi_n^2(f)} \qquad (13\text{-}65)$$

The optimum filter described by (13-65), sometimes referred to as an Eckart filter [3], has the shape of the signal power spectral density if the noise spectrum is white. It is therefore similar to the matched filter described earlier for the known signal in white noise. If the noise is not white, the optimum filter response tends to be reduced in those portions of the spectrum where the noise power is large.

The results above were derived based on the use of a postdetection filter with a rectangular impulse response. Actually, the optimum predetection filter shape is not affected significantly by the choice of postdetection filter provided that the bandwidth of the postdetection filter is small compared with the noise bandwidth. For instance, let $h_2(t)$ have an exponential impulse response with time constant T. It is left to the reader to prove that if the target signal persists for a time much larger than T, and if

$$h_2(t) = \frac{1}{T} \exp\left(\frac{-t}{T}\right), \qquad t \geq 0$$

then

$$\text{SNR}(z) = 2T\beta_n \text{SNR}^2(y), \qquad T\beta_n \gg 1 \qquad (13\text{-}66)$$

The form of the postdetection filter is selected based on expected encounter geometry and dynamics. Because of potentially large variations expected in these parameters, relatively simple forms, such as the rect function or exponential function, are usually appropriate.

Passive narrowband system.   Passive targets such as ships and submarines emit narrowband as well as broadband acoustic signals. The receiver mechanization appropriate for the narrowband components of the target signature consists of a bank of parallel narrowband filters covering the expected frequency range. The output of each filter is envelope-detected, processed by a postdetection filter if necessary, and compared with a Neyman–Pearson threshold. Such a mechanization is shown in Figure 13-9.

The typical passive narrowband receiver forms the parallel filter bank by a discrete Fourier transformation of the input signal. The bandwidth of each filter is the inverse of the time duration of the segment of the input signal used to form the transform. This bandwidth is typically very small compared with the total bandwidth occupied by the target total acoustic signature. If the filter bandwidth is equal to the bandwidth of the target line component, no further performance gain will be realized by postdetection processing in $h_2(t)$. More typically, the product of signal duration and filter bandwidth will exceed unity, in which case the performance of each narrowband channel is governed by equations identical to those for the broadband case.

For the narrowband system it is not unreasonable to assume that the noise spectral density is constant over the band of each individual filter. This results in some simplification of the output signal-to-noise calculation. As an example, assume a complex sinusoidal signal, narrowband filter and noise spectrum as follows:

$$\Psi_s(f) = a^2\delta(f - f_1)$$

$$\Psi_n(f) = \Psi_n(f_1)$$

$$H_1(f) = \text{rect}\left(\frac{f - f_1}{\beta}\right)$$

Then

$$\text{SNR}(y) = \frac{a^2}{\Psi_n(f_1)\displaystyle\int \text{rect}^2[(f - f_0)/\beta]\, df} = \frac{a^2}{\Psi_n(f_1)\beta}$$

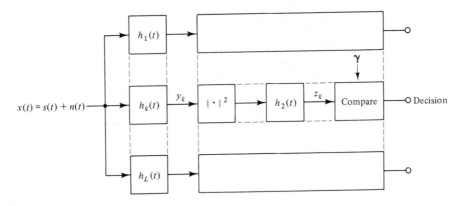

Figure 13-9   Passive narrowband receiver mechanization.

For the rect-function filter, the effective bandwidth with white noise input is equal to β. Assuming that $h_2(t)$ has a rect function impulse response of width T, the output signal-to-noise ratio is

$$\mathrm{SNR}(z) = T\beta_n\,\mathrm{SNR}^2(y)$$

$$= T\beta\frac{a^4}{\Psi_n^2(f_1)\beta^2} = \left(\frac{T}{\beta}\right)\left(\frac{a^2}{\Psi_n(f_1)}\right)^2 \qquad (13\text{-}67)$$

13.2 RECEIVER OPERATING CHARACTERISTICS

The detection performance of a receiving system is calculated by determining the input signal power required to achieve a given probability of detection while maintaining a fixed acceptable probability of false alarm. For this purpose the required signal-to-noise ratio at the input to the envelope detector (SNR(y)) is first determined. The required receiver input signal is then determined from

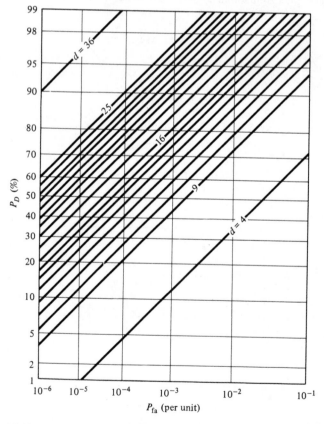

Figure 13-10 Receiver operating characteristic curves for $p_0(z)$ and $p_1(z)$ both Gaussian. SNR (z) = d for indicated P_D and p_{fa}.

knowledge of the linear transfer function from the receiver input to the envelope detector input.

The detection performance may be presented graphically by plotting the detection probability versus false-alarm probability with signal-to-noise as the family parameter. Such a plot is called the *receiver operating characteristic* (ROC) curves.

In general, a series of such curves is required to cover the possible range of time–bandwidth products and postdetection processing configurations. A particularly simple ROC curve results with a mechanization as shown in Figure 13-8 and with $\beta T \gg 1.0$. In this case $p_0(z)$ and $p_1(z)$ are both Gaussian, and there is a simple relationship between SNR(y) and SNR(z), given by (13-61). Therefore, SNR(z) may be used as the family parameter with the result that only one set of ROC curves is required. Such a plot is shown in Figure 13-10. In this figure the parameter d, called the *detection index,* is the value of SNR(z) required for the indicated values of P_D and P_{fa}.

As an example of the use of Figure 13-10, assume a passive broadband receiver with $T\beta = 1000$. We require the value of SNR(y) to achieve a 90% probability of detection with a false-alarm probability of 10^{-4}. From Figure 13-10, $d = $ SNR(z) $= 25$. Therefore, from (13-61) we obtain

$$\text{SNR}(y) = \left(\frac{d}{T\beta}\right)^{1/2} = \left(\frac{25}{1000}\right)^{1/2} = 0.158$$

Receiver operating characteristic curves for more general situations have been prepared by a number of workers in the field [4–6]. The curves prepared by Robertson [6] are based on a linear envelope detector and the postdetection summation of a number of independent samples. Whalen [1] includes a full set

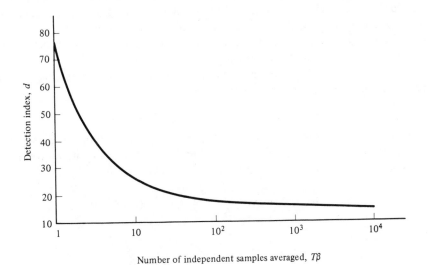

Number of independent samples averaged, $T\beta$

Figure 13-11   Detection index versus $T\beta$ for $P_D = 50\%$ and $p_{fa} = 10^{-4}$.

of the Robertson curves covering the range from 1 to 8192 independent samples. The number of independent samples is equivalent to the time–bandwidth product used in the preceding section. The family parameter for the Robertson curves is the signal-to-noise ratio at the detector input.

For large time–bandwidth products the results obtained using curves such as the Robertson curves are nearly identical to those obtained from Figure 13-10. In Figure 13-11 the detection index is plotted as a function of the $T\beta$ product for $P_D = 50\%$ and $P_{fa} = 10^{-4}$. For $T\beta \geq 32$, the error in computing SNR(y) caused by using Figure 13-10 rather than Figure 13-11 is less than 1 dB.

13.3 ESTIMATION OF SIGNAL PARAMETERS

In addition to providing for the detection of target signals, the receiver system is often required to provide estimates of specific target parameters. For instance, we may be interested in the target range, bearing, speed, and size. These target parameters relate to target signal parameters such as time of arrival, relative time of arrival at two separated points, frequency shift, and received signal strength.

As with the detection problem, the received waveform includes the desired target signal in combination with noise. We therefore desire to choose for an estimate of a given parameter that value that maximizes the conditional probability density function for the parameter, given the received waveform. Let x be the received signal and α the parameter to be measured. Then

$$p(\alpha \mid x) = \frac{p(x \mid \alpha)p(\alpha)}{p(x)}$$

In most cases of interest $p(\alpha)$, the a priori distribution for α, is unknown and may be assumed nearly uniformly distributed over the region of interest. Also note that $p(x)$ is independent of α. Therefore, maximizing $p(\alpha \mid x)$ is equivalent to maximizing $p(x \mid \alpha)$. The estimate selected in this manner is called a *maximum likelihood estimate* since $p(x \mid \alpha)$ is the conditional likelihood function for x given the value α for the unknown parameter.

To investigate the maximum likelihood estimate of a single unknown parameter α, let the received waveform be

$$x(t) = s(t, \alpha) + n(t), \qquad 0 \leq t \leq T$$

with $s(t, \alpha)$ completely contained in the interval T and with the noise $n(t)$ assumed white and Gaussian. For the maximum likelihood estimate we form the partial derivative of $p(x \mid \alpha)$ with respect to α. Thus, with

$$p(x \mid \alpha) = K \exp\left\{ -\frac{1}{N_0} \int_0^T [x(t) - s(t, \alpha)]^2 \, dt \right\}$$

we get

$$\frac{\partial p(x \mid \alpha)}{\partial \alpha} = \frac{2K}{N_0} \left\{ \int_0^T [x(t) - s(t, \alpha)] \frac{\partial s(t, \alpha)}{\partial \alpha} \, dt \right\} \exp\{\cdot\} \qquad (13\text{-}68)$$

Choose $\tilde{\alpha}$ as the estimate of α such that (13-68) is zero.

$$\int_0^T [x(t) - s(t, \tilde{\alpha})]\frac{\partial s(t, \tilde{\alpha})}{\partial \tilde{\alpha}} dt = 0 \qquad (13\text{-}69)$$

13.3.1 Estimation of Time Delay

Target range is typically determined by measuring the time delay between the envelopes of the transmitted waveform and received echo. For convenience we shall use the complex representations of noise and signal envelope functions. Assume that the real noise at the receiver input is white with spectral density $N_0/2$, and that the real signal spectrum is centered at a carrier frequency f_0 with a bandwidth small compared with f_0. The complex envelope of the total waveform at the input is

$$x(t) = \eta(t) + \mu(t - \tau)$$

where, in accordance with the methods developed in Section 6.10, we define

$$\eta(t) = [n(t) + j\hat{n}(t)]\exp(-j2\pi f_0 t) = \text{noise envelope}$$

$$\mu(t) = [s(t) + j\hat{s}(t)]\exp(-j2\pi f_0 t) = \text{signal envelope}$$

and τ is the time location of the target echo. With no loss in generality, we let $\tau = 0$ in the following discussion.

The autocorrelation function of the real and imaginary parts of $\eta(t)$ will be required. From the defining relationships for the Hilbert transform, the Fourier transform of the real part of $\eta(t)$ is

$$\text{Re}[\eta(t)] = \eta_r(t) \longleftrightarrow \tfrac{1}{2}\{N(f + f_0)[1 + \text{sgn}(f + f_0)]$$
$$+ N(f - f_0)[1 - \text{sgn}(f - f_0)]\} \qquad (13\text{-}70)$$

where $N(f)$ is the spectrum of the real noise $n(t)$. From (13-70) we obtain

$$R_{\eta_r}(\tau) \longleftrightarrow \frac{N_0}{4}[2 + \text{sgn}(f + f_0) - \text{sgn}(f - f_0)]$$

$$= \frac{N_0}{2}\left[1 + \text{rect}\left(\frac{f}{2f_0}\right)\right] \qquad (13\text{-}71)$$

from which

$$R_{\eta_r}(\tau) = \frac{N_0}{2}[\delta(\tau) + 2f_0 \text{ sinc }(2f_0\tau)] \qquad (13\text{-}72)$$

It is easily shown that the autocorrelation function of the imaginary part of $\eta(t)$ is also given by (13-72). Our use of $R_{\eta_r}(\tau)$ and $R_{\eta_i}(\tau)$ is limited to integrals of the products of these functions with functions of the narrowband signal envelope. Therefore, the sinc function in (13-72) will behave approximately as an impulse and $R_{\eta_r}(\tau)$ can be expressed as

$$R_{\eta_r}(\tau) = R_{\eta_i}(\tau) \simeq N_0 \, \delta(\tau) \qquad (13\text{-}73)$$

From (13-69), the estimate of time delay $\tilde{\tau}$ is the solution of

$$\text{Re}\left\{ \int_0^T [x(t) - \mu(t - \tilde{\tau})] \frac{\partial \mu^*(t - \tilde{\tau})}{\delta \tilde{\tau}} \, dt \right\} = 0 \qquad (13\text{-}74)$$

which can also be written as

$$\text{Re}\left\{ \frac{\partial}{\partial \tilde{\tau}} \mathcal{R}_\mu(\tilde{\tau}) + \int_0^T \eta(t) \frac{\partial \mu^*(t - \tilde{\tau})}{\partial \tilde{\tau}} \, dt - \frac{\partial}{\partial \tilde{\tau}} \int_0^T |\mu(t - \tilde{\tau})|^2 \, dt \right\} = 0 \qquad (13\text{-}75)$$

where $\mathcal{R}_\mu(\tilde{\tau})$ is the signal-envelope autocorrelation function. The third term in (13-72) is identically zero.

Expand $\mathcal{R}_\mu(\tilde{\tau})$ as a Taylor series and obtain

$$\text{Re}[\mathcal{R}_\mu(\tilde{\tau})] = \mathcal{R}_\mu(0) + \frac{\mathcal{R}_\mu''(0)}{2} \tau^2 + \cdots$$

Dropping terms of higher order than the second and differentiating, we get

$$\text{Re}\left\{ \frac{\partial}{\partial \tilde{\tau}} \mathcal{R}_\mu(\tilde{\tau}) \right\} = \mathcal{R}_\mu''(0) \, \tilde{\tau} \qquad (13\text{-}76)$$

Substitute (13-76) into (13-75) and obtain for the estimate (remembering that the actual value of τ is zero)

$$\tilde{\tau} = -\frac{1}{\mathcal{R}_\mu''(0)} \text{Re}\left\{ \int_0^T \eta(t) \frac{\partial \mu^*(t - \tilde{\tau})}{\partial \tilde{\tau}} \, dt \right\} \qquad (13\text{-}77)$$

The mean value of $\tilde{\tau}$ is zero (or equal to τ in the general case), indicating that it is an *unbiased* estimate.

The variance of the estimate is the expected value of the square of (13-77). Using the definition of $R_{\eta_r}(\tau)$ from (13-73), the expression for the variance reduces to

$$\text{Var}[\tilde{\tau}] = \left(\frac{1}{\mathcal{R}_\mu''(0)} \right)^2 N_0 \int_0^T \left| \frac{\partial \mu(t - \tilde{\tau})}{\partial \tilde{\tau}} \right|^2 \, dt \qquad (13\text{-}78)$$

From basic Fourier transform relationships and Parseval's theorem, the integral in (13-78) is easily shown to be equal to $\mathcal{R}_\mu''(0)$. Thus

$$\text{Var}[\tilde{\tau}] = \frac{N_0}{\mathcal{R}_\mu''(0)} \qquad (13\text{-}79)$$

We now define a bandwidth measure for the signal envelope in terms of the normalized second moment of its energy spectrum. This bandwidth measure, β_0, is sometimes called the rms bandwidth. Thus, with $\mu(t) \longleftrightarrow M(f)$,

$$\beta_0^2 = \frac{(2\pi)^2 \int f^2 |M(f)|^2 \, df}{\int |M(f)|^2 \, df} = \frac{\mathcal{R}_\mu''(0)}{\mathcal{R}_\mu(0)} \tag{13-80}$$

The mean value, or first moment, of $|M(f)|^2$ is assumed to be zero in the definition given by (13-80).

Recalling that $\mathcal{R}_\mu(0) = 2E$ where E is the energy in the real signal waveform, (13-80) may be substituted in (13-78) to obtain

$$\mathrm{Var}[\bar{\tau}] = \left[\frac{2E}{N_0} \beta_0^2 \right]^{-1} \tag{13-81}$$

It can be shown [1] that no other estimator of the delay results in a smaller variance than that given by (13-81). This minimum variance is called the *Cramer–Rao* bound [7].

13.3.2 Estimation of Frequency Shift

Assume the envelope function of the total received waveform to be

$$x(t) = \mu(t) \exp(j2\pi v t) + \eta(t) \tag{13-82}$$

where $\mu(t)$ is the envelope of a narrowband signal and v is an unknown Doppler frequency shift to be estimated. Again with no loss in generality we may let the actual value of $v = 0$, and obtain the estimate $\tilde{v}$ from the equation

$$\mathrm{Re}\left\{ \int_0^T [x(t) - \mu(t) \exp(j2\pi \tilde{v} t)] \frac{\partial \mu^*(t) \exp(-j2\pi \tilde{v} t)}{\partial \tilde{v}} \, dt \right\} = 0 \tag{13-83}$$

which can be put in the form

$$\mathrm{Re}\left\{ \frac{\partial \mathcal{K}_\mu(\tilde{v})}{\partial \tilde{v}} \right\} = -\mathrm{Re}\left\{ \int_0^T \eta(t) \frac{\partial \mu^*(t) \exp(-j2\pi \tilde{v} t)}{\partial \tilde{v}} \, dt \right\} \tag{13-84}$$

where $\mathcal{K}_\mu(\tilde{v})$ is the frequency-domain autocorrelation function of the spectrum of $\mu(t)$ with respect to the frequency shift $\tilde{v}$.

Proceeding as before, expand $\mathcal{K}_\mu(\tilde{v})$ in a Taylor series, differentiate, and obtain as the desired estimate

$$\tilde{v} = -\left(\frac{1}{\mathcal{K}_\mu''(0)} \right) \mathrm{Re}\left\{ \int_0^T \eta(t) \frac{\partial \mu^*(t) \exp(-j2\pi \tilde{v} t)}{\partial \tilde{v}} \, dt \right\} \tag{13-85}$$

The variance of the estimate is

$$\mathrm{Var}[\tilde{v}] = \left(\frac{1}{\mathcal{K}_\mu''(0)} \right)^2 N_0 \int_0^T (2\pi)^2 t^2 |\mu(t)|^2 \, dt \tag{13-86}$$

We now identify an rms measure of envelope time duration, τ_0, as

$$\tau_0^2 = \frac{(2\pi)^2 \int t^2 |\mu(t)|^2 \, dt}{\int |\mu(t)|^2 \, dt} = \frac{\mathcal{H}_\mu''(0)}{\mathcal{H}_\mu(0)} \tag{13-87}$$

But $\mathcal{H}_\mu(0) = \mathcal{R}_\mu(0) = 2E$, so that the variance may be written as

$$\text{Var}[\tilde{v}] = \left[\frac{2E}{N_0} \tau_0^2 \right]^{-1} \tag{13-88}$$

Notice that the variance in estimating frequency is reduced by increasing the time duration of the signal, whereas the variance in estimating time is reduced by increasing the signal bandwidth.

The receiver mechanization to estimate frequency typically consists of a parallel bank of filters and integrators, similar to that shown in Figure 13-7(a) or Figure 13-9. The center frequency of the filter with the maximum output is the desired estimate.

13.3.3 Estimation of Target Bearing

To estimate time delay accurately we found in Section 13.3.1 that the signal waveform should occupy a considerable extent in the frequency domain. Similarly, to estimate frequency, the time duration of the signal should be large. We shall now demonstrate that to estimate target bearing it is necessary to have spatially separated signal samples so that we may consider the signal to have an appreciable extent as measured in linear space.

For simplicity, consider a narrowband signal and narrowband isotropic noise at frequency f and bandwidth $\beta \ll f$ available over a finite portion of a straight line such as the y-axis. We wish to estimate the bearing ψ of the target relative to a normal to the y-axis.

Let the received envelope per unit length at the origin be $\mu(t)$. The received signal at any other point on the y-axis is then

$$\mu(t, y, \psi) = \mu(t) \exp\left(j \frac{2\pi f y \sin \psi}{c} \right) \tag{13-89}$$

For convenience, we drop the explicit recognition of the time dependence and write the total received waveform per unit length along the y-axis as

$$x(y) = \mu(y, \psi) + \eta(y) \tag{13-90}$$

where $\eta(y)$ is the contribution from the isotropic noise field. Let the actual target bearing ψ be zero and estimate the bearing from the equation

$$\text{Re}\left\{ \int_{-L/2}^{+L/2} [x(y) - \mu(y, \tilde{\psi})] \frac{\partial \mu^*(y, \tilde{\psi})}{\partial \tilde{\psi}} \, dy \right\} = 0 \tag{13-91}$$

or

$$\text{Re}\left\{\int_{-L/2}^{+L/2} \mu(y)\frac{\partial \mu^*(y,\bar{\psi})}{\partial \bar{\psi}}\, dy\right\} = -\text{Re}\left\{\int_{-L/2}^{+L/2} \eta(y)\frac{\partial \mu^*(y,\bar{\psi})}{\partial \bar{\psi}}\, dy\right\} \quad (13\text{-}92)$$

where L defines the portion of the y-axis where the signal is available. The limits of integration in (13-92) may be extended to infinity by introducing the rectangular aperture function

$$g(y) = \text{rect}\left(\frac{y}{L}\right)$$

With this change and performing the indicated differentiation, we obtain

$$\text{Re}\left\{-j\int y|\mu(y)|^2\, g(y)\exp\left(-jky\sin\bar{\psi}\right)dy\right\}$$

$$= \text{Re}\left\{j\int y\eta(y)\mu^*(y)g(y)\exp\left(-jky\sin\bar{\psi}\right)dy\right\} \quad (13\text{-}93)$$

where $k = 2\pi f/c = 2\pi/\lambda$. Equation (13-93) may be further simplified by assuming $L/\lambda \gg 1$ and $\bar{\psi}$ small compared with λ/L. We also assume that the magnitude of the signal envelope is independent of y. Then

$$-|\mu|^2 k\bar{\psi}\int y^2 g(y)\, dy = \text{Re}\left\{j\mu^*\int y\eta(y)g(y)\exp\left(-jky\bar{\psi}\right)dy\right\} \quad (13\text{-}94)$$

and the bearing estimate is

$$\bar{\psi} = -\frac{\text{Re}\left\{j\int y\eta(y)g(y)\exp\left(-jky\bar{\psi}\right)dy\right\}}{\mu k \int y^2 g(y)\, dy} \quad (13\text{-}95)$$

For isotropic noise, the spatial single-frequency autocorrelation function for the real or imaginary part of the envelope is

$$E[\eta_r(y)\eta_r(y')] = E[\eta_i(y)\eta_i(y')] = P_n \text{ sinc }[k(y-y')] \quad (13\text{-}96)$$

With the help of (13-96) the variance of the bearing estimate becomes

$$\text{Var}[\bar{\psi}] = \frac{P_n(\lambda/2)\int y^2 g^2(y)\, dy}{|\mu|^2 k^2 \left(\int y^2 g(y)\, dy\right)^2} \quad (13\text{-}97)$$

Recognizing that for this example $g(y) = g^2(y)$, this can be put in the form

$$\text{Var}[\bar{\psi}] = \frac{P_n \lambda^3}{2|\mu|^2 (2\pi)^2 \int y^2 g^2(y)\, dy} \quad (13\text{-}98)$$

Now define the rms length L_0 of the aperture as

$$L_0^2 = \frac{(2\pi)^2 \int y^2 g^2(y)\, dy}{\int g^2(y)\, dy} \tag{13-99}$$

For the rectangular aperture function $L_0 = \pi L/\sqrt{3}$. The ratio $|\mu|^2/P_n$ is the signal-to-noise ratio which, assuming matched filter temporal processing, is $2E/N_0$. With this substitution the variance becomes

$$\text{Var}[\hat\psi] = \left[\left(\frac{2E}{N_0}\right)\left(\frac{2L}{\lambda}\right)\left(\frac{L_0^2}{\lambda^2}\right) \right]^{-1} \tag{13-100}$$

In (13-100) the factor $2L/\lambda$ is the directivity factor for a uniformly weighted line aperture in an isotropic noise field. The variance of $\hat\psi$ can therefore be expressed in terms of the signal-to-noise ratio at the output of a line array and the effective aperture length. Thus

$$\text{Var}[\hat\psi] = \left[\text{SNR}_o \left(\frac{L_0}{\lambda}\right)^2 \right]^{-1} \tag{13-101}$$

where

$$\text{SNR}_o = \left(\frac{2E}{N_0}\right)\left(\frac{2L}{\lambda}\right)$$

In principle, target bearing can be estimated with a variance given by (13-101) using two separated point sensors provided that their effective separation, as defined by (13-99), is equal to L_0, and provided that the target signal level is increased to give a signal-to-noise ratio equal to SNR_o.

In practice, a noise-free estimate of the received target envelope to use as a reference in (13-91) is seldom available. The split-aperture correlator discussed in Section 11.7 uses the received waveform itself as a reference by dividing the aperture in two and cross-correlating the waveforms received in the two halves. Because the "reference" waveform unavoidably contains noise, the resulting performance does not equal that suggested by the Cramer–Rao bound in (13-100).

To derive the bearing-estimate variance for the split-aperture correlator, consider the schematic mechanization shown in Figure 13-12. Let the signal and noise be narrowband and complex with n_1 and n_2 independent. The true target bearing is assumed to be small relative to the beamwidth of each half-array so that we may ignore the pattern function and write

$$s_1 = a \exp\left(-j\frac{\pi L}{2\lambda}\psi\right)$$

and

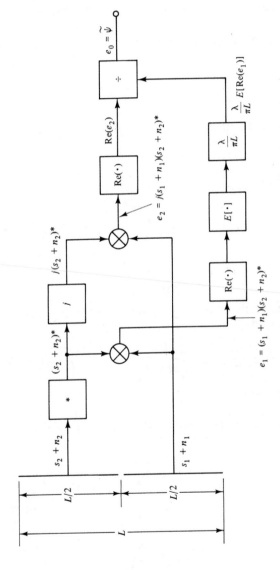

Figure 13-12 Schematic representation of a split-aperture correlator.

$$s_2 = a \exp\left(+j\frac{\pi L}{2\lambda}\psi\right)$$

where the time variable is again suppressed. Assume the signal is available for a time T and the noise bandwidth is $\beta = 1/T$. The real-signal energy is then $E = a^2 T/2$ and the average real-noise power at the input to each half-array is $N_0 \beta = N_0/T$.

At the multiplier outputs we are interested in the real parts of the complex products. Thus

$$\text{Re}[e_1] = \text{Re}[s_1 s_2^* + s_1 n_2^* + s_2^* n_1 + n_1 n_2^*]$$

and

$$\text{Re}[e_2] = \text{Re}[js_1 s_2^* + js_1 n_2^* + js_2^* n_1 + jn_1 n_2^*]$$

But

$$\text{Re}[s_1 s_2^*] = a^2 \cos\left(\frac{\pi L}{\lambda}\psi\right) \simeq a^2$$

$$\text{Re}[js_1 s_2^*] = a^2 \sin\left(\frac{\pi L}{\lambda}\psi\right) \simeq a^2\left(\frac{\pi L}{\lambda}\psi\right)$$

Also

$$E\{\text{Re}[e_1]\} \simeq a^2$$

For large signal-to-noise ratio, we may ignore the noise–noise cross-product term and write

$$e_0 = \frac{\lambda\,\text{Re}[e_2]}{\pi L a^2} = \psi + \frac{\lambda}{\pi L a^2}\{\text{Re}[js_1 n_2^* + js_2^* n_1]\} = \check{\psi} \quad (13\text{-}102)$$

The variance of the estimate is

$$\text{Var}[\check{\psi}] = \left(\frac{\lambda}{\pi L a^2}\right)^2 E\left\{[\text{Re}(js_1 n_2^* + js_2^* n_1)]^2\right\} \quad (13\text{-}103)$$

The expectation in (13-103) is easily shown to be

$$4E[s_{1_r}^2 n_{2_r}^2] = 4E[s_{1_r}^2]E[n_{2_r}^2] = 2a^2 E[n_{2_r}^2]$$

The noise out of each half-array is related to the input noise by the directivity factor. Hence

$$E[n_{2_r}^2] = \frac{N_0}{T}\left(\frac{\lambda}{L}\right)$$

Letting $a^2 = 2E/T$, the variance finally becomes

$$\text{Var}[\check{\psi}] = \left[\left(\frac{2E}{N_0}\right)\left(\frac{\pi^2 L^3}{2\lambda^3}\right)\right]^{-1}$$

But for the total aperture the effective length is $L_0 = \pi L / \sqrt{3}$, so that the variance can be put in the form

$$\text{Var}[\hat{\psi}] = \left[\left(\frac{2E}{N_0} \right) \left(\frac{2L}{\lambda} \right) \left(\frac{L_0^2}{\lambda^2} \right) \frac{3}{4} \right]^{-1} \tag{13-104}$$

Comparing this expression with (13-100), we see that the penalty paid for the lack of a noise-free reference is an increase by a factor of $\frac{4}{3}$ in the variance compared with the minimum possible variance.

PROBLEMS

13.1. Let $x(t)$ be a zero-mean Gaussian noise waveform with variance σ^2. For positive values of x a Neyman–Pearson threshold γ is established for the purpose of deciding on the presence or absence of a signal. Using appropriate tables of integrals for the Gaussian function, determine the required ratio γ/σ for false-alarm probabilities of 10^{-2}, 10^{-3}, and 10^{-4}.

13.2. Repeat Problem 13.1 assuming that x is a two-degree-of-freedom chi-squared variable with a density function

$$p(x) = \frac{1}{\bar{x}} \exp\left(\frac{-x}{\bar{x}} \right) \qquad \text{for } x \geq 0$$

Recall that with this density function $\bar{x} = \sigma_x$.

13.3. Assume $p_0(x)$ and $p_1(x)$ given by

$$p_0(x) = \frac{1}{\sqrt{2\pi}\,\sigma} \exp\left(\frac{-x^2}{2\sigma^2} \right)$$

$$p_1(x) = \frac{1}{\sqrt{2\pi}\,\sigma} \exp\left[\frac{-(x-\mu)^2}{2\sigma^2} \right]$$

Calculate the values of the likelihood ratio λ_0 corresponding to the threshold settings determined in Problem 13.1 assuming in each case that $\mu = \gamma$.

13.4. With a passive broadband receiver mechanization, as shown in Figure 13-8, let $h_2(t)$ be

$$h_2(t) = \frac{1}{T} \exp\left(\frac{-t}{T} \right) \qquad \text{for } t \geq 0$$

With T large compared with the reciprocal of the effective noise bandwidth, derive an expression for the variance of z_0. Compare the result with the variance obtained if $h_2(t)$ is a rectangular function of width T.

13.5. Assume a passive broadband system with $T\beta = 1000$. Calculate the required signal-to-noise ratio in dB at the input to the square-law envelope detector to achieve a 50% P_D with values of P_{fa} equal to 10^{-6}, 10^{-5}, 10^{-4}, and 10^{-3}.

13.6. The bearing estimate variance for a narrowband split-aperture correlator is given in (13-104), assuming a large signal-to-noise ratio at the output of each half array. Demonstrate that the variance with arbitrary signal-to-noise ratio is obtained by multiplication of (13-104) by the factor $[1 + (N_0/2E)(\lambda/L)]$. Write the expression for the variance assuming a signal-to-noise ratio much less than 1.

SUGGESTED READING

1. Whalen, A. D., *Detection of Signals in Noise*. New York: Academic Press, Inc., 1971.

2. Helstrom, C. W., *Statistical Theory of Signal Detection*. Elmsford, N.Y.: Pergamon Press, Inc., 1960.

3. Eckart, C., "Optimal Rectifier Systems for the Detection of Steady Signals," Marine Physical Laboratory of Scripps Institute of Oceanography, La Jolla, Calif., Ref. 52-11, 1952.

4. Marcum, J. I., "A Statistical Theory of Target Detection by Pulsed Radar, Mathematical Appendix," Rand Corporation Report RM-339, July 1948.

5. Fehlner, L. F., "Marcum and Swerling's Data on Target Detection by Pulsed Radar," The Johns Hopkins University, Applied Physics Laboratory, Silver Spring, Md., TG 451, 1962.

6. Robertson, G. H., "Operating Characteristics for a Linear Detector of CW Signals in Narrow-Band Gaussian Noise," *Bell Syst. Tech. J.*, Vol. 46, No. 4 (1967).

7. Cramer, H., *Mathematical Method of Statistics*. Princeton, N.J.: Princeton University Press, 1958.

8. Van Trees, H. L., *Detection, Estimation and Modulation Theory*, Part I. New York: John Wiley & Sons, Inc., 1968.

14 System Performance Analysis: Examples

14.0 INTRODUCTION

Performance analysis of underwater acoustic systems involves consideration of the characteristics of targets, the ocean medium, and the fundamentals of statistical detection and estimation theory. Previous chapters have addressed each of these areas in some detail. We conclude this work with selected examples of system performance analysis intended to demonstrate some of the principles developed earlier. Of course, space does not permit the inclusion of all elements of system design and analysis considered in the previous 13 chapters. It is hoped that the reader will be able to apply the basic principles presented to practical system analysis without benefit of specific examples in every case.

14.1 PASSIVE NARROWBAND DETECTION

As discussed in Section 13.1.4, the passive narrowband detection system is essentially a spectrum analyzer. A typical system block diagram, including a beamformer-spatial filter at the input is shown in Figure 14-1. The target signal is assumed to be a pure sinusoid at a frequency equal to the center frequency of one of the predetection filters. The noise bandwidth of each filter is β and a total of N contiguous filters are used to cover the desired overall bandwidth $N\beta$. For this example, the postdetection processor provides a linear integration for a time T such that $\beta T \geqslant 1.0$.

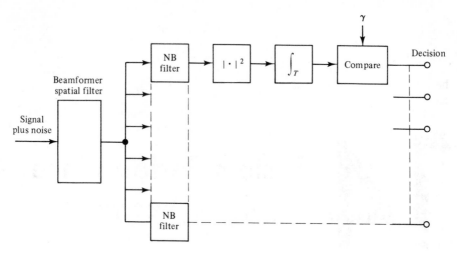

Figure 14-1   Narrowband receiving system.

The output signal-to-noise ratio, SNR(z), for any channel is given by (13-67). We now include the effect of the spatial filter (assuming isotropic noise) and write

$$\text{SNR}(z) = \left(\frac{T}{\beta}\right)\left[\left(\frac{D(f_1)}{\Psi_n(f_1)}\right)P_s\right]^2 \tag{14-1}$$

where $D(f_1)$ is the directivity factor of the spatial filter at the target signal frequency f_1, and $\Psi_n(f_1)$ is the ambient noise power spectral density per hertz. The target power at the system input is P_s.

It is convenient to include the effect of transmission loss from the target to the receiver and write (14-1) in decibel notation. Thus

$$\frac{1}{2}\left[\text{SNR}(z)_{dB} - 10\log\left(\frac{T}{\beta}\right)\right] = (\text{SL} - \text{TL}) - (\text{NL}_s - \text{DI}) \tag{14-2}$$

where   SL = target source level at f_1
       TL = transmission loss at f_1 for target at range r
       NL$_s$ = ambient noise level in 1-Hz band at f_1
       DI = directivity index at f_1

To define system detection performance we now select an acceptable false alarm probability for the multichannel system shown in Figure 14-1. Each channel provides an independent output every T seconds so that the total number of false alarm *opportunities* per second is N/T, where N is the total number of channels. The number of false alarms per second is therefore $(N/T)P_{fa}$, and the reciprocal of this number is called the average false alarm time $\bar{t}_{fa}$. Thus

$$\bar{t}_{fa} = \frac{T}{NP_{fa}} \tag{14-3}$$

or

$$P_{\text{fa}} = \frac{T}{N \bar{t}_{\text{fa}}} \tag{14-4}$$

The false-alarm probability may therefore be determined by first selecting an acceptable false-alarm time.

From the false-alarm probability and a desired detection probability, the *required value d* of SNR (z) is selected from Figure 13-10 (because $\beta T \gg 1.0$, we assume Gaussian output statistics). Substitute $10 \log d = $ SNR $(z)_{\text{dB}}$ in (14-2) and solve for the transmission loss corresponding to the selected values of P_{fa} and P_D.

$$\text{TL} = \text{SL} - \text{NL}_s + \text{DI} - 5 \log d + 5 \log \left(\frac{T}{\beta} \right) \tag{14-5}$$

The last two terms on the right in (14-5) correspond to the required ratio of *total signal power to noise power per hertz* at the input to the envelope detector. This ratio expressed in decibels is the *narrowband detection threshold*, DT_{Hz}. Thus

$$\text{DT}_{\text{Hz}} = 5 \log d - 5 \log \left(\frac{T}{\beta} \right) \tag{14-6}$$

As a numerical example, assume the following signal and system parameters:

Target signal   pure sinusoid at $f_1 = 500$ Hz with source level SL $= 125$ dB//μPa @ 1 m.
Ambient noise level   NL$_s = +65$ dB//μPa in 1-Hz band at 500 Hz. Corresponds approximately to sea state 3.
Spatial filter   horizontal line aperture with $L = 100$ ft.
Predetection processor   bank of contiguous narrowband filters with $\beta = 0.5$ Hz covering a 500-Hz band centered at 500 Hz.
Postdetection processor   linear integrators with $T = 100$ sec.
Desired system $\bar{t}_{\text{fa}}$ 1000 sec.

We shall now calculate the probability of detection as a function of range first assuming a simple spherical spreading transmission loss, and then a transmission loss characteristic as shown in Figure 5-15.

We require first the false-alarm probability corresponding to the desired $\bar{t}_{\text{fa}}$. Using (14-4), we obtain

$$P_{\text{fa}} = \frac{T}{N \bar{t}_{\text{fa}}} = 10^{-4}$$

With this value of P_{fa}, Figure 13-10 is used to tabulate the detection index and detection threshold for several values of P_D. For example:

P_D (%)	d	$5 \log d$	DT_{Hz} (dB)
10	6.2	4.0	−7.5
50	14	5.7	−5.8
90	36	7.8	−3.7

The directivity index for the horizontal line aperture is obtained from the approximate relationship given by (11-23). Hence, at 500 Hz ($\lambda = 10$ ft)

$$DI \simeq 10 \log \left(\frac{2L}{\lambda} \right) = 10 \log (20) = 13 \text{ dB}$$

The allowed transmission loss is now obtained from

$$TL = SL - NL_s + DI - DT_{Hz}$$
$$= +73 - DT_{Hz} \tag{14-7}$$

For the three selected values of P_D, the transmission loss along with the corresponding range with spherical spreading is

P_D(%)	TL (dB)	$r = $ antilog (TL/20) (m)
10	80.5	10,600
50	78.8	8,710
90	76.7	6,840

With the transmission-loss characteristic in Figure 5-15, the detection performance for this example is limited approximately to the range defining the transition from the direct-path mode to the bottom-bounce mode of propagation. The detection probability versus range for the two transmission loss models is plotted in Figure 14-2.

The transmission loss given by (14-7), when calculated for a P_D of 50%, is sometimes called the system *figure of merit* (FOM). Note, however, that the FOM includes target and environmental characteristics, and is therefore not a unique constant characterizing system capability in general.

It is instructive to determine the sensitivity of the calculated detection range to the choice of false alarm probability. For a detection probability of 50% let $P_{fa} = 10^{-5}$ and 10^{-6}, corresponding to $t_{fa} = 10^4$ and 10^5 sec. From Figure 13-10, the detection index takes on the values 18 and 23, from which the detection ranges, assuming spherical spreading, are

P_{fa}	r (m)
10^{-5}	8180
10^{-6}	7690

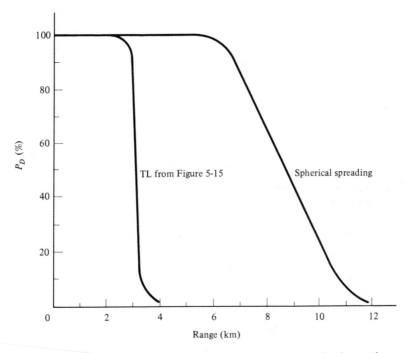

Figure 14-2 Probability of detection versus range for narrowband example.

In this example, reducing the false-alarm probability by two orders of magnitude results in a calculated detection range reduction of only about 12%. In general, it will be found that the calculated detection range is not very sensitive to the choice of P_{fa}.

14.2 PASSIVE BROADBAND DETECTION

One channel, representing one beam of what may be a multibeam system, of a passive broadband receiving system is shown in block diagram form in Figure 14-3. The spectral densities at the spatial filter input are

Received signal: $\Psi_{s_r}(f)$
Ambient noise: $\Psi_n(f)$

The spatial filter is assumed to have unity gain for a target on the MRA. Therefore, the signal and noise spectral densities at the input to the square-law envelope detector are

Signal: $\Psi_{s_r}(f)|H(f)|^2$
Noise: $\Psi_n(f)|H(f)|^2/A(f)$

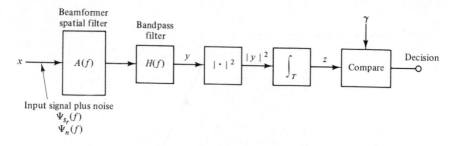

Figure 14-3   One channel of a passive broadband receiving system.

where $A(f)$ is the spatial filter array gain.

The value of SNR(y) is given by

$$\text{SNR}(y) = \frac{\int \Psi_{s_r}(f)|H(f)|^2\, df}{\int [\Psi_n(f)|H(f)|^2/A(f)]\, df} \tag{14-8}$$

The output signal-to-noise ratio SNR(z) is, from (13-61),

$$\text{SNR}(z) = T\beta_n \text{SNR}^2(y)$$

where β_n is the effective noise bandwidth at the input to the envelope detector. The value of β_n may be calculated from the noise spectral density at y. Thus

$$\beta_n = \frac{\left[\displaystyle\int_0^\infty \Psi_n'\, df\right]^2}{\displaystyle\int_0^\infty \Psi_n'^2\, df} \tag{14-9}$$

where the argument of Ψ_n' has been dropped for convenience and

$$\Psi_n' = \frac{\Psi_n|H|^2}{A}$$

Substitution of (14-8) and (14-9) in the equation for SNR(z) gives

$$\text{SNR}(z) = T\frac{\left[\displaystyle\int_0^\infty \Psi_n'\, df\right]^2}{\displaystyle\int_0^\infty \Psi_n'^2\, df} \cdot \frac{\left[\displaystyle\int_0^\infty \Psi_{s_r}|H|^2\, df\right]^2}{\left[\displaystyle\int_0^\infty \Psi_n'\, df\right]^2} \tag{14-10}$$

Assuming $T\beta_n$ is large compared to unity, the ROC curves of Figure 13-10 may be used to determine the required detection index for selected values at P_D and P_{fa}. We then write

$$\text{req'd SNR}(z) = d = T\beta_n \text{SNR}^2(y)$$

or

$$\text{req'd SNR}(y) = \left(\frac{d}{T\beta_n}\right)^{1/2} \tag{14-11}$$

Converting to decibel notation, we wish to determine the range r that satisfies the equality

$$SL_T(r) - NL_T = DT \tag{14-12}$$

where

$$SL_T(r) = \text{total received signal level at range } r$$

$$= 10 \log\left[\int \Psi_{s_r}|H|^2 \, df\right] \tag{14-13}$$

$$NL_T = \text{total noise level at input to envelope detector}$$

$$= 10 \log\left[\int \Psi_n' \, df\right] = 10 \log\left[\int (\Psi_n|H|^2/A) \, df\right] \tag{14-14}$$

$$DT = \text{broadband detection threshold}$$

$$= 5 \log d - 5 \log T\beta_n \tag{14-15}$$

Notice that the broadband detection threshold differs from the narrowband definition. This is the result of the common practice of writing the narrowband equation in terms of the noise spectral level in a 1-Hz band rather then using the total noise level in the filter band.

As discussed in Section 11.6, the shape of the received target-signal spectrum may, in general, be a function of range. That is,

$$\Psi_{s_r}(f) = \Psi_s(f)T(r,f)$$

where $\Psi_s(f)$ is the target spectral density at a range of 1 m and $T(r,f)$ is the transmission loss at range r as a function of frequency. If the transmission loss frequency dependence cannot be neglected, an iterative process must be used to find the detection range r that results in equality for (14-12). If the frequency dependence can be neglected over the frequency band of interest, the transmission loss term may be removed from the integral in (14-13) to obtain

$$SL_T(r) = 10 \log\left[T(r)\int \Psi_s|H|^2 \, df\right] = SL_T - TL(r) \tag{14-16}$$

Substitution of (14-16) in (14-12) permits an explicit solution for the transmission loss corresponding to the selected values of P_D and P_{fa}. Thus

$$TL(r) = SL_T - NL_T - DT \tag{14-17}$$

The detection range, or equivalently the allowed transmission loss, is increased by increasing the integration time T (provided that T does not exceed

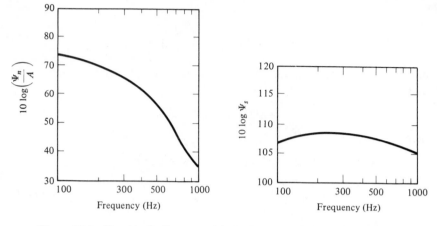

Figure 14-4 Signal and noise spectral levels for broadband detection example.

the duration of the target signal), or by increasing the array gain. The bandpass filter transfer function $H(f)$ affects all three terms on the right side of (14-17). Performance is optimized by selecting $H(f)$ in accordance with (13-65). Including the effect of the array gain, the optimum filter shape is given by

$$|H(f)|^2 = \frac{\Psi_{s_r}(f)A^2(f)}{\Psi_n^2(f)} \qquad (14\text{-}18)$$

Note that, in general, the optimum filter shape is a function of range because of the effect of range on the shape of the target spectrum. However, in most applications the approximate detection range may be anticipated and the filter designed to be optimum at that range. The following example will demonstrate that great precision is not warranted in selecting the filter shape.

Assume an ambient noise spectral level at the output of the spatial filter and a target spectral level as given by Figure 14-4. For $P_D = 50\%$, $P_{fa} = 10^{-4}$, and $T = 100$ sec, we wish to determine the following:

1. Detection range assuming a rectangular bandpass filter with the upper corner at 1000 Hz and the lower corner at 900, 800, 700, 600, or 500 Hz

2. The shape of the optimum filter

3. The detection range using the filter from part 2

Spherical spreading, independent of frequency, will be assumed in each case.

The integrations required for the solution of equation (14-17) may be accomplished using simple numerical methods as described in Section 2.5.3. Divide the spectral densities in Figure 14-4 into 100-Hz bands and tabulate the intensity at the center of each band. As an example, the signal power in some band f_1 to f_2 is

$$\int_{f_1}^{f_2} \Psi_s |H|^2 \, df \simeq (100 \text{ Hz}) \sum_i \Psi_{s_i} |H_i|^2 \qquad \text{for all } i \text{ in } [f_1, f_2]$$

or

$$SL_T = 10 \log (100) + 10 \log \left(\sum_i \Psi_{s_i} |H_i|^2 \right)$$

With the rectangular filter, H_i may be taken as unity and dropped from these expressions. The following tabulation for the rectangular filters defined in part 1 of this example serves to illustrate the method:

f (Hz)	SL_s (dB)	Ψ_{s_i}	$(100) \sum_i \Psi_{s_i}$	SL_T (dB)	Band (Hz)
950	105.2	3.31×10^{10}	3.31×10^{12}	125.2	1000–900
850	105.8	$3.8 \ \times 10^{10}$	7.11×10^{12}	128.5	1000–800
750	106.2	4.17×10^{10}	11.28×10^{12}	130.5	1000–700
650	106.8	4.78×10^{10}	16.06×10^{12}	132	1000–600
550	107.3	5.37×10^{10}	21.43×10^{12}	133.3	1000–500

The intensity level is first tabulated at the center of each rectangular element. The third column converts the spectral level in dB to a numeric ratio. The fourth column provides the cumulative sum to give the signal power in the required bands. Finally, the fifth column converts the power sum to a band level in dB.

To complete the calculation for part 1 we require the noise band levels and the effective bandwidths for each of the rectangular filters. These are obtained from the following approximate expressions:

$$NL_T \simeq 10 \log \left[100 \sum_i \frac{\Psi_{n_i} |H_i|^2}{A_i} \right]$$

$$\beta_n \simeq 100 \left[\sum_i \frac{\Psi_{n_i} |H_i|^2}{A_i} \right]^2 \left[\sum_i \left(\frac{\Psi_{n_i} |H_i|^2}{A_i} \right)^2 \right]^{-1}$$

with the following results:

Band	NL_T (dB)	β_n (Hz)
1000–900	+ 55	100
1000–800	59.8	180
1000–700	64	223
1000–600	69.5	183
1000–500	76.1	158

Notice that the effective noise bandwidth is not equal to the rectangular filter bandwidth except for the first 100-Hz band centered at 950 Hz. The rapid falloff of the noise spectrum between 500 and 1000 Hz ensures that the noise bandwidth is, in general, less than the filter bandwidth.

With $d = 14$, for the required P_D and P_{fa}, and $T = 100$ sec, we may now calculate detection threshold, FOM and detection range for each assumed filter. The results are as follows:

Band	$SL_T - NL_T$	DT	FOM	r (m)
1000–900	70.2	−14.3	84.5	16,788
1000–800	68.7	−15.5	84.2	16,218
1000–700	66.5	−16	82.5	13,335
1000–600	62.5	−15.6	78.1	8,035
1000–500	57.2	−15.3	72.5	4,217

In this example the best detection range is obtained with a 100-Hz filter at the upper end of the band. Because of the spectral shapes assumed for signal and noise, this is the region that provides the best signal-to-noise ratio. Notice that the best detection threshold is obtained with the 1000 to 700-Hz band, but the increased noise resulting from including the lower-frequency region results in a reduced detection range.

For part 2 of the example, we use (14-18) to define the optimum filter. Normalizing the peak value to unity, we obtain the following:

f	$|H/H_{max}|^2$
950	$1.0 = 0$ dB
850	$0.29 = -5.3$ dB
750	$0.05 = -12.7$ dB
650	$0.0036 = -24.4$ dB
550	$0.0001 = -40$ dB

The optimum filter response peaks sharply at the upper end of the band, supporting the result obtained in part 1.

For part 3, it is convenient to substitute the optimum filter expression in the integrals involved in (14-17) and simplify to obtain

$$\text{FOM} = \text{TL}(r) = 5 \log \left[\int \left(\frac{\Psi_s A}{\Psi_n} \right)^2 df \right] - 5 \log d + 5 \log T \qquad (14\text{-}19)$$

Notice that the assumption of an optimum filter reduces to one the number of integrals that must be evaluated.

Numerical evaluation of (14-19) gives

$$FOM = 85.25 \text{ dB}$$
$$r = 18,300 \text{ m}$$

Comparison with the results in part 1 verifies that the performance with the optimum filter is indeed better than those obtained with the rectangular filters. Note, however, that the optimum FOM is only 0.75 dB better than that for the rectangular filter in the band 1000 to 900 Hz. This indicates that careful consideration of the signal and noise characteristics may lead to a simple filter configuration that will perform in a near-optimum manner. Determination of the optimum filter shape is a simple procedure and the results can be used as a guide to the selection of a simple filter form as an approximation to the optimum. An extremely close match between the ideal and actual filter shapes is seldom justified by the performance improvement gained relative to the added cost and complexity.

14.3 PASSIVE BROADBAND ANGLE TRACKING

In Chapters 11 and 13 the split-aperture correlator was discussed as a possible mechanization for narrowband bearing-angle estimation. As shown in Figure 13-12, a 90° phase shift at the center frequency of the narrowband signal is inserted in one signal channel so that the correlator output is responsive to the electrical phase difference between the two half-array signals. The electrical phase difference in the narrowband case is proportional to the target bearing relative to the array MRA.

A similar result is obtained with a broadband target signal by introducing a 90° phase shift at all frequencies over the frequency band of interest. This is equivalent to a Hilbert transformation of the signal in question. That is, in Section 6.10 it was shown that given a signal $x(t)$ and its Fourier transform $X(f)$, the Hilbert transform of $x(t)$ is defined by

$$\hat{x}(t) \longleftrightarrow -j \, \text{sgn}\,(f)X(f)$$

The broadband split-aperture correlator using a Hilbert transformation is shown schematically in Figure 14-5. It is left as an exercise in the problems section to demonstrate that a practical implementation of this system exists.

Signal and noise are assumed to be independent ZMG processes with n_1 independent of n_2, and s_1 and s_2 differing only by a time delay caused by target bearing relative to the array MRA. The target bearing angle is assumed small relative to the beamwidth of the array, so that small-angle approximations are appropriate for all trigonometric functions of the bearing angle.

The integrator outputs in Figure 14-5 are

$$z_1 = \int_T \hat{s}_1 s_2 \, dt + \int_T (\hat{n}_1 s_2 + n_2 \hat{s}_1 + \hat{n}_1 n_2) \, dt$$

$$z_2 = \int_T s_1 s_2 \, dt + \int_T (n_1 s_2 + n_2 s_1 + n_1 n_2) \, dt$$

(14-20)

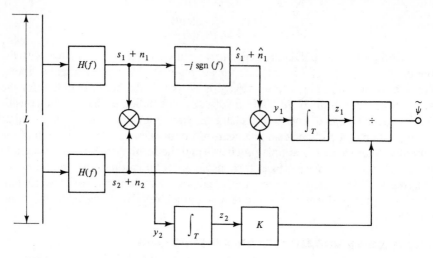

Figure 14-5 Broadband split-aperture correlator with Hilbert transformation for bearing estimation.

Assume that the signal-to-noise ratios at the outputs of the filters, $H(f)$, are small, but that T is sufficiently large so that we may write

$$z_1 \simeq \int_T \hat{s}_1 s_2 \, dt + \int_T \hat{n}_1 n_2 \, dt$$

$$z_2 \simeq \int_T s_1 s_2 \, dt$$

For small bearing angles, s_1 is nearly identical to s_2, so that $z_2 \simeq TE[s^2]$.

The output signal z_1/Kz_2 is the bearing estimate, and its expected value is

$$E[\tilde{\psi}] = \frac{E[\hat{s}_1 s_2]}{KE[s^2]} \tag{14-21}$$

Assuming that the noise effective bandwidth is large compared with $1/T$, the variance of the estimate is

$$\text{Var}[\tilde{\psi}] = \frac{1}{K^2 TE^2[s^2]} \int_{-\infty}^{+\infty} R_n^2(\tau) \, d\tau \tag{14-22}$$

which can be put in the form

$$\text{Var}[\tilde{\psi}] = \left[2K^2 T\beta_n \frac{R_s^2(0)}{R_n^2(0)} \right]^{-1} \tag{14-23}$$

where β_n is the effective noise envelope bandwidth.

Let the target signal be a plane wave originating in the azimuth plane at a bearing angle ψ relative to the array MRA. The signals s_1 and s_2 and their Fourier transforms are (with the small-angle approximations)

$$s_1(t, \psi) \longleftrightarrow G(\psi, f) H(f) S(f) \exp\left(j\frac{\pi f L \psi}{2c}\right)$$

$$\simeq G(\psi, f) H(f) S(f)\left(1 + j\frac{\pi f L \psi}{2c}\right)$$

$$s_2(t, \psi) \longleftrightarrow G(\psi, f) H(f) S(f) \exp\left(-j\frac{\pi f L \psi}{2c}\right) \qquad (14\text{-}24)$$

$$\simeq G(\psi, f) H(f) S(f)\left(1 - j\frac{\pi f L \psi}{2c}\right)$$

where $G(\psi, f)$ = half-array pattern function when centered at the origin
$S(f)$ = Fourier transform of plane wave signal at array input with $\psi = 0$

The Fourier transform of $\hat{s}_1(t, \psi)$ is

$$\hat{s}_1(t, \psi) \longleftrightarrow -j \,\mathrm{sgn}(f) G(\psi, f) H(f) S(f) \exp\left(j\frac{\pi f L \psi}{2c}\right) \qquad (14\text{-}25)$$

from which, with the help of Parseval's theorem,

$$E\left[\int_T \hat{s}_1(t) s_2(t)\, dt\right] = TE[\hat{s}_1 s_2]$$

$$= -jT \int_{-\infty}^{+\infty} \mathrm{sgn}(f) |G|^2 |H|^2 \Psi_s \left(1 + j\frac{\pi f L \psi}{c}\right) df \qquad (14\text{-}26)$$

where $\Psi_s(f) = \overline{|S(f)|^2}$. The imaginary part of (14-26) is zero, so that we may write

$$E[\hat{s}_1 s_2] = \frac{2\pi L \psi}{c} \int_0^\infty f |G|^2 |H|^2 \Psi_s\, df \qquad (14\text{-}27)$$

The integral in (14-27) is the centroid of the one-sided spectral densities of s_1 and s_2. When properly normalized this can be interpreted as the "mean frequency" of the broadband spectrum. Thus, define

$$\bar{f} = \frac{\displaystyle\int_0^\infty f |G|^2 |H|^2 \Psi_s\, df}{\displaystyle\int_0^\infty |G|^2 |H|^2 \Psi_s\, df} \qquad (14\text{-}28)$$

The denominator in (14-28) is $\frac{1}{2} R_s(0) = \frac{1}{2} E[s^2]$. Substitution of (14-28) in (14-27) gives

$$E[\hat{s}_1 s_2] = \frac{\pi L \bar{f} \psi}{c} R_s(0) = \frac{\pi L \psi}{\bar{\lambda}} E[s^2] \qquad (14\text{-}29)$$

where $\bar{\lambda} = c/\bar{f}$ is the mean wavelength.

Now let $K = \pi L/\bar{\lambda}$ in (14-21) and the expected value of the estimate equals the true bearing. Substitution for K in (14-23) gives for the variance

$$\mathrm{Var}[\hat{\psi}] = \left[2\left(\frac{\pi L}{\bar{\lambda}}\right)^2 T\beta_n\, \mathrm{SNR}^2 \right]^{-1} \tag{14-30}$$

where SNR is the signal-to-noise ratio of each half-array signal measured at the output of the predetection filter.

Note that selection of the constant K implies that the signal mean frequency is known. In general, an error in the estimate of $\bar{f}$ will result in an error in the expected value of the bearing estimate unless the true bearing is zero. A closed-loop null-steering system, as shown in Figure 11-23, may be used to avoid this problem.

We shall now determine the form of the predetection filter to minimize the variance of the broadband bearing estimator. Starting with (14-22) it is easy to show that the variance is minimized if we maximize the function

$$\frac{\left[\int_0^\infty f\Psi_s|H|^2\, df\right]^2}{\int_{-\infty}^{+\infty} (\Psi_n')^2|H|^4\, df} \tag{14-31}$$

where Ψ_n' is the noise spectral density at the outputs of the half-arrays. By multiplying and dividing the integrand in the numerator by Ψ_n' and using the Schwarz inequality, we obtain

$$\frac{\left[\int_0^\infty f\Psi_s|H|^2\, df\right]^2}{\int_{-\infty}^{+\infty} (\Psi_n')^2|H|^4\, df} \le \frac{1}{2}\int_0^\infty f^2\left(\frac{\Psi_s}{\Psi_n'}\right)^2 df \tag{14-32}$$

The inequality in (14-32) becomes an equality if

$$|H(f)|^2 = \frac{f\Psi_s(f)}{\Psi_n'^2(f)} \tag{14-33}$$

The optimum filter for angle estimation differs only by the factor f from the optimum detection filter described by (13-64).

As a numerical example let the signal and noise spectral densities at the output of each half-array have the shape given in Figure 14-4. The transmission loss is assumed independent of frequency and an optimum filter as defined by (14-33) is used in each channel of the split-aperture system. The bearing estimate variance in this case can be put in the form

$$\mathrm{Var}[\hat{\psi}] = \frac{1}{2T}\left(\frac{c}{\pi L}\right)^2\left[T^2(r)\int_0^\infty f^2\left(\frac{\Psi_s}{\Psi_n'}\right)^2 df\right]^{-1} \tag{14-34}$$

where $T(r)$ is the transmission factor at range r and Ψ_s is the signal spectral density at $r = 1$. We are interested in the range at which the standard deviation of the bearing estimate, $\sigma_{\tilde{\psi}}$, is reduced to an acceptable level. Converting (14-34) to decibel form and solving for transmission loss, we obtain

$$\text{TL} = 10 \log \sigma_{\tilde{\psi}} + 5 \log (2T) - 10 \log \left(\frac{c}{\pi L} \right) + 5 \log \left[\int f^2 \left(\frac{\Psi_s}{\Psi_n'} \right)^2 df \right]$$

(14-35)

Let the total array length be 100 ft, the integration time $T = 10$ sec, and the sound speed $c = 5000$ ft/sec. If the acceptable bearing-estimate standard deviation is 0.01 rads, the transmission loss must then not exceed

$$\text{TL} = -25.5 + 5 \log \left[\int f^2 \left(\frac{\Psi_s}{\Psi_n'} \right)^2 df \right]$$

(14-36)

From the spectral densities in Figure 14-4, we tabulate

f	$f^2 \left(\dfrac{\Psi_s}{\Psi_n'} \right)^2 (\Delta f)$
950	$98.8 \ \times 10^{20}$
850	$26.3 \ \times 10^{20}$
750	$4.1 \ \times 10^{20}$
650	$0.24 \ \times 10^{20}$
550	0.008×10^{20}
	$129.5 \ \ \times 10^{20} = \Delta f \sum f^2 \left(\dfrac{\Psi_s}{\Psi_n'} \right)^2$

and obtain

$$\text{TL} = -25.5 + 5 \log (129.5 \times 10^{20}) = 85 \text{ dB}$$

(14-37)

In this example the mean frequency is very nearly the upper frequency of the band, or 1000 Hz. The beamwidth is therefore approximately $\lambda / L = 0.05$ rad and the selected standard deviation of the bearing estimate is one-fifth of the beamwidth of the full array. Notice that the TL given by (14-37) is about the same as obtained for detection in Section 14.2 using an optimum detection filter. Actually, for a fair comparison the ambient noise spectral density for the bearing estimate calculation should have been doubled because this was assumed to be the noise level at the half-array output. This would have reduced the allowed transmission loss by 3 dB to a value of 82 dB.

The calculated accuracy of the bearing estimate in this example is based on a single plane wave target in isotropic noise. The presence of plane wave signals at other bearings and of nonisotropic noise fields may in practice result in increased errors in the bearing estimate unless the ratio of desired signal power to total interference is large.

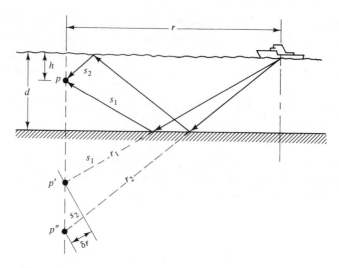

Figure 14-6   Geometry for simple two-path passive ranging.

14.4 PASSIVE MULTIPATH RANGING

A passive sonar system may determine range to a target by observing the target from two vantage points—that is, by triangulation. A closely related technique takes advantage of the complex nature of the transmission channel to achieve a result similar to that obtained by observing the target from multiple receiver locations.

Consider the geometry in Figure 14-6. Two possible propagation paths from a surface-ship target to a submerged receiver are shown. If the target is beyond the direct-path range, these two paths may be the only paths of consequence. For simplicity in the following analysis, refraction effects are neglected, the bottom-bounce loss is assumed equal for the two paths, and the surface reflection loss is assumed zero. Hence the received signals have essentially equal amplitude and differ only in relative time delay caused by the difference in transmission path lengths. The received signals, s_1 and s_2, at the receiver location p are the same as signals that would be received by direct transmission at the separated image points p' and p'' (aside from a sign change resulting from the surface reflection of s_2).

If $s_1 = s(t)$, then $s_2 = s(t - \Delta)$, where $\Delta = \delta r / c$. If the receiver depth h and water depth d are small compared to the horizontal range r, the relationship between horizontal range and path length difference, δr, is easily shown to be

$$ r = \frac{4hd}{\delta r} $$

Thus, if d and h are known and δr can be measured, the target range can be estimated. As a numerical example, let

$$d = 5000 \text{ ft}$$

$$h = 200 \text{ ft}$$

$$r = 50{,}000 \text{ ft}$$

then $\delta r = 80$ ft and the time delay of s_2 relative to s_1 is approximately 16 msec (for $c = 5000$ ft/sec).

Figure 14-7 provides a functional mechanization for determining the time delay Δ, and thereby δr, associated with the two propagation paths in Figure 14-6. Actually, the mechanization shown simply provides an estimate of the received signal autocorrelation function, from which the relative time delay can be estimated. In this discussion we restrict our attention to establishing the signal-to-noise ratio of the appropriate components of $y(\tau, t)$. The receiving sensor is assumed to have an omnidirectional response in the vertical plane so that the individual paths are not resolved in angle.

The output of the integrator is an estimate of the autocorrelation function of the received signal. Thus, with $T \gg \tau$,

$$y(t, \tau) = \tilde{R}_x(\tau) = \frac{1}{T} \int_t^{t+T} x(t')x(t' - \tau)\, dt' \tag{14-38}$$

and the expected value of $y(t, \tau)$ is

$$\overline{y(t, \tau)} = R_x(\tau) = R_{s_1}(\tau) + R_{s_2}(\tau) + R_n(\tau) + R_{s_1 s_2}(\tau) + R_{s_2 s_1}(\tau) \tag{14-39}$$

Because s_1 and s_2 differ only by the relative time delay, $R_{s_1}(\tau) = R_{s_2}(\tau)$. Also, $R_{s_1 s_2}(\tau) = R_s(\tau - \Delta) = R_{s_2 s_1}(-\tau)$. Hence

$$R_x(\tau) = 2R_s(\tau) + R_n(\tau) + R_s(\tau - \Delta) + R_s(\tau + \Delta) \tag{14-40}$$

The component parts of $R_x(\tau)$ are sketched in Figure 14-8 for τ-positive, assuming that the widths of the main lobes of the correlation functions are narrow compared with the relative delay Δ.

In order to estimate delay it is first necessary to detect the presence of the cross-correlation peak in the vicinity of $\tau = \Delta$. To this end, we now derive the signal-to-noise ratio at $\tau = \Delta$, defined as

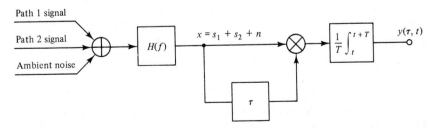

Figure 14-7   Autocorrelator for multipath passive range measurement.

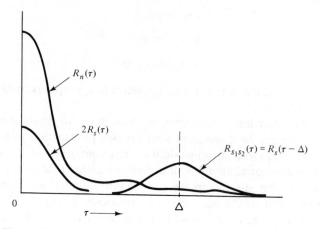

Figure 14-8   Component parts of $R_x(\tau)$ with two propagation paths.

$$\text{SNR}(y) = \frac{\{E[y_{s+n}(\Delta)] - E[y_n(\Delta)]\}^2}{\text{Var}[y_n(\Delta)]} \tag{14-41}$$

For $R_s(0) \ll R_n(0)$ and for Δ large compared with the width of the correlation functions, we have

$$E[y_{s+n}(\Delta)] \simeq R_s(0) + R_n(\Delta)$$

and

$$E[y_n(\Delta)] = R_n(\Delta)$$

from which the numerator in (14-41) is $R_s^2(0)$.

To obtain the variance with noise only, we require

$$E[y_n^2(\Delta)] = \frac{1}{T^2} \iint \text{rect}\left(\frac{t}{T}\right) \text{rect}\left(\frac{t'}{T}\right) \overline{n(t)n(t')n(t - \Delta)n(t' - \Delta)}\, dt\, dt' \tag{14-42}$$

Assuming that the noise is ZMG, the expectation of the fourfold product in (14-42) may be evaluated using the relationship given in (9-140). Thus

$$\overline{n(t)n(t')(n(t - \Delta)n(t' - \Delta)} = \overline{[n(t)n(t')]}\,\overline{[n(t - \Delta)n(t' - \Delta)]}$$
$$+ \overline{[n(t)n(t-\Delta)]}\,\overline{[n(t')n(t' - \Delta)]}$$
$$+ \overline{[n(t)n(t' - \Delta)]}\,\overline{[n(t')n(t - \Delta)]}$$
$$= R_n^2(t - t') + R_n^2(\Delta)$$
$$+ R_n(t - t' + \Delta)R_n(t' - t + \Delta) \tag{14-43}$$

With the assumed conditions, the third term in (14-43) will vanish on integration. Now let $t - t' = t''$ and obtain

$$E[y_n^2(\Delta)] = \frac{1}{T^2} \iint \text{rect}\left(\frac{t}{T}\right) \text{rect}\left(\frac{t - t''}{T}\right) [R_n^2(t'') + R_n^2(\Delta)] \, dt \, dt''$$

Integration with respect to t gives

$$E[y_n^2(\Delta)] = R_n^2(\Delta) + \frac{1}{T} \int_{-T}^{+T} \left(1 - \frac{|t''|}{T}\right) R_n^2(t'') \, dt''$$

For T large compared with the width of $R_n(t'')$ this is approximately

$$E[y_n^2(\Delta)] = R_n^2(\Delta) + \frac{1}{T} \int_{-\infty}^{+\infty} R_n^2(t'') \, dt''$$

from which

$$\text{Var}[y_n(\Delta)] = \frac{1}{T} \int_{-\infty}^{+\infty} R_n^2(t'') \, dt'' \tag{14-44}$$

Using the definition of effective noise envelope bandwidth, we have

$$\text{Var}[y_n(\Delta)] = \frac{R_n^2(0)}{2T\beta_n} \tag{14-45}$$

The value of SNR(y) is therefore

$$\text{SNR}(y) = 2T\beta_n \left[\frac{R_s^2(0)}{R_n^2(0)}\right] = \frac{T\beta_n}{2} \text{SNR}^2(x) \tag{14-46}$$

Where SNR(x) is the signal-to-noise ratio at the filter output. Note that (14-46) is identical in form to the output signal-to-noise ratio given by (13-61) for a broadband detection system. The value of SNR(y) is maximized by the use of the optimum filter defined in (13-65).

For (14-45) to be valid, we required that the relative delay, Δ, be large compared with the width of the signal and noise autocorrelation functions. This is equivalent to requiring that Δ be large compared with the inverse of the signal and noise effective envelope bandwidths. In the numerical example considered, with $\Delta = 16$ msec, the bandwidth requirements are

$$\beta_n, \beta_s \gg \frac{10^3}{16} \simeq 62 \text{ Hz}$$

Hence signal and noise bandwidths of 400 Hz or more may be required for successful passive ranging with the geometry given.

Assuming a reasonable signal-to-noise ratio, the accuracy of the relative path-delay estimate is obtained from an expression of the form given in (13-81). That is, the variance estimate is inversely related to the signal-to-noise ratio and the signal-envelope bandwidth measure, β_0.

In practice, the accuracy of the results obtained with multipath ranging is very dependent on the actual propagation characteristics that exist, and on the ability to accurately predict the multipath structure. In general, the multipath structure may be much more complex than shown in Figure 14-6, often resulting in situations where passive multipath ranging is impractical.

14.5 ACTIVE SYSTEM DETECTION PERFORMANCE

Assume an active pulsed sonar system with a baffled circular array providing both the transmit and receiving aperture. A partial listing of system and target characteristics is as follows:

System:

Array	10-ft-diameter baffled circular aperture
Transmit frequency	5 kHz
Pulse shape	0.1 sec rectangular pulse
Source level	$+220$ dB$//\mu$Pa at 1 m

Target:

Target strength	TS $= +10$ dB
Relative velocity	± 40 knots $= \pm 66.8$ ft/sec

Environment:

Ambient noise	$+60$ dB$//\mu$Pa in a 1 Hz band
Surface reverberation	$S_s = -40$ dB

We shall now calculate detection range for $P_D = 50\%$ and $P_{fa} = 10^{-4}$ assuming the following performance limitations:

1. Detection limited by ambient noise with zero relative target velocity
2. Ambient-noise limited with nonzero relative target velocity
3. Surface-reverberation limited with zero relative target-to-reverberation velocity
4. Surface-reverberation limited with nonzero target-to-reverberation velocity

14.5.1 Ambient-Noise Limited: Zero Relative Velocity

For this case, assume that the predetection filter is centered at the frequency of the transmitted signal with a bandwidth equal to the inverse of the pulse width. Because the bandwidth is narrow compared with the center frequency, the noise spectral density is essentially constant over the receiver bandwidth. The transmission loss may therefore be calculated by an equation similar to that used for the passive system. That is,

$$TL = \tfrac{1}{2}(SL + TS - NL_s + DI - DT_{Hz}) \qquad (14\text{-}47)$$

Where TL is the *one-way* transmission loss and the factor of $\tfrac{1}{2}$ results from the required *two-way* transmission for the active system. The detection performance will be calculated based on a single-pulse transmission. Therefore, the narrow-band detection threshold is

$$DT_{Hz} = 5 \log d - 5 \log\left(\frac{t_p}{\beta}\right) \qquad (14\text{-}48)$$

where t_p is the rectangular pulse width and β is the predetection filter bandwidth. With $\beta = 1/t_p$ this becomes

$$DT_{Hz} = 5 \log d - 5 \log(t_p^2) \qquad (14\text{-}49)$$

For this system the product of pulse width and noise bandwidth is unity. Therefore, we may not assume Gaussian statistics in selecting the detection index d. For the given values of $P_D = 50\%$ and $P_{fa} = 10^{-4}$, Figure 13-11 may be used to arrive at the value of $d = 75$ for $t_p\beta = 1.0$. With this value of d, DT_{Hz} for this example is approximately $+19$ dB.

The directivity index for the circular aperture is obtained from (11-84), which gives

$$DI = 20 \log\left(\frac{\pi D}{\lambda}\right) = 20 \log\left(\frac{10\pi}{1}\right) \simeq 30 \text{ dB}$$

The one-way transmission loss corresponding to the selected values of P_D and P_{fa} is obtained by substitution in (14-47). Thus

$$TL = \tfrac{1}{2}(220 + 10 - 60 + 30 - 19) = 90.5 \text{ dB}$$

Assuming spherical spreading with absorption loss of 0.5 dB/km (see Figure 5-10), the resulting detection range is the solution of

$$90.5 = 20 \log(r) + 0.5(r \times 10^{-3})$$

with r in meters. The result is $r = 15{,}000$ m.

14.5.2 Ambient-Noise Limited: Nonzero Relative Velocity

With a possible target relative velocity of ± 40 knots, the received signal may be shifted in frequency relative to the transmitted frequency by the Doppler effect. The total frequency range for this example is

$$f_{d_{max}} = \pm\frac{2V}{\lambda} = \pm\frac{2(66.8)}{1} \simeq \pm 134 \text{ Hz}$$

As discussed in Section 13.1.4, either a single wideband filter may be used to accommodate the expected Doppler shift, or a bank of contiguous narrowband filters may be used each of which is matched to the pulse bandwidth. First consider the performance with a single wideband filter with $\beta = 268$ Hz. With this approach, the time–bandwidth product is $t_p\beta = 26.8$. A postdetection filter is assumed with a bandwidth matched to the rectangular envelope bandwidth.

From Figure 13-11 the required detection index is approximately 20, so that the detection threshold is

$$DT_{Hz} = 5\log(20) - 5\log\left(\frac{0.1}{268}\right) \approx +24 \text{ dB}$$

Note that the increased noise bandwidth results in about a 5 dB increase in the required detection threshold compared with the zero-Doppler case.

The transmission loss now becomes

$$TL = \tfrac{1}{2}(220 + 10 - 60 + 30 - 24) = 88 \text{ dB}$$

from which the detection range for spherical spreading with absorption loss is $r = 12,500$ m. This represents a 17% reduction in range compared with the previous example.

If a bank of narrowband filters, each matched to the pulse bandwidth, is used, the detection performance will be very nearly equal to that calculated for the zero relative velocity case. The number of parallel channels required is approximately 27, which would result in a slight increase in detection threshold to maintain the same false-alarm characteristics as with the single narrowband channel.

14.5.3 Reverberation-Limited: Zero Target-to-Reverberation Velocity

Let the transmit and receive pattern functions be oriented parallel to the ocean surface. Assume that the received reverberation signal exceeds ambient noise, and for simplicity we assume that surface reverberation is the dominant reverberation component.

The reverberation spectral density will have the characteristics discussed in Section 12.3.1 with the central peak shifted in frequency relative to the transmitted signal by an amount related to platform speed and beam orientation relative to the platform velocity vector. In reverberation-limited conditions it is common practice to shift the transmitted frequency to maintain the centroid of the reverberation spectrum at a fixed frequency. This permits the use of fixed filter designs in the receiving system.

If the target is stationary relative to the scatterers causing the reverberation, the received target spectral density will be essentially identical in form to the reverberation spectral density. In this case the ratio of received target echo to reverberation in decibel form is given by (12-20). Thus

$$EL - RL_s = TS - S_s - \frac{TL}{2} - 10 \log \left(\frac{ct_p \gamma_B}{2} \right)$$

where RL_s is the received surface reverberation level and γ_B is a measure of two-way azimuth beamwidth. From Table 8.1 in Urick [1] we obtain

$$10 \log \gamma_B = 10 \log \left(\frac{\lambda}{\pi D} \right) + 6.9$$

which for the 10-ft-diameter array at 5 kHz gives $\gamma_B = 0.156$ rad.

The detection performance for the reverberation-limited case is obtained by equating the received echo-to-reverberation level to a detection threshold and solving for transmission loss. Thus

$$TL = 2 \left[TS - S_s - 10 \log \left(\frac{ct_p \gamma_B}{2} \right) - DT \right] \qquad (14\text{-}50)$$

Because the target-signal and reverberation have the same spectral shape, no processing advantage results in this case from a judicious selection of the predetection filter shape (of course, a predetection filter will be used to minimize the effect of ambient noise). Furthermore, postdetection filtering of the signal received from a single-pulse transmission will not result in a processing gain because the product of target-signal duration and reverberation bandwidth is essentially unity. Therefore, the detection threshold is a function only of the detection index. Thus

$$DT = 5 \log d$$

where d must be obtained from ROC curves based on a unity time–bandwidth product. For this example, we may use Figure 13-11, from which $d = 75$ and $DT = +9$ dB.

From (14-50) we now obtain

$$TL = 2(10 + 40 - 11 - 9) = +60 \text{ dB}$$

Assuming spherical spreading, the resulting detection range is limited to 1000 m. Notice that absorption loss is not considered in this example because absorption loss affects reverberation and target signal equally. Absorption loss should be considered when comparing reverberation level and ambient noise to distinguish between reverberation-limited and ambient-noise-limited conditions.

From (14-50), the only parameters available to the system designer to improve detection performance are the projected pulse width and the azimuth beamwidth. Performance is improved by reducing either of these parameters *provided* that the spatial dimensions of the pulse in the water are not reduced below the spatial dimensions of the target.

Note that the calculated performance is quite sensitive to the assumptions concerning target strength and reverberation scattering strength, both of which parameters are typically not known with great precision. Thus an error of 10 dB in either of these terms results in a 20 dB error in calculated transmission loss.

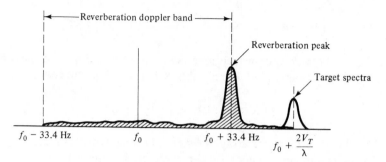

Figure 14-9 Reverberation and target spectra for a bow-on encounter and a closing target.

14.5.4 Reverberation-Limited: Nonzero Target-to-Reverberation Velocity

Let the sonar platform be moving at a speed of 10 knots with the array beam pattern aligned with the platform velocity vector. The target is assumed aligned with the array MRA with a closing relative velocity in excess of 10 knots. The received spectral densities for reverberation and target signal are sketched in Figure 14-9.

The reverberation peak in this case has the maximum Doppler shift of $+2V/\lambda = 33.4$ Hz, where V is the platform speed. Some reverberation energy is spread over the range $f_0 \pm 33.4$ Hz as a result of the nonzero side-lobe response of the array. The reverberation energy above the frequency $f_0 + 33.4$ Hz is the result of the shape of the transmitted spectrum.

For the rectangular pulse waveform, the main-beam reverberation spectral density is approximately

$$\Psi_R(f) = \Psi_{max} \operatorname{sinc}^2[(f - f_0 - f_d)t_p] \tag{14-51}$$

This is sketched in Figure 14-10 together with the envelope of the peaks of the reverberation side lobes, and a target spectrum centered at a higher Doppler frequency.

Now let the target relative closing speed be 20 knots resulting in a Doppler shift of 66.8 Hz. The target spectrum is therefore centered at a frequency $f' = 33.4$ Hz above the center of the main-beam reverberation spectrum. The peak reverberation levels at the target frequency will be on the order of

$$10 \log \Psi_R = 10 \log \Psi_m - 20 \log (\pi f' t_p)$$

$$= 10 \log \Psi_m - 20 \log \left(\frac{33.4\pi}{10}\right)$$

$$= 10 \log \Psi_m - 20 \text{ dB}$$

Thus, to a first approximation a filter with bandwidth $1/t_p$ centered at the target

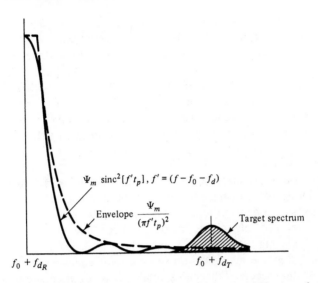

$\Psi_m \, \mathrm{sinc}^2[f't_p], \, f' = (f - f_0 - f_d)$

Envelope $\dfrac{\Psi_m}{(\pi f' t_p)^2}$

Target spectrum

$f_0 + f_{d_R}$

$f_0 + f_{d_T}$

Figure 14-10   Details of main-beam reverberation spectra with rectangular transmitted pulse envelope.

signal frequency will provide an echo-to-reverberation ratio 20 dB better than if the target spectrum coincides with the reverberation spectrum. Let the reverberation reduction factor in dB be R and modify (14-50) to give

$$TL = 2\left[TS - S_s + R - 10 \log\left(\frac{ct_p \gamma_B}{2}\right) - DT \right] \qquad (14\text{-}52)$$

where for the rectangular pulse envelope

$$R = 20 \log\left[(f_{d_t} - f_{d_R})\pi t_p\right] \qquad (14\text{-}53)$$

f_{d_t} = target Doppler frequency

f_{d_R} = main-beam reverberation Doppler frequency

Substitution of the numerical values for this example gives

$$TL = 2(10 + 40 + 20 - 11 - 9) = 100 \text{ dB}$$

Note that this value of transmission loss is actually 12 dB better than calculated for the ambient-noise-limited case. This means that for this relative target velocity, ambient noise rather than reverberation will limit the detection performance. If the target closing speed is reduced to 15 knots, the transmission loss is reduced to 88 dB, so that the reverberation-limited and noise-limited ranges will be about the same.

From (14-52) and (14-53) it is clear that performance with a high-Doppler target is improved by *increasing* the pulse width. This is in contrast with the situation where the target Doppler frequency is the same as the reverberation Doppler frequency. With the high-Doppler target, the increase in reverberation area caused by an increase in pulse width is more than offset by the reduction in reverberation spectral side-lobe level at the target frequency.

With a rectangular pulse envelope, the spectral side-lobe level decreases inversely with the square of frequency. Because of the direct relationship between spectral side-lobe level and detection performance with a high-Doppler target, shapes other than the simple rectangular envelopes are often used to achieve improved performance. Techniques identical to those discussed in Section 11.3 for the reduction of spatial side lobes for array pattern functions are used to select the shape of the time-domain envelope function.

In this example, if the target relative closing velocity is less than 10 knots, the reverberation contribution of the array spatial side lobes should be considered when calculating performance. If the array is used for both transmission and reception, the side-lobe reverberation level is reduced proportional to the fourth power of the ratio of side lobe to main-lobe amplitude. As an example, let the relative target closing velocity be 5 knots. The received target frequency will be about 17 Hz below the main-beam reverberation frequency, and must compete with reverberation received through spatial side lobes at an angle off the MRA determined from

$$\psi = \cos^{-1}\left(\frac{\lambda f_d}{2V}\right) = \cos^{-1}\left[\frac{(1)(17)}{(2)(10)}\right]$$

$$\approx 32°$$

The circular array normalized pattern function is given by

$$G(\psi) = \frac{2J_1[(\pi D/\lambda)\sin\psi]}{(\pi D/\lambda)\sin\psi}$$

Where $J_1(\bullet)$ is a first-order Bessel function of the first kind. With $D = 10$ ft and $\lambda = 1$ ft, the peak side-lobe level in the vicinity of $\psi = 32°$ is about 0.13. Therefore, the reverberation reduction factor relative to the main lobe is

$$R = -10 \log[G^4(\psi)] = 35 \text{ dB}$$

Assuming a rectangular pulse envelope, the spatial side-lobe reverberation contribution will in this case be considerably less than the pulse spectral side-lobe contribution. However, if the envelope shape is selected to minimize the spectral side lobes, the spatial side-lobe contribution may become appreciable.

An omnidirectional transmit pattern is sometimes used in combination with a directional receiving pattern. The reverberation contribution from the side-lobe regions is then reduced as the square of the pattern function rather than as the fourth power. Thus, in the example above, the reduction factor would be 17.5 dB relative to the main lobe, rather than 35 dB. If only the receiver pattern is directional, it may be desirable to reduce the spatial side lobes by array shading techniques to improve performance.

PROBLEMS

14.1. A narrowband receiving system forms 1000 contiguous parallel channels, each with a bandwidth of 0.5 Hz. The postdetection integration time is 100 sec for each channel, and the desired false-alarm time is $\frac{1}{2}$ hr. Calculate the narrowband detection threshold for a 90% detection probability.

14.2. Repeat Problem 14.1 for false-alarm times of 15 min and 2 hr.

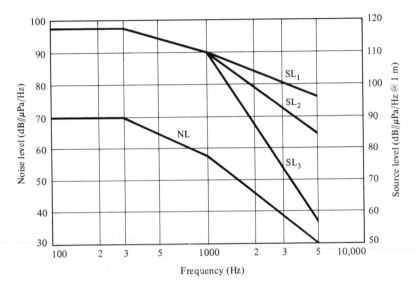

Figure P14-1

14.3. In Figure P14-1 are given the noise spectral level at the spatial filter output, and three alternative source spectral levels for use in calculating performance of a broadband receiving system. Determine the optimum broadband filter shape covering the frequency range from 100 to 5000 Hz for each of the source spectra.

14.4. Calculate the FOM for the broadband system for each of the three target spectra given in Problem 14.3. Assume the use of the optimum filters, a postdetection integration time of 300 sec, $P_D = 50\%$ and $P_{\text{fa}} = 10^{-4}$.

14.5. Repeat Problem 14.4 assuming that the filters have upper cutoff frequencies of 3000 Hz, 2000 Hz, and 1000 Hz. Based on these results, is the use of the full 5000-Hz bandwidth justified for each of the target spectra?

14.6. In Figure 14-5 a broadband bearing estimation mechanization is shown using a Hilbert transformation. Demonstrate that the mechanization shown in Figure P14-2 is a practical implementation of the Hilbert transformation technique. For simplicity, assume that the target and noise signals, s and n, are bandlimited signals with effective carrier frequency f_0, and the filters $H_2(f)$ are rectangular baseband filters with bandwidth equal to the signal and noise bandwidth.

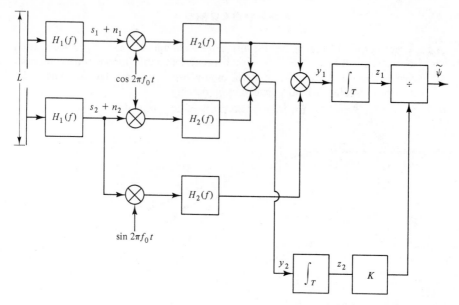

Figure P14-2

SUGGESTED READING

1. Urick, R. J., *Principles of Underwater Sound for Engineers,* 2nd ed. New York: McGraw-Hill Book Company, 1975, Chaps. 12 and 13.
2. "Principles of Sonar Installation," Naval Underwater Systems Center, Technical Document 6059, 1980.
3. Stewart, J. L., "Active Acoustic Signal Processing" (Unclassified), *U.S. Navy J. Underwater Acoust.,* Vol. 21, No. 4 (Oct. 1971).
4. Stocklin, P. L., "Passive Acoustic Signal Processing—The State of the Art" (Unclassified), *U.S. Navy J. Underwater Acoust.,* Vol. 21, No. 4 (Oct. 1971).
5. Winder, A. A., "Sonar System Technology," *IEEE Trans. Sonics, Ultrasonics,* Vol. SU-22, No. 5 (Sept. 1975).
6. Knight, W. S., R. G. Pridham, and S. M. Kay, "Digital Signal Processing for Sonar," *Proc. IEEE,* Vol. 69, No. 11 (Nov. 1981).
7. Gong, F. K., and J. S. Davis, "Evaluation of Target Motion Analysis in a Multipath Environment," Naval Underwater Systems Center, Technical Report 4814, Mar. 1976.
8. Lindgren, A. G., and K. F. Gong, "Measurement of Relative Delay between Signals Propagating in a Multipath Environment," IEEE International Conference on Engineering in the Ocean Environment, 1974.
9. Nuttall, A. H., "Signal Processing in Reverberation—A Summary of Performance Capability," Naval Underwater Systems Center, Technical Memo, TC-173-72, 1972.

Index